Renewable Energy and AI for Sustainable Development

Electronic device usage has increased considerably in the past two decades. System configurations are continuously requiring upgrades; existing systems often become obsolete in a matter of 2–3 years. Green computing is the complete effective management of design, manufacture, use, and disposal, involving as little environmental impact as possible. This book intends to explore new and innovative ways of conserving energy, effective e-waste management, and renewable energy sources to harness and nurture a sustainable eco-friendly environment.

This book:

- Highlights innovative principles and practices using effective e-waste management and disposal
- Explores artificial intelligence based sustainable models
- Discovers alternative sources and mechanisms for minimizing environmental hazards
- Highlights successful case studies in alternative sources of energy
- Presents solid illustrations, mathematical equations, as well as practical in-the-field applications
- Serves as a one-stop reference guide to stakeholders in the domain of green computing, e-waste management, renewable energy alternatives, green transformational leadership comprising theory concepts, practice and case studies
- Explores cutting-edge technologies like internet of energy and artificial intelligence, especially the role of machine learning and deep learning in renewable energy and creating a sustainable ecosystem
- Explores futuristic trends in renewable energy

This book aims to address the increasing interest in reducing the environmental impact of energy as well as its further development and will act as a useful reference for engineers, architects, and technicians interested in and working with energy systems; scientists and engineers in developing countries; industries, manufacturers, inventors, universities, researchers, and interested consultants to explain the foundation to advanced concepts and research trends in the domain of renewable energy and sustainable computing.

The content coverage of the book is organized in the form of 11 clear and thorough chapters providing a comprehensive view of the global renewable energy scenario, as well as how science and technology can play a vital role in renewable energy.

Innovations in Intelligent Internet of Everything (IoE)
Series Editor: Fadi Al-Turjman

Computational Intelligence in Healthcare: Applications, Challenges, and Management
Meenu Gupta, Shakeel Ahmed, Rakesh Kumar, Chadi Altrjman

Blockchain, IOT and AI technologies for Supply Chain Management
Priyanka Chawla, Adarsh Kumar, Anand Nayyar and Mohd Naved

Renewable Energy and AI for Sustainable Development
Editors: Sailesh Iyer, Anand Nayyar, Mohd Naved and Fadi Al-Turjman

For more information about the series, please visit: https://www.routledge.com/ Innovations-in-Intelligent-Internet-of-Everything-IoE/book-series/IOE

Renewable Energy and AI for Sustainable Development

Edited by
Sailesh Iyer
Anand Nayyar
Mohd Naved
Fadi Al-Turjman

CRC Press
Taylor & Francis Group
Boca Raton London New York

CRC Press is an imprint of the
Taylor & Francis Group, an **informa** business

Designed cover image: Author

First edition published 2024
by CRC Press
2385 Executive Center Drive, Suite 320, Boca Raton, FL 33431

and by CRC Press
4 Park Square, Milton Park, Abingdon, Oxon, OX14 4RN

CRC Press is an imprint of Taylor & Francis Group, LLC

ISBN: 9781032439495 (hbk)
ISBN: 9781032439501 (pbk)
ISBN: 9781003369554 (ebk)

DOI: 10.1201/9781003369554

Typeset in Times
by KnowledgeWorks Global Ltd.

Contents

Preface

Electronic devices usage has increased considerably in the past two decades. System configurations are continuously requiring upgrades, as existing systems become obsolete in a matter of 2–3 years. Inefficient disposal of obsolete IT devices generates huge e-waste, harming the environment and leading to various issues like climate change and higher CO_2 emissions. The current primary environmental concern is lack of effective waste management including e-waste disposal mechanisms.

Green computing is the complete effective management of design, manufacture, use and disposal of electronics, involving as little environmental impact as possible. This book explores new and innovative ways of conserving energy, effective e-waste management and renewable energy sources to harness and nurture a sustainable eco-friendly environment.

The book encapsulates new trends and technologies for renewable energies, policies and strategies for renewable energies, smart grids, batteries, control techniques for renewable energies, hybrid renewable energies, renewable energy research and applications for industries, applications of renewable energies, artificial intelligence and machine learning studies for renewable energies and renewable energy systems in smart cities. Both theoretical and applied aspects are addressed in this book. Many illustrations and mathematical equations as well as practical on-the-field applications are incorporated. This book aims to contribute to the increasing interest in reducing the environmental impact of energy as well as its further development.

The book will serve as a practical guide to stakeholders in the domains of green computing, e-waste management, renewable energy alternatives, green transformational leadership; the book also explores how cutting-edge technologies like internet of energy, artificial intelligence especially machine learning and deep learning can play a vital role in renewable energy and creating a sustainable ecosystem. The book comprises 11 chapters covering diverse aspects of renewable energy and artificial intelligence for sustainable development.

Chapter 1 titled "Green and Clean Energy: Current Global Scenario" highlights the importance of solar energy, current status as well as future growth perspectives. In addition, the chapter discusses the global energy investment in terms of technology and also stresses on impact of COVID-19 on energy sector. **Chapter 2** titled "An Overview of Global Renewable Energy Resources: Present Scenario, Policies, and Future Perspectives" focuses on the Global Energy crisis and the various types of renewable energy resources, the barriers to the development of renewable energy generation and the policies to promote renewable energy generation around the world. This chapter also discusses the benefits and future perspectives of generating electricity from renewable energy resources. **Chapter 3** titled "Plastic Waste Conversion: A New Sustainable Energy Model in the Circular Economy Era" highlights the role of plastic in Circular Economy from manufacturing to the final recycling model, including the plastic waste conversion models deployed like waste to new plastic, waste to fuel and chemicals and waste to fertilizer. In addition, the chapter also delves on the principle of CE to develop or formulate a strategy for the byproducts generated during waste to energy conversion. **Chapter 4** titled "Layout Planning of a Small-Scale Manufacturing Industry Using an Integrated Approach based on AHP-PSI" illustrates five various layout alternatives and the evaluation process carried out with the structured framework consisting preference selection index approach (PSI) and analytic hierarchical process (AHP) approach. In addition, the chapter proposes a novel integrated approach based on AHP-PSI for layout planning for small-scale manufacturing units. **Chapter 5** titled "ATC Enhancement Due to Charging/Discharging of BESS in Wind Power Integrated Systems" proposes an effective methodology to improve the available transfer capacity value during off-peak hours. In addition, the chapter also calculates ATC values using proposed charging/discharging of battery energy storage system (BESS) for enhancing the ATC during congestion conditions. **Chapter 6** titled "Concentrated Solar Power: A Promising Sustainable Energy Option" elaborates on concentrated solar thermal power, disadvantages of PV systems, various types of concentrated solar thermal power, thermal storage and Brayton cycle-based power block. **Chapter 7** titled "Efficient and Effective Techniques for Intensification of Renewable Energy (Wind) Using Deep Learning Models" discusses the various advancements of deep learning and the scenarios of how AI and ML can elevate and enhance the potential, production and conservation of renewable energy. In addition, feedforward and LSTM networks are demonstrated and compared on predefined parameters. **Chapter 8** titled "Machine Learning for Renewable Energy Applications" provides an insight into the classification of machine language techniques and the procedures that include pre-processing of data, and possible potential opportunities for the effective deployment of machine learning techniques for prediction of energy from the renewable sources. **Chapter 9** titled "Effective Contribution of Green Human Resources Practices on Environmental Sustainability in the Era of Industry 4.0: Evidence from India" highlights a study of the association between the green transformational leadership of the leaders of the diverse manufacturing industry (chemical, cement, and medicine) and green environmental behavior (GEB) of the employees of the respective companies. The chapter also aims to analyze the mediating role of green human resource management policies and their implementation. **Chapter 10** titled "Recent Developments in Waste Valorization: An Overview of Indian and Worldwide Perspectives" highlights the status of waste valorization research in India and highlights a comprehensive depository of recently developed different innovative and fruitful strategies of waste valorization especially in Indian subcontinents. **Chapter 11** titled "Eco-Friendly Cities and Villages with Sustainability: Futuristic Perspectives" describes the challenges involved, principles followed in different existing green cities and villages, their impact on environment, physical and mental health of the people residing there, economic benefits and overall growth achieved with emerging trends and best practices performed for green cities and villages.

The book will be of great use to engineers, architects, and technicians interested in and working with energy systems; scientists and engineers in developing countries; industries, manufacturers, inventors, universities, researchers, and interested consultants to explain the foundation to advanced concepts and research trends in the domain of renewable energy and sustainable computing.

Sailesh Iyer
Anand Nayyar
Mohd Naved
Fadi Al-Turjman

About the Editors

Dr. Sailesh Iyer holds a Ph.D. degree (computer science) and is currently serving as a professor at Rai University, Ahmedabad. He has more than 23 years of experience in academics, industry, and corporate training out of which 19 years are in core academics. He has patents to his credit, various SCIE and Scopus publications and is involved as an editor for various book projects with IGI Global (USA), Taylor and Francis (UK), and Bentham Science (UAE). A hardcore academician and administrator, he has excelled in corporate training, and delivered expert talks in various AICTE-sponsored STTPs, ATAL FDPs, reputed universities, government-organized workshops, and orientation and refresher courses organized by HRDC and Gujarat University. Research contributions include reputed publications, Track Chair at *ICDLAIR 2020* (Springer Italy), *icSoftComp 2020, IEMIS 2020* (Springer), *ICRITO 2020* (IEEE), *ARISE-2021, FTSE-2021*; Dr. Iyer is also a TPC member of various reputed international and national conferences, a reviewer of International Journals like *Multimedia Tools and Applications* (Springer), *Journal of Computer Science* (Scopus Indexed), *International Journal of Big Data Analytics in Healthcare* (IGI Global), *Journal of Renewable Energy and Environment*, and an editor in various other journals. In addition to guiding research scholars as supervisor, Dr. Iyer has also been invited as a judge for various events, an examiner for reputed universities, he is a lifetime member of Computer Society of India (CSI), and he also served as managing committee (MC) member, CSI Ahmedabad Chapter from 2018 to 2020.

Dr. Anand Nayyar received his Ph.D. degree (computer science) from Desh Bhagat University in 2017, in the area of wireless sensor networks, swarm intelligence, and network simulation. He is currently working in the School of Computer Science, Duy Tan University, Da Nang, Vietnam as assistant professor, scientist, vice-chairman (research), and director – IoT and Intelligent Systems Lab. Dr. Nayyar is a certified professional with more than 125 professional certificates from CISCO, Microsoft, Amazon, EC-Council, Oracle, Google, Beingcert, EXIN, GAQM, Cyberoam and many more. He has published more than 150 research papers in various high-quality web of science, ISI-SCI/SCIE/SSCI Impact Factor Journals and Scopus/ESCI indexed Journals, more than 70 papers in international conferences indexed with Springer, IEEE and ACM Digital Library, and more than 40 book chapters in various SCOPUS, WEB OF SCIENCE indexed books with Springer, CRC Press, Wiley, IET, and Elsevier. Dr. Nayyar has 10000+ citations, H-Index: 52 and I-Index: 180. He is a member of more than 60 Associations as Senior and Life Member including IEEE and ACM. He has authored/co-authored and edited more than 50 books of computer science, and is associated with more than 500 International Conferences as Program Committee/Chair/Advisory Board/Review Board member. He has 18 Australian Patents, 4 German Patents, 1 Japanese Patent, 11 Indian Design and Utility Patents, 3 Indian Copyrights, and 2 Canadian Copyrights to his credit in the areas of wireless communications, artificial intelligence, cloud computing, IoT, and image processing. He has been awarded more than 38 awards for teaching and research – *Young Scientist, Best Scientist, Young Researcher Award, Outstanding Researcher Award, Excellence in Teaching*, and many more. He is listed in top 2% of scientists as per Stanford University (2020, 2021). He is acting as associate editor for *Wireless Networks* (Springer), *Computer Communications* (Elsevier), *International Journal of Sensor Networks* (IJSNET) (Inderscience), *Frontiers in Computer Science, PeerJ Computer Science, Human Centric Computing and Information Sciences* (HCIS), *IET-Quantum Communications, IET Wireless Sensor Systems, IET Networks, IJDST, IJISP, IJCINI*, and *IJGC*. He is acting as editor-in-chief of IGI-Global, and a US journal titled *International Journal of Smart Vehicles and Smart Transportation* (IJSVST). He has reviewed more than 2000 articles for diverse Web of Science and Scopus Indexed Journals. He is currently researching in the areas of wireless sensor networks, internet of things, swarm intelligence, cloud computing, artificial intelligence, drones, blockchain, cyber security, network simulation, big data, and wireless communications.

Dr. Mohd Naved is a machine learning consultant and researcher, currently teaching in Amity University (Noida), India for various degree programs in analytics and machine learning. He has been actively engaged in academic research on various topics in management as well as on 21st-century technologies. He has published more than 30 research papers in reputed journals (SCI/Scopus Indexed). He has 16 patents in AI/ML and actively engages in commercialization of innovative products developed at the university level. His interview has been published in various national and international magazines. A former data scientist, he is an alumnus of Delhi University. He holds a Ph.D. from Noida International University.

Prof. Dr. Fadi Al-Turjman received his Ph.D. degree in computer science from Queen's University, Canada, in 2011. He is the associate dean for research and the founding director of the International Research Center for AI and IoT at Near East University, Nicosia, Cyprus. Prof. Al-Turjman is the head of the Artificial Intelligence Engineering Department, and a leading authority in the areas of smart/intelligent IoT systems, wireless, mobile networks' architectures, protocols, deployments, and performance evaluation in artificial intelligence of things (AIoT).

His publication history spans over 400 SCI/E publications, in addition to numerous keynotes and plenary talks at flagship venues. He has authored and edited more than 40 books about cognition, security, and wireless sensor networks' deployments in smart IoT environments, which have been published by well-reputed publishers such as Taylor and Francis, Elsevier, IET and Springer. He has received several recognitions and best paper awards at top international conferences. He also received the prestigious *Best Research Paper Award* from Elsevier *Computer Communications Journal* for the period 2015–2018, in addition to the *Top Researcher Award* for 2018 at Antalya Bilim University, Turkey. Prof. Al-Turjman has led a number of international symposia and workshops in flagship communication society conferences. Currently, he serves as book series editor and the lead guest/associate editor for several top tier journals, including the *IEEE Communications Surveys and Tutorials* (IF 23.9) and the *Elsevier Sustainable Cities and Society* (IF 7.8), in addition to organizing international conferences and symposiums on the most up-to-date research topics in AI and IoT.

Contributors

Tazim Ahmed
Industrial and Production Engineering
Jessore University of Science and
    Technology
Bangladesh

Kavita Arora
Department of Computer Applications
Manav Rachna International Institute of
    Research & Studies
Faridabad, Haryana, India

Dhanasekaran Arumugam
Center for Energy Research
Department of Mechanical Engineering
Chennai Institute of Technology
Chennai, Tamil Nadu, India

Tarana Afrin Chandel
Department of Electronics and
    Communication Engineering
Integral University
Lucknow, Uttar Pradesh, India

Raviprakash Chandra
Department of Engineering and
    Physical Sciences
Institute of Advanced Research
Gandhinagar, India

Aditya Dharaiya
Department of Geology
Savitribai Phule Pune University
Pune, Maharashtra, India
&
Wildlife and Conservation Biology
    Research Foundation
Patan, Gujarat, India

A. Gayathri
Department of EEE
Sri Krishna College of Technology
Coimbatore, Tamil Nadu, India

Anil B. Ghubade
School of Mechanical Engineering
Lovely Professional University
Phagwara, India

Atif Iqbal
Department of electrical
    Engineering
Qatar University
Doha, Qatar

Shirazul Islam
Department of Electrical Engineering
Qatar University
Doha, Qatar

Sailesh Iyer
Department of Computer Science and
    Engineering
Rai School of Engineering
Rai University
Gujarat, India

Vishnupriyan Jegadeesan
Center for Energy Research
Department of Electrical and
    Electronics Engineering
Chennai Institute of Technology
Chennai, Tamil Nadu, India

Ibrahim Denka Kariyama
Department of Agricultural
    Engineering
Dr. Hilla Limann Technical
    University
Wa, Upper West Ghana, Ghana

Preeti Kashyap
Department of Biological Science and
    Biotechnology
Institute of Advanced Research
Gandhinagar, India

M.A. Mallick
Department of Electrical Engineering
Integral University
Lucknow, Uttar Pradesh, India

V. Manimegalai
Department of EEE
Sri Krishna College of Technology
Coimbatore, Tamil Nadu, India

V. Mohanapriya
Department of EEE
Bannari Amman Institute of
    Technology Erode
Tamil Nadu, India

V. Sathya Moorthi
Department of Business
    Administration
Kalasalingam Academy of Research
    and Education
Virudhunagar, Tamil Nadu, India

Soumitra Mukhopadhyay
Doosan Power Systems (I) Pvt. Ltd.
Kolkata, West Bengal, India

Aishvarya Narain
Department of Electrical Engineering
United College of Engineering and
    Research
Prayagraj, Uttar Pradesh, India

Egharevba Godshelp Osas
Industrial Chemistry Programme
Department of Physical Sciences
Landmark University
Omu-Aran, Kwara State, Nigeria

Mariya Ouaissa
Computer Science and Networks
Moulay Ismail University Meknes
Morocco

P. Pandiyan
Department of EEE

KPR Institute of Engineering and
    Technology
Coimbatore, Tamil Nadu, India

Richa Parmar
Solar Water Pump Laboratory
National Institute of Solar Energy
Gurugram, Haryana, India

Rushika Patel
Wildlife and Conservation Biology
    Research Foundation
Patan, Gujarat, India
&
Gujarat Biotechnology Research Centre
Department of Science and Technology
Government of Gujarat
Gandhinagar, Gujarat, India

Shital Patel
Department of Pharmacy
Bharat Institute of Technology
Mangalpally Ibrahimpatnam
Telangana, India

Ajay John Paul
School of Mechanical Engineering
Kyungpook National University
Daegu, Gyeongbuk Province, South
    Korea

Dipesh Popli
Mechanical Engineering Department
GD Goenka University
Gurgaon, India

V. Rukkumani
Department of EIE
Sri Ramakrishna Engineering College
Coimbatore, Tamil Nadu, India

Kunal Shah
Department of Anesthesiology
University of Texas MD
Anderson Cancer Center
Houston, Texas, United States

Parveen Sharma
School of Mechanical Engineering
Lovely Professional University
Phagwara, India

Bhima Sridevi
Department of Pharmacy
Bharat Institute of Technology
 Mangalpally Ibrahimpatnam
Telangana, India

Christopher Stephen
Department of Mechanical Engineering
Vel Tech Rangarajan Dr. Sagunthala
 R&D Institute of Science and
 Technology
Chennai, Tamil Nadu, India
&

Teaching Associateship for Research
 Excellence (TARE) Fellow under
 SERB
National Institute of Solar Energy
 (NISE)
Gurugram, Haryana, India

Dolly Vadaviya
Department of Biological Science and
 Biotechnology
Institute of Advanced Research
Gandhinagar, India

Gajendra Singh Vishwakarma
Department of Biological Science and
 Biotechnology
Institute of Advanced Research
Gandhinagar, India

1 Green and Clean Energy
Current Global Scenario

Tarana Afrin Chandel[1], M.A. Mallick[2], Atif Iqbal[3], and Shirazul Islam[3]
[1]Department of Electronics and Communication Engineering, Integral University, Lucknow, Uttar Pradesh, India
[2]Department of Electrical Engineering, Integral University, Lucknow, Uttar Pradesh, India
[3]Department of Electrical Engineering, Qatar University, Doha, Qatar

CONTENTS

DOI: 10.1201/9781003369554-1

1.1   INTRODUCTION

In many countries with the revolution in industry, renewable energy (RE) became the major source energy demand over fossil fuel to fulfil their primary requirement. A total of 75% carbon emission, which was due to fossil fuel and RE, shows major effect in global climatic change. This carbon emission due to fossil fuels affected 5 million premature deaths per year. To overcome these problems, it is necessary to reduce CO_2 emissions and sip pollutant. These objectives can be accomplished with different resources and implementing it with technologies and by making policies. Two major goals can fulfil the transformation pathway. The first goal is to achieve qualitative and quantitative framework   for the recognition of policies, estimate transition towards energy system and uplift sustainability development. The second goal is to promote the integration of varied energy problems and main-tain global balance [1]. The international policies designed will help us to enhance social-economic developments, energy generation and distribution globally. Thus, it requires switching towards nuclear and renewable innovations globally. Solar, wind, waves, tides, geothermal and hydropower are the sources of primary energy in renewable technologies. About 11% of global primary energy came from renew-able energy resources in 2019 [2]. Figure 1.1 shows the contribution of energy gen-eration with different sources in renewable technologies since 1965 till 2020 [2].

Figure 1.2 shows changes in energy generation with different energy source along with time from the year 1985 till 2020 [2]. The generation of energy via hydropower has increased from 2000 TWh in 1985 to 4000 TWh in 2020. Apart from this, solar and wind energy generation has grown rapidly from the year 2010 onwards as per World in Data Journal. Sustainable development is also a key point towards green and clean energy. It includes social, economic and environment developments. Sustainable development motivates everyone to safeguard, preserve and boost the resources. United Nations approved 2030 agenda of developing a new roadmap of sustainable develop-ment having sustainable development goals [3]. This goal was to protect and safeguard human on the planet by providing employment, food, water, sanitization and energy. At a social level, sustainability can bring developments of people, regions and cultures, bringing equality in life, healthcare and education worldwide. Economical sustainabil-ity is dependent on the economic progress. This can be achieved by investing in and developing green businesses. Developing green industry businesses and reducing car-bon dioxide emission is an approach towards environmentally sustainable development.

Renewable energy generation, World

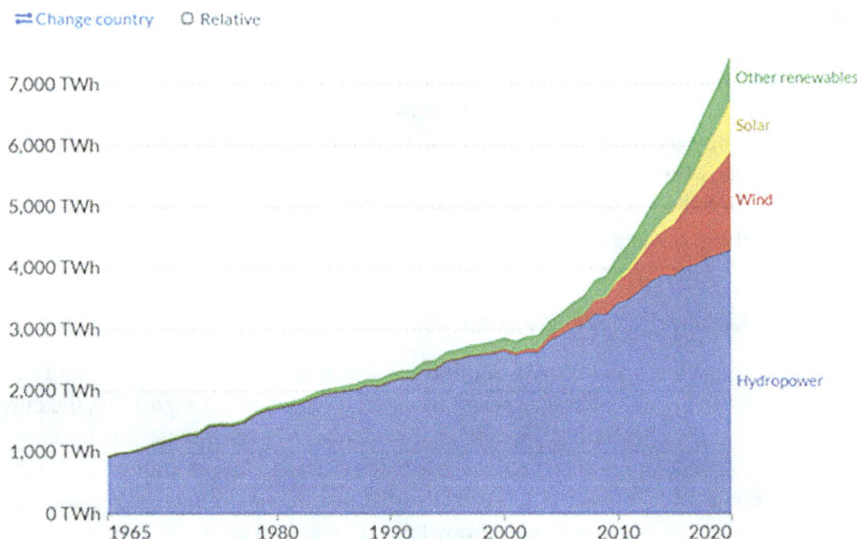

FIGURE 1.1 Primary energy generation with different energy sources.

Modern renewable energy generation by source, World

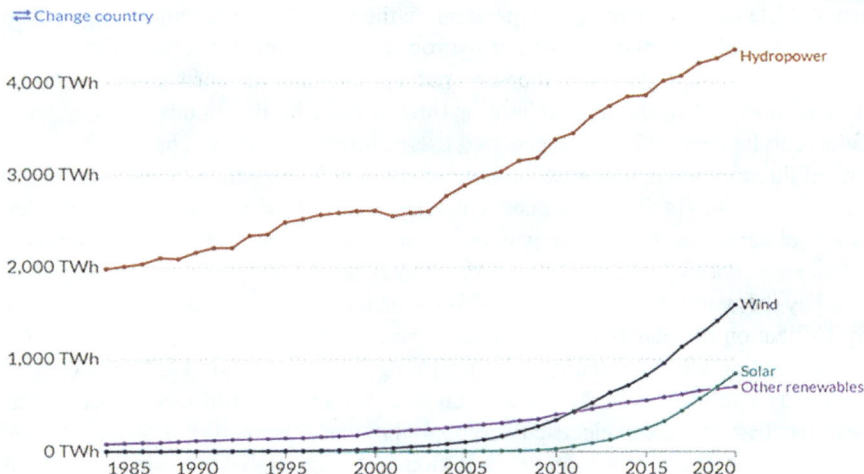

FIGURE 1.2 Change in renewable energy generation with time.

1.1.1 OBJECTIVES OF THE CHAPTER

This chapter has the following objectives:

- To study the importance of solar energy, current status of solar energy, International Solar Alliance (ISA), growth in RE and also includes international climate policies, energy transformation towards RE, leaving behind fossil fuel;
- Discuss global energy investment in terms of technology;
- And, to study the impact of COVID-19 on energy.

1.1.2 ORGANIZATION OF THE CHAPTER

This chapter is organized as follows: Section 1.2 discusses the role of solar energy as well as outlines the growth of solar energy worldwide. Section 1.3 highlights the current status of solar photovoltaic (PV) capacity – worldwide, region-wise and country-wise. Section 1.4 enlightens international climate policies like international efforts for climate change and Nationally Determined Contributions. Section 1.5 illustrates International Solar Alliance in terms of power system transformation, power system operator survey towards global energy transformation and status of power system transformation. Section 1.6 discusses global energy investment in terms of investment by technology and investment by region. Section 1.7 focusses on employment in renewable energy. Section 1.8 stresses on the impact of COVID-19 on energy and Section 1.9 concludes the chapter with future scope.

1.2 ROLE OF SOLAR ENERGY

Solar energy is abundant and inexhaustible in nature. It is a green and clean energy source. All life on this universe is dependent on the sun [4]. The radiant energy of sun provides heat and light by the fusion of hydrogen into helium at its core [4] known as solar radiation. A total of 50% of the solar radiation reaches the earth's surface, while the rest is absorbed in the atmosphere or thrown back by the clouds [4]. The solar radiation can be captured and turned into useful forms of energy. The sun energy is generated due to nuclear fusion at the core of the sun. The earth receives this energy in the form of sunlight. This sun energy is beneficial in production of electricity due to photo-voltaic effect in the same way as green plant use sun energy to produce their food known as photosynthesis. The sun can contribute 1,000 times more energy than required by the world; however, only 0.02% of the total energy is being used currently [5, 6]. Utilization of solar radiation can be done for lightening the homes, buildings, street and heating the surroundings by generating electricity from it. The major concern towards solar energy is the climatic change. Production of electricity with solar radiation is clean and green electricity compared to that generated from fossil fuel, i.e. free from air, water and environment pollution with zero global warming and threat free to public health. According to the US Department of Energy, the amount of solar radiation reaching the earth's surface is more than enough to fulfil the requirement of energy globally for a period of one year [4]. The amount of power generated by solar radiation and stored within half month is equal to the sum of energy stored by coal, oil and natural gases in all planets. Once the electric power energy is harnessed for solar radiation, fuels are free.

1.2.1 GROWTH OF SOLAR ENERGY WORLDWIDE

Growth of solar energy worldwide with PV techniques was exponential from 1992 to 2018 as shown in Figure 1.3. This duration was called as solar PV era which evolved from small scale and spread all over to larger market as a mainstream electricity source. The solar PV system was potentially recognized by the subsidy programmes regarding tariffs. This was implemented by many governments to provide economic incentives for investments. For many years, Japan and European countries took the advantage of the subsidy and made progress in the economy. In 2011, the United States and China made the Solar Roof project for enhancing PV deployment with five-year plan energy generation. The deployment of PV system via bottom up market strategies is very powerful. This strategy concentrate towards technologies for providing rooftop energy services i.e. helpful for house hold goods and for commercial process technology in United States via 2030. The market of utility scale plant came into existence in the early 21st century with generation and distribution applications. The growth of PV worldwide is shown in Figure 1.4. Till the year 2015, 30 countries generated electric power and distributed it at a low cost of electricity, which is quite less than or equal to the energy supplied from the grid. The United State will be on top in solar market. Next will be Japan and then Germany. At Present China is the world largest manufacturer of solar PV panels in manufacturing PV module and

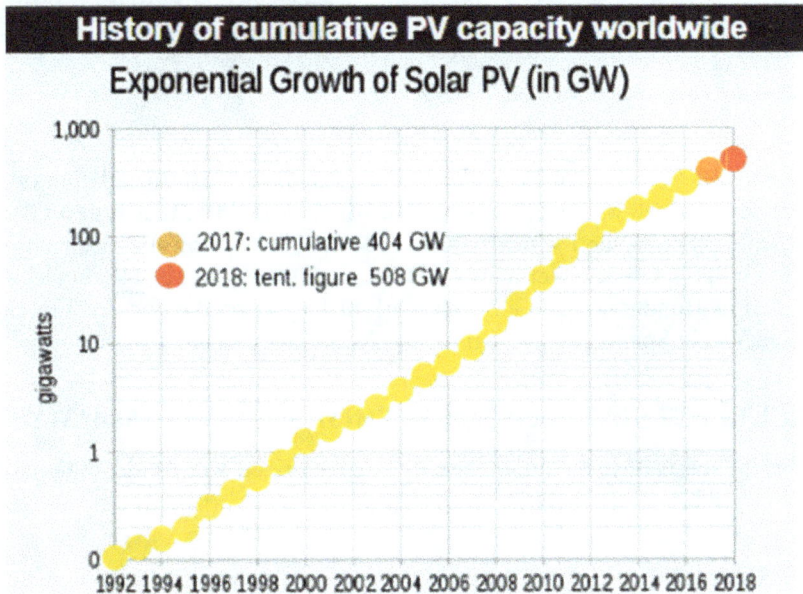

FIGURE 1.3 Exponential growth in solar photovoltaic energy.

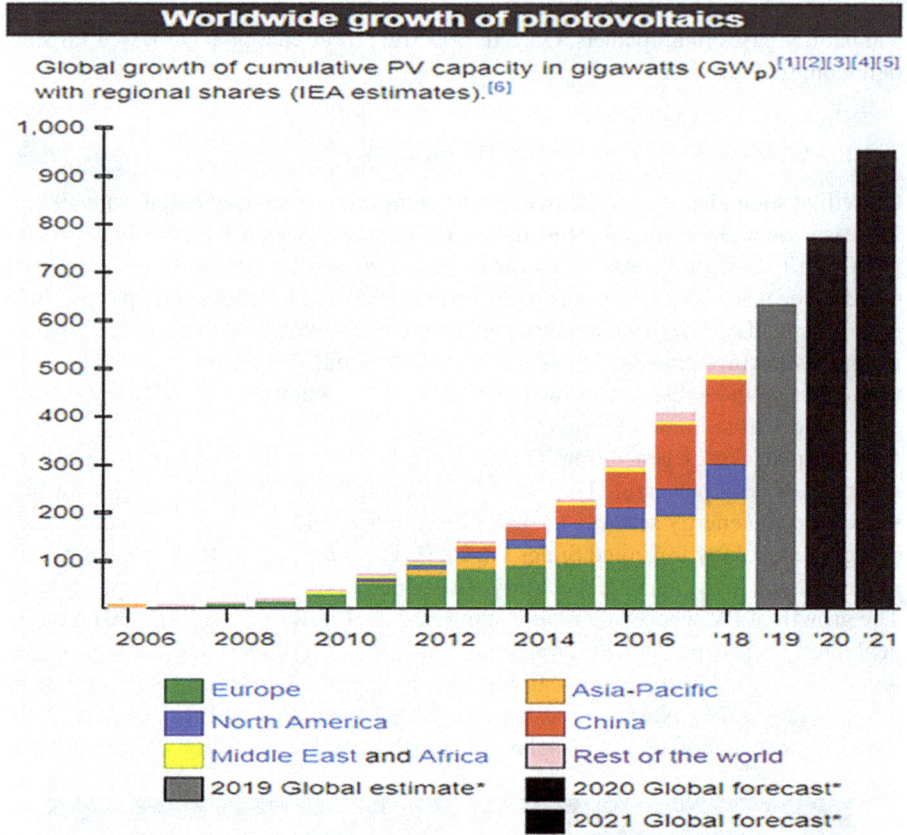

FIGURE 1.4   Growth of photovoltaic energy worldwide.

is also a leading country in electricity generation since 1950. Installation of solar PV globally reached about 512 GW in 2015 including 35%, that is 180 GW was utility scale plant. Solar PV power provided 7%–8% of domestic electricity consumption in Italy, Greece, Germany and Chile in 2018. A total of 14% of electricity generation in Honduras made a bench mark of largest production, 11% in United Kingdom, 4% in Spain and, approximately, 2.55% in India and China.

1.3   CURRENT STATUS OF SOLAR PHOTOVOLTAIC CAPACITY

The current status of installation capacity is categorized worldwide, regional-wise and country-wise.

1.3.1   WORLDWIDE

PV installation capacity was enhanced by 95 GW in the year 2017. The total installation capacity crossed 401 GW the same year, supplying 2.1% of the total power

TABLE 1.1

Photovoltaic Stations with the Maximum Installation Capacity across the World

Year	Name of Photovoltaic Power Station	Country	Capacity (MW)
1982	Lugo	United States	1
1985	Carrisa Plain	United States	5.6
2005	Bavaria Solarpark (Mühlhausen)	Germany	6.3
2006	Erlasee Solar Park	Germany	11.4
2008	Olmedilla Photovoltaic Park	Spain	60
2010	Sarnia Photovoltaic Power Plant	Canada	97
2011	Huanghe Hydropower Golmud Solar Park	China	200
2012	Agua Caliente Solar Project	United States	290
2014	Topaz Solar Farm	United States	550
2015	Longyangxia Dam Solar Park	China	850
2016	Tengger Desert Solar Park	China	1547
2019	Pavagada Solar Park	India	2050
2020	Bhadla Solar Park	India	2245

consumption globally [7, 8]. Table 1.1 shows the PV stations with largest capacity worldwide from 1982 to 2020 with its generation capacity [9].

1.3.2 REGIONAL

Among regions, Asia is one of the rapidly growing regions with 75% of the global installation capacity. In 2017, China itself is producing more than half of the worldwide consumption. Total installation capacity of the Asian region is 401 GW [7]. Europe declined to 28% global capacity, America 19% and Middle East 2%. Installation capacity of the European Union was twice or more than China and 25% more than the United States.

However, with respect to per capita installation, the European Union has more than twice the capacity compared to China and 25% more than the United States [10]. European fulfilled 3.5% electricity demand and 7% peak electricity demand by solar PV power in 2014 [9].

1.3.3 COUNTRIES AND TERRITORIES

Solar power plant was installed as an alternative conventional energy source in many countries as well as territories. Table 1.2 shows the different countries and territories with their solar PV capacity.

China is the world's leading country in the generation of solar power with an installation capacity more than 200 GW at the end of the year 2019 [11]. It also has the biggest market for solar PV and solar thermal energy. In the past five years, more than half of PV was marketed by China with the help of suitable policy makers,

TABLE 1.2

Solar Photovoltaic Capacity (MW) Installed in Different Countries and Territories [9]

Country	2016		2017		2018		2019		2020	
	New	Total	New	Total	New	Total	New	Total	New	Total
China	34,540	78,070	53,000	131,000	45,000	175,018	30,100	204,700	49,655	254,355
European Union	101,433		107,150	8,300	115,234	16,000	134,129	18,788	152,917	101,433
The United States	14,730	40,300	10,600	51,000	10,600	53,184	13,300	60,682	14,890	75,572
Japan	8,600	42,750	7,000	49,000	6,500	55,500	7,000	63,000	4,000	67,000
Germany	1,520	41,220	1,800	42,000	3,000	45,930	3,900	49,200	4,583	53,783
India	3,970	9,010	9,100	18,300	10,800	26,869	9,900	35,089	4,122	39,211
Italy	373	19,279	409	19,700		20,120	600	20,800	800	21,600
Australia	839	5,900	1,250	7,200	3,800	11,300	3,700	15,928	1,699	17,627
Vietnam		6		9		106	4,800	5,695	10,909	16,504
South Korea	850	4,350	1,200	5,600	2,000	7,862	3,100	11,200	3,375	14,575
Spain		4,669		4,688		4,707		8,711	5,378	14,089
United Kingdom	1,970	11,630	900	12,700		13,108	233	13,346	177	13,563
France	559	7,130	875	8,000		9,483	900	9,900	1,833	11,733
The Netherlands	525	2,100	853	2,900	1,300	4,150		6,725	3,488	10,213
Brazil			900	1,100		2,413	2,138	4,595	3,145	7,881

industries and the subsidies by the government help in reducing the solar power cost. In the field of solar water heating, China is accountable for generating 70% total world energy requirement capacity. China has its future mission to hit the target of solar capacity of 1300 GW by the end of 2050 [11].

Japan is on second position in solar market, expanding solar power since 1990s. Installing 8 GW solar powers in 2017, it has reached over 50 GW cumulative installation capacities by the end of the same year [9]. Now the country is supplying 2.5% of electricity demand annually.

India holds third position in the world in solar power. In 2017, India made a record of 9255 MW with another project of 9627 MW which was under development [11]. National Solar Mission was launched in 2010 to produce 20 GW power till 2022 under the action plan on Climatic Change. This target was achieved in 2018 only, that is four years before the target date. After achieving 20 GW, Prime Minister Shri Narendra Modi was motivated and enhanced the solar capacity to 100 GW and overall renewable capacity to 175 GW by 2022.

1.4 INTERNATIONAL CLIMATE POLICIES

Climate change is a major global challenge for the 21st century. The earth's surface temperature is continuously increasing due to the use of fossil fuels releasing CO_2 and other gasses into the environment resulting in global warming. The Intergovernmental Panel on Climate Change (IPCC) reported the finding of the global climate change research in the Fourth Assessment Report held in 2007 [12, 13]. During its Fifth Assessment Report in 2014, the conclusion of the research was the same as the previous report, that is "to keep global warming below 2°C, emissions of carbon dioxide (CO2) and other greenhouse gases (GHGs) must be halved by 2050".

1.4.1 INTERNATIONAL EFFORTS FOR CLIMATE CHANGE

In 1992, an agreement was made by the international community under UN Framework Convention on Climatic Change (UNFCCC) to preserve the environmental concentration of CO_2 and other greenhouse gases at a certain magnitude to prevent further dangerous interference with the climate system [12]. In their next conference, UNFCCC made emphasis on reducing greenhouse gas emission for industrialized countries. This agreement was reinforced in 2005 and the undertaking was committed to being effective till 2020.

1.4.2 FRAMEWORK CONVENTION ON CLIMATE CHANGE

Conference of the Parties (COP21) in 2015, Paris made an agreement with three major goals for all countries [12]. First was to limit global warming to less than 2°C as compared ideally to 1.5°C at the pre-industrial level. This goal would lower the risks on health issues and impacts of climate change on the environment. Second goal was to adapt the climatic change and lower the emission in order to prevent food production from risk. Last was to align the finance flow and ensure the reduction of

Cumulative energy-related carbon emissions (Gt CO_2)

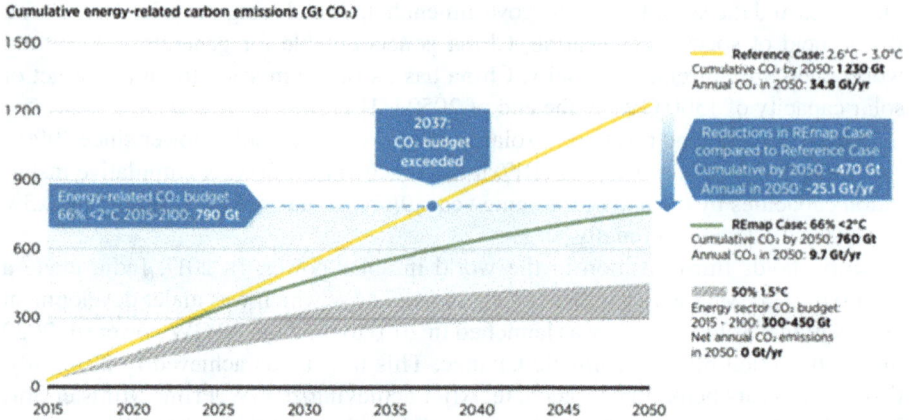

FIGURE 1.5 Budget till 2050 for reducing CO_2 emissions globally [14].

greenhouse gas emission and more durability towards climate change. This agreement was enforced on November 4, 2016 [12, 13] national climate policy. Germany and EU maintained the above-mentioned agreement limiting the global warming below 2°C with pre-industrial times. The German government forced for organizing annual Petersburg Climate Dialogue [13]. In 2009, Federal Chancellor, Angela Merkel, launched the Petersburg Climate Dialogue which reveals the climate negotiation in Copenhagen [13]. This was in-between climate summit, bringing environment ministers together from developing, developed and newly industrializing countries for open discussion. This open discussion was to achieve faster progress in international climate negotiation. The goal to achieve environment temperature rise below 2°C was a big issue [14]. Overall emissions are to be reduced further by 470 Gigaton (Gt) till 2050 [14] compared to current and planned policies (business-as-usual) to hit the target. The mission to reduce CO_2 below 2°C related to energy globally budget is shown in Figure 1.5 [14].

1.4.3 NATIONALLY DETERMINED CONTRIBUTIONS

Australia submitted an emission reduction policy which was known as NDC under the Paris Agreement Act [15]. The first NDC is available on the UNFCCC platform which includes the following points [15]:

  i. 2015 NDC includes commitment for the reduction of emissions from 26% to 28% below that was in 2005, by the end of 2030.
 ii. 2020 NDC includes confirmation of the target set in 2015, and amendment of new plans and operations.
iii. 2021 NDC includes bond for 100% free carbon emission by the end of 2050.

By the end of 2025, Austria will submit next NDC to UNFCCC which will include the target set till 2030 [13].

1.5   INTERNATIONAL SOLAR ALLIANCE

ISA was launched during Paris Climate Conference, 2015, to upgrade solar technologies especially for those countries having high solar resources but under-developed electricity access [15]. Since Austria was the founder of ISA, it promised to share its expertise regarding solar technology and generation of power via providing training and webinars [15]. The founder of ISA also promised to provide library of tools and resources for policy development through Clean Energy Solution Centre [15].

1.5.1   POWER SYSTEM TRANSFORMATION

Many countries in the world are moving towards the electricity system through renewable energy and other low carbon emission energy resources to fulfil the economic, environment and reliability goals [14]. In achieving this goal government, research and educational institutes are facing challenges in gaining knowledge and applying this elegant knowledge to transform the power system [16, 17]. To overcome these challenges and barriers towards different countries requires a global consortium, that is Global Power System Transformation (G-PST) Consortium [16, 17]. The G-PST Consortium provides coordinated, holistic support and knowledge infusion for system operators pursuing clean energy transitions, including performing cutting-edge research; providing implementation support for world-class engineering and operational solutions; supporting workforce development; building and disseminating open-access data and tools; and accelerating localized technology adoption, standards development and testing programs [16, 17]. There are five areas where this Consortium will take action while reinforcing the following initiatives:

i. **System Operator Research and Peer Learning and Presentation**
   Application-based research should be done to produce best solutions to the system operations and developing awareness through learning.
ii. **System Operator Technical Assistance**
   Training should be given to technical staff through workshop.
iii. **Foundational Workforce Development**
   Curriculum at university level should be changed to create manifold workforce for future and developing technical skills to the system operators.
iv. **Localized Technology Adoption Support**
   Every country should adapt modern innovations of power system through testing programs and skilled development workshops.
v. **Open Data and Tools**
   With the help of tools and data, various plans, operations and real-time systems can be monitored.

This transition took place with speed globally with the increase in flexible renewable energy (FRE), inverter-based resources (IBRs) and distributed energy resources (DERs).

1.5.2 POWER SYSTEM OPERATOR SURVEY TOWARDS GLOBAL ENERGY TRANSFORMATION

IEEE SA and the IEEE Power and Energy Society are working together under the Global Power Systems Transformation Consortium (G-PSTC) to evaluate their opinions towards next-generation energy (e.g. renewables, energy storage, DER, energy efficiency and grid modernization, etc.) to fulfil their basic needs [18]. The survey was done in two parts [18]:

i. Priorities towards system operator technical strategy.
ii. Enabling key standards and their uses.

In this survey, 39 responses were about the query of the system operation with 78 respondents. The representation of the global respondents was: the United States (16 responses, i.e. 41.9%), Asia Pacific (14 responses, i.e. 35.9%), Europe (5 responses, i.e. 12.8%), Canada (2 responses, i.e. 5.1%) and Latin America (2 responses, i.e. 5.1%) [18].

1.5.3 STATUS OF POWER SYSTEM TRANSFORMATION

Transformation of power system means power system flexibility. In other words, we can say that energy transition to boost system flexibility is not simple. Power system flexibility encompasses all relevant characteristics of a power system that facilitates the reliable and cost-effective management of variability and uncertainty in both supply and demand. It is a lively process, occurring in different places with different installation capacities and at different speeds globally. This change in power system has diverse pilots, that is driving new technologies, policy making goals, social economics change, financial goals and business development techniques [19]. For customer satisfaction it is required to design qualitative and quantitative policies which can help them in implementing legal document and infrastructure for developing the utility resources. This exercise will be beneficial toward rapid power transformation and will retain for decades. Investment is flowing increasingly not just towards new generation technologies like renewable energy and cleaner conventional generation but also towards an ecosystem of smarter grids, energy efficiency technologies, demand-side flexibility, storage, electric vehicles and integrated heating and cooling systems [19]. These changes can be seen globally today. Crystal clear vision of transformation of the RE power system is still limited.

Alteration in power system transformation is visible in 11 areas [19], which include Environmental Stewardship, Transmission Systems, Distribution Systems, Transmission-Distribution System Interface, Finance, Markets, Pricing, and Cost Allocation, Static and Dynamic Load, Flexible Generation, Integration with Heating and Cooling and Integration with Transport, Energy Storage, Microgrid [19].

1.5.3.1 Environmental Stewardship

Environmental steward is the step taken by the individual or group with the motive to protect and take care of the environment from the environmental hazard. Keeping in view about the climatic change, quality of environmental gas/air, water death, ecosystem challenges along with low carbon emission technologies, a new boarder electricity plan towards energy sector is set [20]. These electricity planning process and policy plans will help us to proceed towards renewable power energy sector and fulfil the standards and regulation of the environmental goals [21].

1.5.3.2 Transmission Systems

With an emerging trend in the field of renewable energy, both wind and solar energy move parallel with the growth of transmission infrastructure. The installation location of these energy sites is usually far away from consumers or transmission sections. RE can provide better use of generation and transmission resources with well-organized new transmission lines [22]. Reactive and proactive approaches are the two best ways for expansion of transmission section. Reactive approach relates to the commitment of RE projects while proactive approach relates to the purpose of guiding the efficient growth of the power system [22].

1.5.3.3 Distribution Systems

The distribution system is the last stage of an electric power system. It is located near or inside a village, town or industrial area [23] and distributes electricity to the consumer received from the transmission system [23]. The distribution network is fully capable of balancing the generation and control over flexible and dynamic load. It is also engaged in two-way power flow, that is generation distribution and storage. The distribution network interacts and keeps control on both side operators, that is distribution system operator and transmission system operator [19]. Distribution system operation will monitor and collect the data and model their distribution system for further use.

1.5.3.4 Transmission–Distribution System Interface

The electric transmission and distribution systems are two different power systems at different locations far away from each other. Generation of electricity connects the transmission system for transporting the electricity to the distribution system for the customer to avail the electricity. The transmission–distribution interface becomes a junction of economic value, giving rise to a new electric market and technical control electric power system [23]. Relationship of distribution connected generators, PV and others to the wholesale market, and the data transfers and coordination [24].

1.5.3.5 Finance, Markets, Pricing and Cost Allocation

Finance is an important part of power system function. Market value, low cost of the electricity is evolving in support of renewable power system transformation. Innovation emerged in four different modes. These are as follows [19]:

i. Upcoming with new financial investment policy into power sector.
ii. Investing finance according to priority areas.
iii. Involving smart technologies to enhance power system efficiency and reducing carbon emission.
iv. Developing new market strategies to maintain the latest system efficiencies and providing power system flexibility.

In the past, power sector investments were dependent on the return of investments (ROI), set by the regulators. Finance gained by ROI by the governments and third-party project developers were having sufficient finance to invest on the expansion of power sector. Thus, new markets and financial innovations pop up with investments in clean and green energy market [19].

1.5.3.6 Static and Dynamic Load

Static load is power efficiency and dynamic load is future power demand. Both the loads are economical to utilize and control it. Various innovations are used to make the load more efficient and dynamic. These technologies make use of grid services and operation. Smart meters are used for fixing new prices of electricity for customer satisfaction. This results in load adjustments or time management response to system condition. These adjustments take initiative at customers' end or as signal price automated response. Thus, opening the sources and pathways for system planning, operation and investment [19].

1.5.3.7 Flexible Power Generation

Before renewable energy era, power system planning and operations were based on three types of energy generations, that is base-load, intermediate and peak [25]. These operations were planned according to the utilization of power plant and its cost throughout the year. The conventional base-load energy generators were coal which fulfilled the fixed amount of energy, considered as fixed source power system. Power generated from the conventional sources easily fulfils the electricity demand, while the solar and wind energy is dependent upon the weather conditions and is highly variable. The energy generation with low tariff is with wind and solar energies, but these power plants are never considered as base-load depending upon their duration, rather operate for some hours of the year and resemble as intermediate energy generation power plants. Figure 1.6 [19] shows the impact of conventional and transformed power systems.

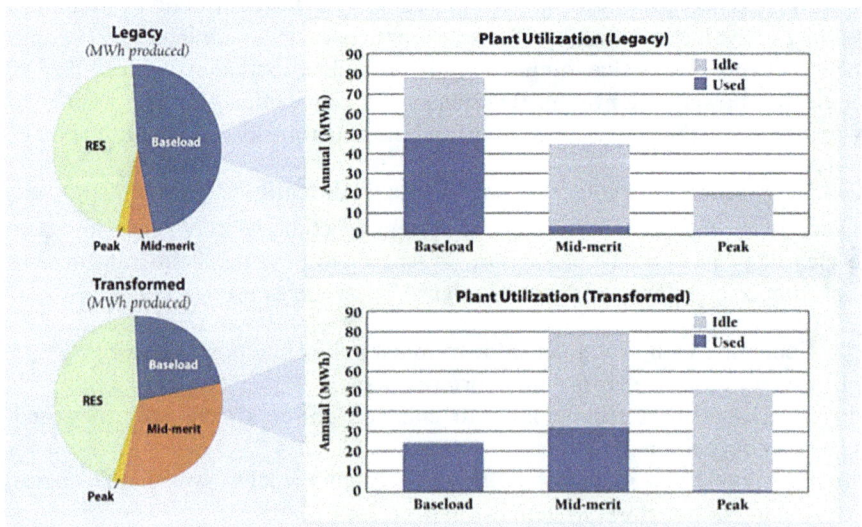

FIGURE 1.6   Impact of conventional and transformed power systems.

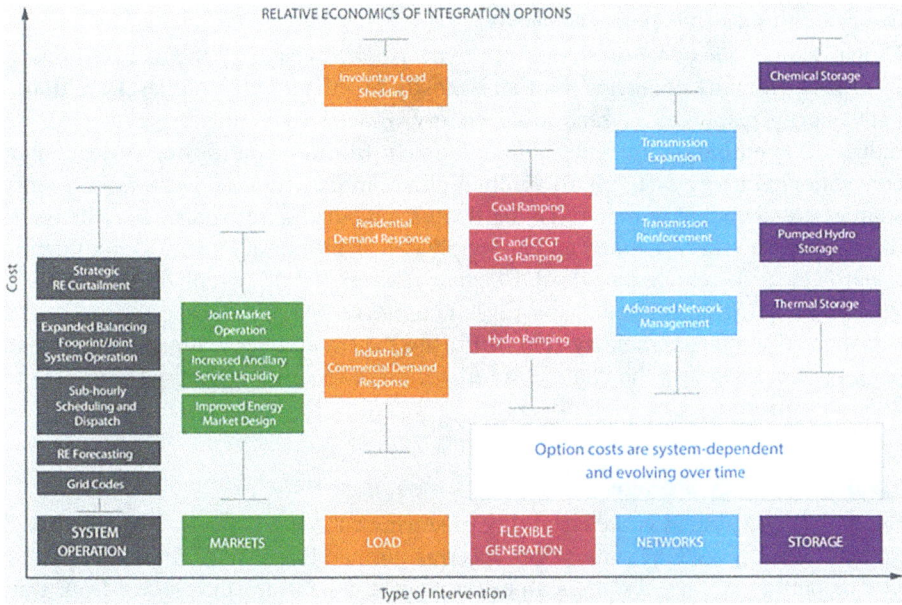

RELATIVE ECONOMICS OF INTEGRATION OPTIONS

		Involuntary Load Shedding			Chemical Storage
				Transmission Expansion	
		Residential Demand Response	Coal Ramping	Transmission Reinforcement	Pumped Hydro Storage
Strategic RE Curtailment			CT and CCGT Gas Ramping		
	Joint Market Operation			Advanced Network Management	Thermal Storage
Expanded Balancing Footprint/Joint System Operation	Increased Ancillary Service Liquidity	Industrial & Commercial Demand Response	Hydro Ramping		
Sub-hourly Scheduling and Dispatch	Improved Energy Market Design				
RE Forecasting			Option costs are system-dependent and evolving over time		
Grid Codes					
SYSTEM OPERATION	MARKETS	LOAD	FLEXIBLE GENERATION	NETWORKS	STORAGE

Type of Intervention

FIGURE 1.7 Flexibility in power system of variable renewable energy [26].

The weather conditions for renewable energies (solar and wind) are very challenging for making a balanced supply of electricity and its necessary demand. But flexible renewable energies have many economic and environmental advantages in the coming days, as shown in Figure 1.5. Interconnect grid will ease the problem of flexible demand. Interconnects in grids can also reduce the wholesale electricity price. One example of France is given, that is during 2016 and 2017, nuclear capacity was very less to fulfil the electricity need of the citizen. The country imported energy from Britain, Germany and their surrounding neighbours, reducing the need of gas power plant. This showed that interconnect is a big advantage and many nations are willing to adopt this technology. The flexibility in power system of variable renewable energy (VRE) is shown in Figure 1.7 [26].

1.5.3.8 Integration with Heating and Cooling

Energy system integration refers to associating numerous energy carriers and storage solutions with each other for strong, authentic and well-planned energy system. This involves heating, cooling, transport and export of electricity and gas. Strong benefits are involved with increased system flexibility with dispatch-able loads and stored energy. This flexibility allows adding more renewable energy with low cost and lesser peak system load. Broad integration with heating and cooling options contains electric power with RE power, renewable-based gases (including "green" hydrogen), use of sustainable bioenergy and the direct use of solar and geothermal heat. The use of green energy will make the future of the energy system free from carbon emission.

1.5.3.9 Integration with Transport

Globally, only 4% of RE is shared in the transport division by the end of the year 2015 [14]. In many countries' biofuels, mostly bioethanol and biodiesel are dominant energy resources in renewable energy. Electricity is the biggest option to reduce CO_2 emissions in renewable energy. In transforming power system, non-conventional energy system will surely reduce carbon emission. Renewable energy sources are abundant in nature having fluctuation outputs, so balance is to be made in between electricity demand and its supply [19]. Transportation sectors are moving towards electric vehicle or hybrid. Electric vehicles (EV) are replacing oil in light vehicle reducing carbon emission [27]. These EVs provide storage to integrate a transportation management system, fulfilling the time of generation with the load [27]. Figure 1.8 shows the change in utilization of renewable energy from 2015 to 2050 in transport sector globally [14].

1.5.3.10 Energy Storage

Energy storage system is the capability of storing energy and power and are able to maintain energy for long time before it is transformed into useful work.

The battery storage system is an example of it which can store power from watt to mega-watt.

The main advantages of energy storage system are availability of energy on time, storing large amount of energy and for transportation.

The current transformation of flexible power system across the world is a starting opportunity for many countries towards VRE, providing electricity services at the reduced cost along with the challenges towards climatic changes [28].

Excessive use of renewable energy is leading towards access to electricity, and storage is the major factor rising towards the renewable energy power system. Energy storage is necessary for integrating FRE, that is solar PV and wind or other sources of flexible power system in the weak grid areas [29]. Energy storage has changed the

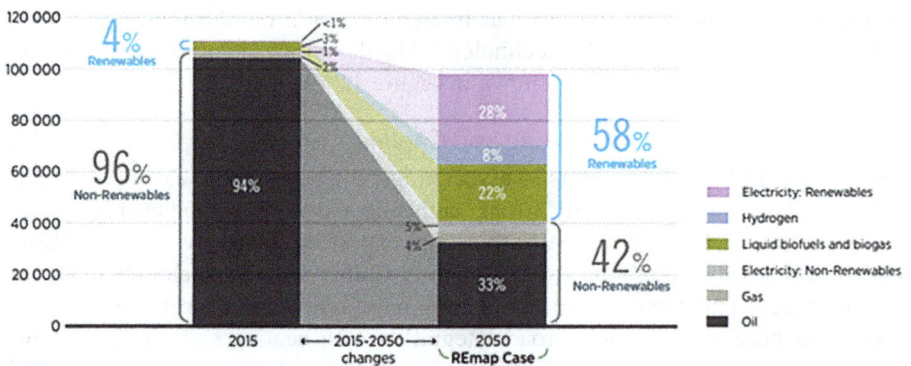

FIGURE 1.8 Renewable energy consumption in transport sector globally from 2015 to 2050.

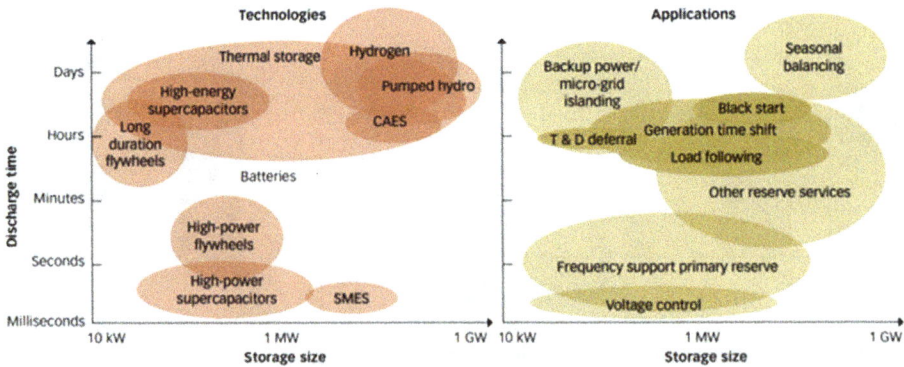

FIGURE 1.9   Energy storage technology with its storing capacity and applications.

universal use of electricity at the international level, reducing CO_2 in the environment. This results in excess of electricity in remote and rural areas [30]. The total storage capacity worldwide is approximately 200 GWh in 2019. The main storage technology used is pumped hydropower (mechanical), electrochemical (batteries) and thermal [29]. In hydropower pump, water is pumped, heated and the steam of hydroelectric generator is stored and released when required. A total of 91% was the electric storage capacity worldwide in 2019 using pumped hydro technology, 5% using electrochemical batteries and 3% with thermal storage technology, while hydrogen and compressed air technology were at lower portion [30]. Pumped hydropower energy system was used in power grids while thermal storage is widely used for cooling and heating buildings, thus reducing carbon emissions. Among all energy storage technologies, electrochemical energy storage is fast growing in the market of renewable energy but is still dominated by electrical vehicles as a consumer application. Figure 1.9 shows different energy storage technologies with its storage capacity and their applications [31]. Battery-based solutions are modular, easy to change, fast respond and reduced cost.

Energy storage at large scale is at grid level energy storage system. This is a necessary step taken for stabilizing the power generation, supply, distribution and its utilization at the customer end [31]. Small storage energy management is easy and modular giving rapid response towards flexible RE installation and its utility. Many other types of battery technologies are used; these are lead–acid, nickel–cadmium, nickel–metal hydride batteries. With the decreasing costs of lithium-ion batteries (LIB) [32], they are used as stationary Battery Energy Storage Systems (BESSs), especially for mobile applications [31]. According to IEA's World Energy Outlook 2020, under the Sustainable Development Scenario, the battery storage capacity can enhance from 6 GW in 2019 to 550 GW by the end of the year 2040 [31, 33].

According to IRENA 2019 report, by the end of 2030, the installation cost of the flow battery storage system will reduce by 66%, high temperature batteries by 50%–60%, flywheel by 35% and compressed air energy storage by 17% [31] as shown in Figure 1.10.

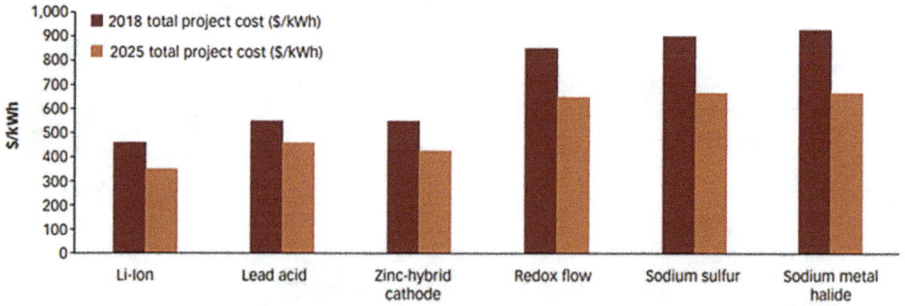

FIGURE 1.10   Energy storage technology and installation cost from 2018 to 2025 [30].

1.5.3.11 Microgrid

Microgrid is a grid-connected setup for flexible energy resources installed within the locality or outside for reliability, resilience and operational economics [34]. It is also referred to as a local group of interconnected loads and distribution within a specific region defining electrical boundaries. Microgrid can get connected or disconnected from the main grid as per the requirements and economic conditions. Renewable energy sources have fulfilled the requirement of energy demand, challenges of climatic changes and contribution towards the sustainable energy development. The integration of flexible energy system is carried out and the distribution in electricity is done through microgrid [33], exchanging information between the consumers and generation-distribution centres. In other words, the energy management system of microgrid maintains the information and controls the generation-distribution system and supplies electricity at a low operational cost [35].

In future, microgrid technology will be used globally, especially in Asia and the Pacific region and North America. Their installation capacity will grow five times from 2018 to 2027 [34], as shown in Figure 1.11. Microgrid is aiming towards infrastructure reliability, investments and achieving green and clean environment free from carbon emission. Residence of the community is the primary customer and

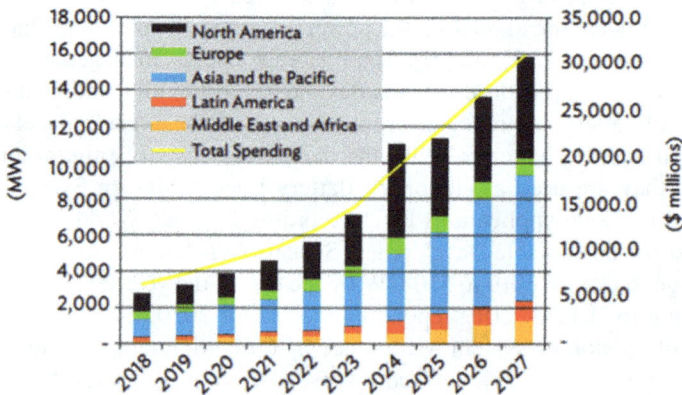

FIGURE 1.11   Installation capacity of microgrid from 2018 to 2027 [34].

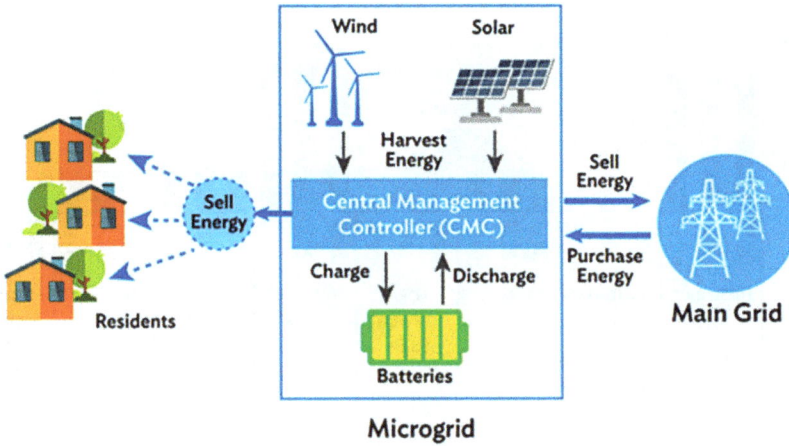

FIGURE 1.12   Microgrid operation [34].

their requirements are fulfilled with affordable cost and security. An overview of operation of the microgrid [34] is shown in Figure 1.12. A microgrid is a group of interconnected loads and DERs within clearly defined electrical boundaries that acts as a single controllable entity with respect to the grid. A microgrid can connect and disconnect from the grid to enable it to operate in both grid-connected or island mode. Advantages of microgrid are reliability, security, reduced electricity cost, clean and green energy which helps to maintain zero carbon emission, can help deploy more zero-emissions energy sources, make use of waste heat, reduce energy lost through transmission lines, help manage power supply and demand, and improve grid resilience to extreme weather.

1.6   GLOBAL ENERGY INVESTMENT

Global energy investment has grown to 1.9 trillion USD in 2021 [36, 37]. A 10% jump in the renewable energy investment in 2021 since 2020, getting out from the crisis level of COVID-19 pandemic [35], shifting towards RE electricity and moving far away from conventional fuel production. According to IEA report, renewable energy has attracted 70% of the total $530 million global energy investment in 2021 for new power generation capacity [36, 37]. Investments are done depending on the present capital spend in energy supply capacity (fuel production, power generation and energy infrastructure) and energy end-use and efficiency sectors (buildings, transport and industry) [36]. New enhanced technology in solar and wind has shown rapid improvements towards electricity generation, resulting four times better electricity production rather than same amount of dollars spend 10 years ago. Figure 1.13 shows global energy supply investment in different sectors of energy [36].

Apart from remarkable growth renewable energy, China has tremendous growth in coal-based power production in December 2020 [37]. Overcoming the risk and hindrances towards finance, climatic finance played an important role in bridging the financial gap and moving towards investment from private sectors to renewable

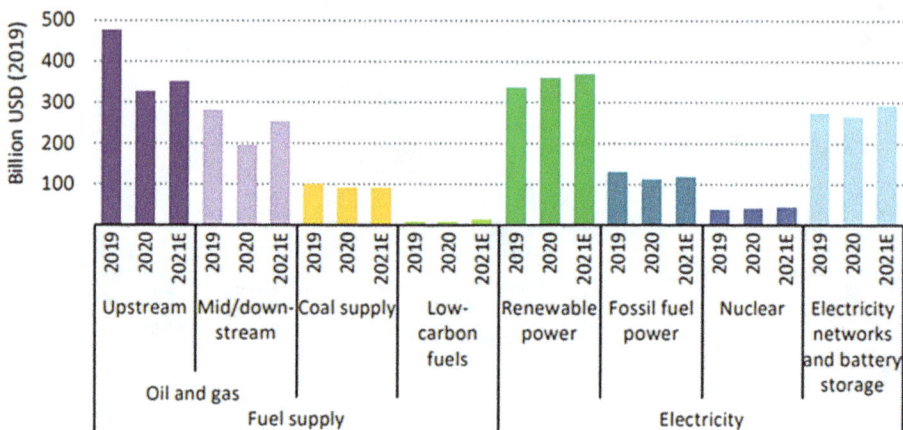

FIGURE 1.13   Global energy investment in different energy sectors [36].

energy power sectors [38]. Figure 1.13 shows annual investments in renewable energy using different technologies [38, 39].

1.6.1   INVESTMENT BY TECHNOLOGY

Solar PV and wind onshore were superior in renewable energy sector in the year 2017 and 2018, committing of spending 77% of the total finance [39, 40]. Huge investment in RE technologies is due to their appropriate policies, improvement in manufacturing technologies, progress in short time period and competition in the market. Investments by technology is shown in Figure 1.14 [38, 39]. From 2014 to

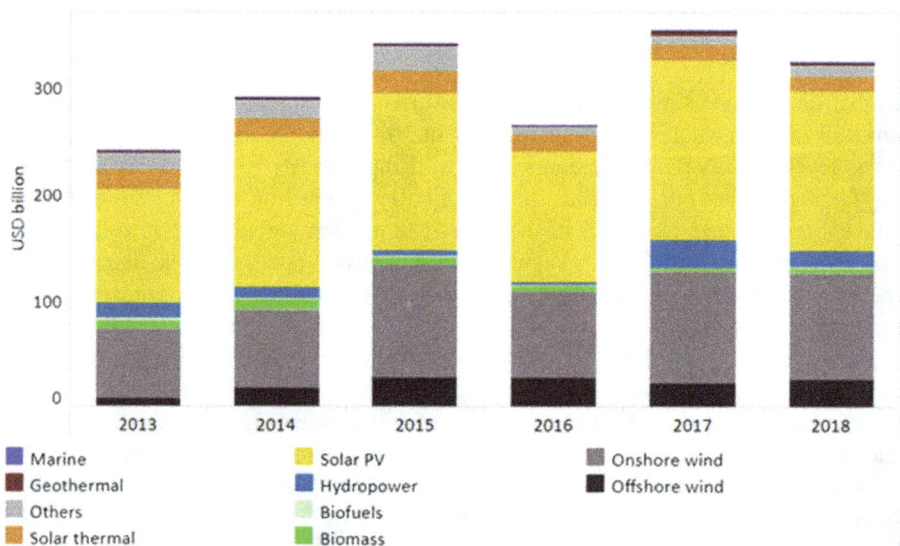

FIGURE 1.14   Investments by technology [38, 39].

2018, investment in offshore wind was alluring and having an average growth of USD 21 billion per year globally and played an important role in reducing CO_2 emissions in the environment.

1.6.2 INVESTMENT BY REGION

China was on the top in renewable energy (solar PV and wind) market, having 93% investment in renewable energy from 2013 to 2018, while Asia and the Pacific region invested approximately 32% of global energy finance commitment in the financial year 2017–2018 [39]. Continuous investment in renewable energy in the United States boosted up the number of countries of the Organization for Economic Co-operation and Development (OECD) located in the Americas [39]. Combining all regions in the financial year 2017–2018 enhanced 22% of global investment in RE with an amount of USD 82 billion peak in 2018. Western Europe also invested USD 51 billion in 2017–2018 [39], becoming 15% of the total investment in the RE sector. In 2017–2018, Asia has a down fall of 53% in renewable energy investment as compared to the year 2015–2016, which may be due to decrease in solar energy investment in Japan.

1.7 EMPLOYMENT IN RENEWABLE ENERGY

Global financial system is dependent on the strength of energy generation and distribution. Everyday life is completely dependent on energy, based on un-replenishable resources, mainly fossil fuels. These conventional energy resources are used at a faster rate than being developed, contributing to CO_2 emissions into the environment. Our dependency on electricity to deliver power for our homes, offices, hospitals and organizations requires an essential shift towards renewable and sustainable power. This transition of energy towards renewable energy power ought to produce new jobs, imparting millions of people global with access to sustainable electricity, in parallel, decreasing CO_2 emissions from the environment. Jobs are the foundation of financial and social improvement. To analyze the exact employment due to productive use of electricity is not an easy task. A survey was done and a review of 50 studies effecting the production of electricity was observed. The analysis of this study shows an increase in 25% in employment, 30% increase in remuneration and 7% increase in school enrolment [40]. They permit humans to earn an income and paint their life out of poverty to reap better livelihoods. Eighty one percent, that is, 163 million of the 200 million human beings are unemployed globally and 1.4 billion are self-employed but are at risk [41]. Accordingly, the advent of satisfactory job is a priority goal for growing countries and a fundamental factor of overseas help. However, creating new jobs is a complex mission, with considerable number of investment and sector-specific factors affecting the final success of policies and plans. NDCs have achieved huge attention from countries because of the 2016 Paris Agreement, upholding 189 countries till 2020 [42].

NDCs have committed and laid the foundation of maintaining the global temperature below 2°C or further below up to 1.5°C to diminish climatic change. Reducing GHG emissions through improving renewable strength era could also make contributions to job introduction and become a using force of countries' financial growth. Climate mitigation effort should be advantageous locally in addition to the worldwide blessings of the GHG emissions reductions and consequently, might be the catalyst for setting greater formidable NDC goals. Reducing GHG emissions through improving renewable strength era can also contribute in introducing new jobs and becoming a dynamic strength of countries' financial growth. According to the International Energy Agency (IEA), the RE area contributes to 42% of worldwide electricity-related CO_2 emissions and, simultaneously, RE electricity era is becoming inexpensive compared to coal and natural gas [41]. Under these circumstances, it is becoming essential to explore and estimate the RE employment opportunities of various decarbonization pathways and RE goals underneath the NDCs.

Transformative approach influenced the labour market and reshaped globally the sectors including movement towards climate trade, digitalization and globalization [43–45]. Energy transformation had an important role towards societies and economic challenges globally. Furthermore, the COVID-19 crisis is certain to exacerbate inequality, both inside and across nations [46, 47]. Crises of COVID-19 require new approaches towards these challenges and require immediate response from policy makers and private energy sectors investments [43].

The renewable power sector has hired 12 million people in 2020 [48]. Employment has grown continuously globally in solar PV, bioenergy, hydropower, and wind strength industries had been the most important employers. According to IRENA's 2021, renewable energy employment globally [49] is estimated as shown in Figure 1.15. There are many factors affecting the employment trends. These factors include the speed at which renewable energy equipment are manufactured, installed and are used [49], as shown in Figure 1.16. Costs of solar and wind are reducing gradually, with proper investment moving towards larger job placements. Further investment in RE enhances new job opportunities and growing labour production.

Policy advisers and aid remain indispensable for setting up universal renewable energy roadmaps, riding ambition and encouraging the adoption of crystal-clear steady policies for feed-in tariffs, auctions, tax incentives, subsidies, permitting

FIGURE 1.15 Global employment in renewable energy sector.

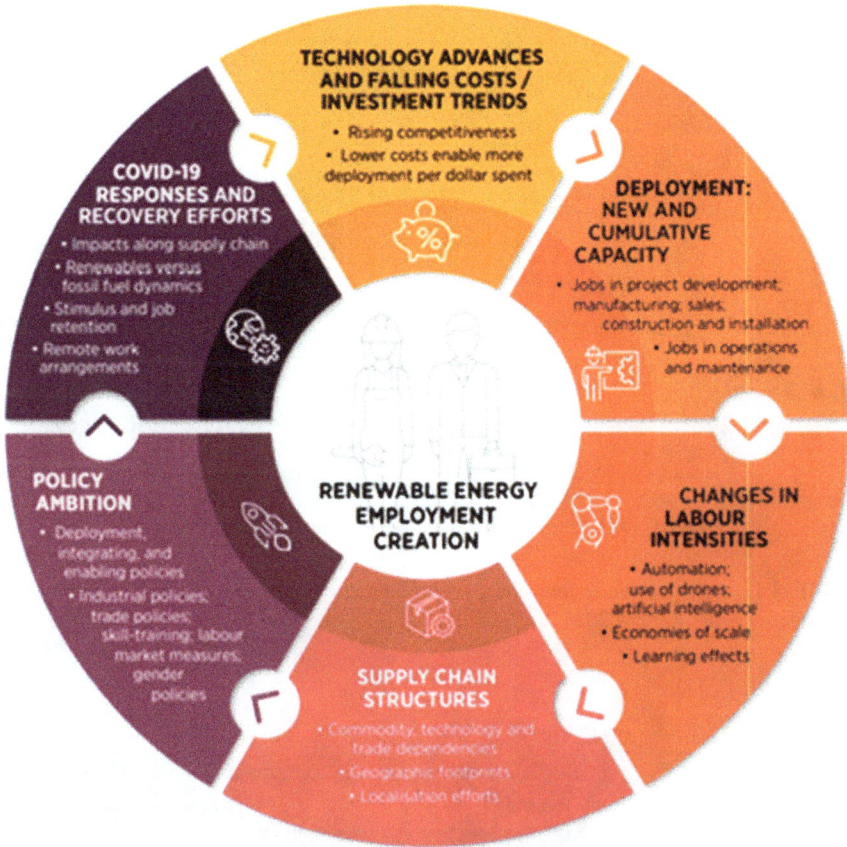

FIGURE 1.16 Factors influencing renewable energy employment [49].

techniques and other regulations [49]. The topographical footstep of renewable energy employment depends on technical policies, industrialized policies, skilful leader's consumer requirements and on the vigorousness of national and regional installation market, strength of production and distribution in the individual countries. Producing skilled team of engineers, scientists, project managers, technicians, electricians, welders, pipefitters, truck drivers, crane operators and many more are also necessary.

Figure 1.17 shows the continuous growth towards employment in RE sector and is expected to reach 42 million jobs by the end of 2050. The IEA has developed a sustainable recovery plan for generation and grid infrastructure energy efficiency buildings and industry sectors; manufacturing of vehicles and other transport measures as well as fuel production, renewables, recycling and innovation [50], as shown in Figure 1.18. Transition towards renewable energy will reduce jobs in fossil fuel sector while enhancing job in RE increasing over 15 million jobs worldwide by 2030 [50].

The planning and transformation energy scenario regarding jobs in energy sector is shown in Figure 1.19.

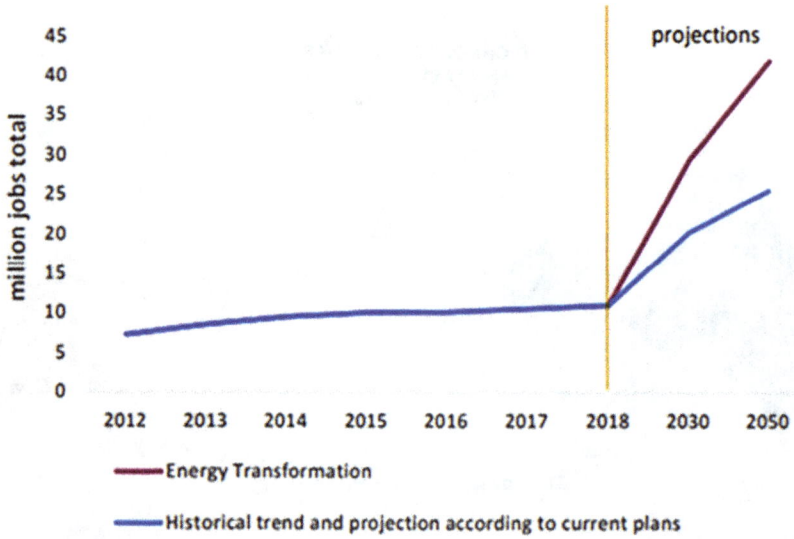

FIGURE 1.17   Global renewable energy employment estimated till 2018 and projected employments from 2030 to 2050.

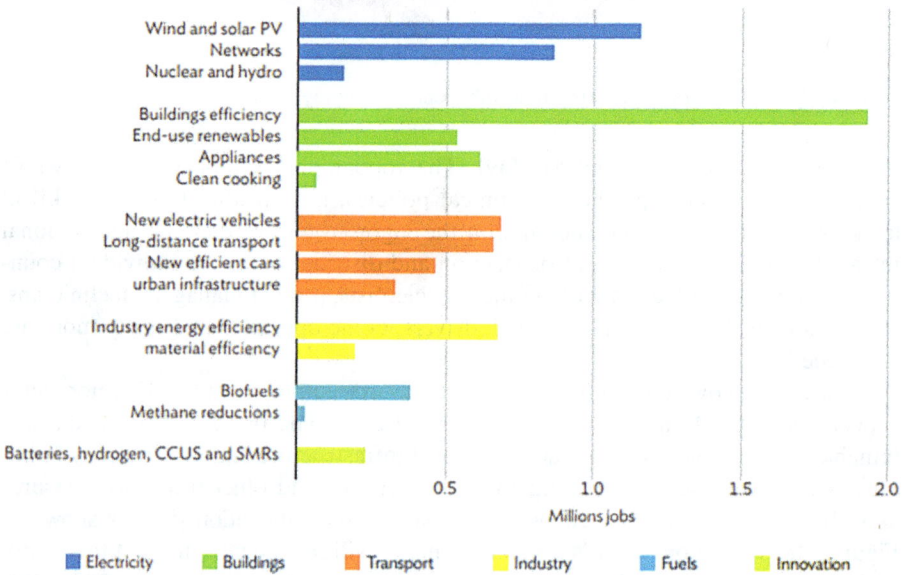

FIGURE 1.18   International Energy Agency's recovery plan of annual construction and manufacturing jobs [50].

Jobs in millions (2020)

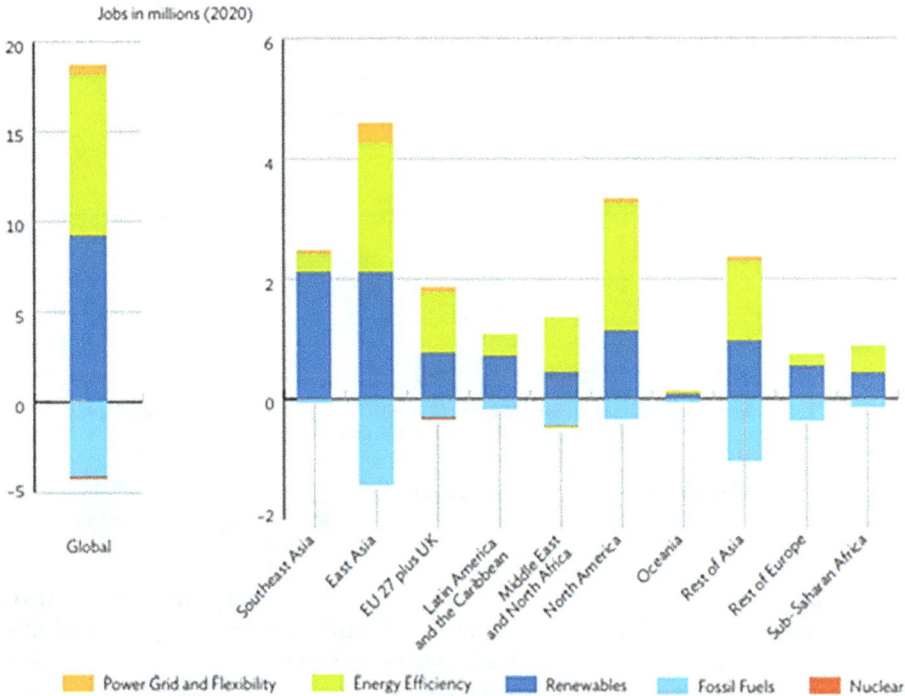

FIGURE 1.19   Jobs in energy sector in 2030 (planning and transforming energy scenario) [48].

1.8   IMPACT OF COVID-19 ON ENERGY

In the starting year of 2020, COVID-19 broke out at the international level leading to the death of many people and affecting the economy worldwide. Statistical analysis of COVID-19 was done by Johns Hopkins University on June 28, 2021 and declared 181,102,393 people infected and 3,923,132 people passed away globally [45]. Three countries namely the United States, India and Brazil had largest number of infected people and deaths globally. In order to stop the spreading of the virus, many countries made restrictions. Educational institutes were closed, partial or full lockdowns and work from residence was advised. These restrictions affected transportation, catering, entertainment, medical care, manufacturing, actual property and many other components. It is easy to portray that the foundation of the global economic system is shattered for most of 2020 and 2021 [49–50]. Worldwide population and energy sectors are the biggest sufferers during the COVID-19 pandemic.

Employment, together with inside the power sector, has been deeply affected by repeated lockdowns and different restrictions which placed stress on deliver chains and constrained financial activity. Many scholars and researchers have studied and discussed the influence of COVID-19 pandemic on energy market related to energy

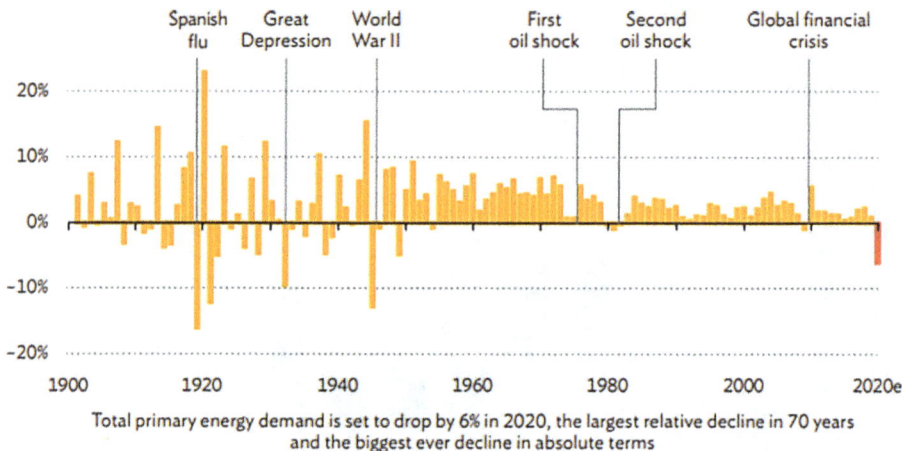

Total primary energy demand is set to drop by 6% in 2020, the largest relative decline in 70 years and the biggest ever decline in absolute terms

FIGURE 1.20 Change in energy demand globally.

consumption, policies, climate and the surrounding environment. According to IEA, in Asia and developing Africa, approximately 110 million people were economically affected and were unable to pay the basic electricity service.

Apart from severe global health and economic crisis during COVID-19, it opens the doors for a green, clean and sustainable energy. The IEA has declared 6% decline in primary energy demand globally by the influence of COVID-19 pandemic [50] as shown in Figure 1.20 and also 7% reduction in global carbon emissions in 2020 [50] as shown in Figure 1.21 (indicated with bold red line), but at a particular financial cost. The decline in energy demand in 2020 is the biggest decline during the past seven decades [51]. These financial losses can be recovered with the help of funds, investments, quality energy infrastructure and producing new job at the same time.

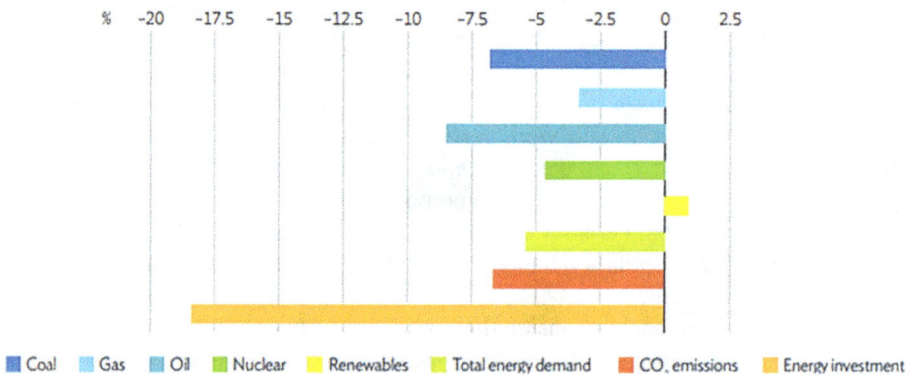

FIGURE 1.21 Reduction in carbon emission in 2020.

1.9 CONCLUSION AND FUTURE SCOPE

COVID-19 has shattered the global economy. To mitigate and to adapt to global climatic change, several plans and policies should be implemented to achieve low carbon emission. RE resources such as wind, solar, water and biomass generate energy with pollution-free environment. Stakeholders can grab the advantage of the emerging technology and new policies in creating innovative devices for energy applications. Investments should be done in generation, transmission and distribution of renewable energy power and creating jobs for millions of people for the recovery at the same time.

Local employment, better health, job opportunities, job creation, consumer choice, improvement of life standard, social bonds creation, income development, demographic impacts, social bonds creation and community development can be achieved by the proper usage of renewable energy system. Along with the benefits of renewable energy resources, these are complex to install and are local, environmental and condition sensitive. Their forecasting, execution and planning require more consideration and knowledge as compared to other projects.

Renewable energy will progress in the coming days, cutting down fossil fuels energy and reducing carbon emissions. Many specialists said that the environment and overall human society will achieve transition towards 100% RE. There exists an amazing consent across all regions that renewable energy power will dominate all other traditional energy in the future. Above 85% of the experts interviewed said that the renewable power share will double by 2050. This is a tremendous change and holds the importance of renewable power industries as principal factors for decarbonization of the energy sector. In renewable energy, it is predicted that solar energy installation capacity will grow by 600 GW globally by 2024 and total renewable electricity is predicted to grow by 1200 GW by 2024, which is equivalent to the total electricity capacity of the United States.

Fatih Birol, executive director, IEA, said that it is a crucial time for renewable energy as solar and wind being the heart of transformation of RE occurring globally. This transformation is also important for efforts to tackle greenhouse gas emissions, reduce air pollution, expanding energy access and making green and clean environment. In the coming decade, there will be an increasing trend of the renewable energy sector particularly in the transportation sector, in which the solar energy-based systems and biofuel energy will increase its share in the RE market. Other applications of solar PV technology will be in traction, charging stations, petrol pumps and roof top systems in industrial and residential installations. The RE share will increase its share in space applications, robotics and medical applications.

REFERENCES

1. Keywan Riahi et al. (2012), Chapter 17: Global Energy Assessment. Energy Pathways for Sustainable Development, Cambridge University Press, http://pure.iiasa.ac.at/id/eprint/10065/1/GEA%20Chapter%2017%20Energy%20Pathways%20for%20Sustainable%20Development.pdf
2. Ritchie Hannah, Roser Max, Pablo Rosado (2020), Renewable Energy, Our World in Data, https://ourworldindata.org/renewable-energy#renewable-energy-generation

3. Acciona (2020), Sustainable Development, https://www.acciona.com/sustainable-development/

4. Venture Solar (16 February 2017), The Importance of Solar Energy, https://venturesolar.com/blog/the-importance-of-solar-energy/

5. Manton, Amanda (2015) "Solar Energy: A Renewable Resource with Global Importance," ESSAI: Vol. 13, Article 26. Available at: https://dc.cod.edu/essai/vol13/iss1/26

6. M Alam et al. (2013), Solar Energy: Trends and Enabling Technologies. Renewable and Sustainable Energy Reviews 19: 555–564.

7. Growth of photovoltaic (20 November 2021), https://en.wikipedia.org/wiki/Growth_of_photovoltaics

8. Whiteman Adrian, IRENA (2021), Renewable Capacity Statistics 2021, International Renewable Energy Agency (IRENA), Abu Dhabi, https://www.irena.org/-/media/Files/IRENA/Agency/Publication/2021/Apr/IRENA_RE_Capacity_Statistics_2021.pdf

9. Snapshot of Global PV 1992–2014 (30 March 2015), International Energy Agency — Photovoltaic Power Systems Programme, 30 March 2015. Archived from the original on 30 March 2015, www.iea-pvps.org/index.php?id=32

10. International Energy Agency (7 December 2021) https://en.wikipedia.org/wiki/International_Energy_Agency

11. Solar Power by Country (8 January 2022), https://en.wikipedia.org/wiki/Solar_power_by_country

12. Federal Ministry of the Environment, Nature Conservation, Nuclear Safety and Consumer Protection, International Climate Change, Accessed at: https://www.bmuv.de/fileadmin/Daten_BMU/Pools/Broschueren/fortschrittsbericht_anpassung_klimawandel_en_bf.pdf

13. Global Climate Change Policies, European Environment Agency (10 January 2022), https://www.eea.europa.eu/themes/climate-change-adaptation/global-climate-change-policies

14. Adnan Z. Amin (2018), Global Energy Transformation: A Roadmap to 2050, https://www.irena.org/-/media/Files/IRENA/Agency/Publication/2019/Apr/IRENA_Global_Energy_Transformation_2019.pdf?rev=6ea97044a1274c6c8ffe4a116ab17b8f

15. Australian Government, Department of Foreign Affairs and Trade (2011), International Cooperation on Climate Change, https://www.dfat.gov.au/international-relations/themes/climate-change/international-cooperation-on-climate-change

16. Global Power System Transformation Consortium (September 2020), https://globalpst.org/wp-content/uploads/GPST_Fact_Sheet.pdf

17. M. O'Malley et al. (2021), Enabling Power System Transformation Globally: A System Operator Research Agenda for Bulk Power issue, IEEE Power & Energy Magazine, DOI: 10.1109/MPE.2021.3104078

18. IEEE Standards Association and IEEE Power and Energy Society (11 January 2021), Power System Operator Survey Toward Global Energy Transformation Summary Report, https://resourcecenter.ieee-pes.org/publications/white-papers/PES_TP_WP_GPSTC_011121.html

19. Mackay Miller et al. (May 2015), Status Report on Power System Transformation: A 21st Century Power Partnership Report, National Renewable Energy Laboratory, https://www.nrel.gov/docs/fy15osti/63366.pdf

20. Nathan J. Bennett et al. (29 December 2017), Environmental Stewardship: A Conceptual Review and Analytical Framework, Environmental Management, 61, 597–614. https://doi.org/10.1007/s00267-017-0993-2

21. Sachi Arakawa et al. (16 March 2018), Environmental Stewardship, In: Handbook of Engaged Sustainability, Springer, DOI: 10.1007/978-3-319-53121-2_37-1

22. Marcelino Madrigal, Steven Stoft (2012), Transmission Expansion for Renewable Energy Scale-Up, International Bank for Reconstruction and Development/The World Bank, DOI: 10.1596/978-0-8213-9598-1

23. Amirhossein Sajadi et al (2016), Transmission System Planning for Integration of Renewable Electricity Generation Units, November 2016, Conference: IEEE Energy Tech. At: Cleveland, Ohio, USA, https://www.researchgate.net/publication/310480331_ Transmission_System_Planning_for_Integration_of_Renewable_Electricity_ Generation_Units

24. The Transmission-Distribution Interface (25 June, 2018), Presented by the Electricity Advisory Committee, https://www.energy.gov/sites/default/files/2018/06/f53/EAC_ Transmission-Distribution%20Interface%20%28June%202018%29.pdf

25. Yony Yen (1 November, 2019), Flexibility (Power System), Renewable Energy Power Plants, https://energypedia.info/wiki/Flexibility_(Power_System)#Renewable_Energy_ Power_Plant

26. J. Cochran, M. Milligan, J. Katz (May 2015), Sources of Operational Flexibility, National Renewable Energy Laboratory, https://www.nrel.gov/docs/fy15osti/ 63039.pdf

27. Henrik Lunda, Willett Kempton (September 2008), Integration of Renewable Energy Into the Transport and Electricity Sectors Through V2G, Energy Policy, 36(9):3578–3587.

28. Kapil Thukral, Priyantha Wijayatunga, Susumu Yoneoka (November 2017), Increasing Penetration of Variable Renewable Energy: Lessons for Asia and the Pacific, Asian Development Bank, https://www.adb.org/sites/default/files/evaluation-document/382641/ files/ied-wp-variable-renewable-energy.pdf

29. Xiayue Fan et al. (8 January 2020), Battery Technologies for Grid Level Large Scale Electrical Energy Storage, Transactions of Tianjin University, 26:92–103, https://doi. org/10.1007/s12209-019-00231-w

30. Energy Storage Grand Challenge Energy Storage Market Report (2020), U.S. Department of Energy, Technical Report, NREL/TP-5400-78461, DOE/GO-102020-5497, December 2020.

31. Daniel Kucevic et al. (April 2020), Standard Battery Energy Storage System Profiles: Analysis of Various Applications for Stationary Energy Storage Systems Using a Holistic Simulation Framework, Journal of Energy Storage, 28, 101077.

32. World Energy Outlook (2020), https://iea.blob.core.windows.net/assets/80d64d90-dc17-4a52-b41f-b14c9be1b995/WEO2020_ES.PDF

33. Ahmed S. Eldessouky, Hossam A. Gabbar (2015), Micro Grid Renewables Dynamic and Static Performance Optimization Using Genetic Algorithm, 2015 IEEE International Conference on Smart Energy Grid Engineering (SEGE), DOI: 10.1109/ SEGE.2015.7324596

34. Kookie Triviño (2020), On Microgrids for Power Quality and Connectivity (e-book), Asian Development Bank, DOI: http://dx.doi.org/10.22617/TIM200182-2

35. Yimy E. García Vera et al. (2019), Energy Management in Microgrids with Renewable Energy Sources: A Literature Review, Applied Science, 9, 3854, DOI: 10.3390/ app9183854

36. World Energy Investment (2021), https://iea.blob.core.windows.net/assets/5e6b3821-bb8f-4df4-a88b-e891cd8251e3/WorldEnergyInvestment2021.pdf

37. Samrat Sengupta (4 June 2021) Down to Earth News, https://www.downtoearth.org. in/blog/climate-change/renewable-energy-estimated-to-attract-70-global-energy-investment-in-2021-iea-77263#:~:text=Published%3A%20Friday%2004%20June% 202021,of%20traditional%20fossil%20fuel%20production

38. IRENA – International Renewable Energy Agency (2020), Finance & Investment, https://www.irena.org/financeinvestment

39. IRENA – International Renewable Energy Agency (2020), Global Landscape of Renewable Energy Finance, https://www.irena.org/-/media/Files/IRENA/Agency/ Publication/2020/Nov/IRENA_CPI_Global_finance_2020.pdf

40. Alexander Ochs and Dean Gioutsos (October 2017), The Employment Effects of Renewable Energy Development Assistance, EUEI Policy Brief, https://www.research-gate.net/publication/321398329_The_Employment_Effects_of_Renewable_Energy_Development_Assistance

41. Stelios Grafakos et al. (June 2020), Employment Assessment of Renewable Energy: Power Sector Pathways Compatible with NDCs and National Energy Plans, Global Green Growth Institute, Seoul. https://gggi.org/employment-assessment-of-renewable-energy-power-sector-pathways-compatible-with-ndcs-and-national-energy-plans/

42. Andreas Klemmer (2019), Green Jobs and Renewable Energy: Low Carbon, High Employment, International Labour Organization, https://www.ilo.org/wcmsp5/groups/public/---ed_emp/---emp_ent/documents/publication/wcms_250690.pdf

43. Veronika. Czako (2020), Employment in Energy Sector, Status Report 2020, JRC Science for Policy Report, Luxembourg: Publications Office of the European Union, DOI: 10.2760/95180

44. Francesco La Camera (2019), ©IRENA 2019, Global Energy Transformation: A Roadmap to 2050, https://www.irena.org/-/media/Files/IRENA/Agency/Publication/2019/Apr/IRENA_Global_Energy_Transformation_2019.pdf

45 Francesco La Camera, Guy Ryder (2022), © IRENA 2022, Renewable Energy and Jobs Annual Review 2022, ISBN: 978-92-9260-364-9.

46. Anh Tuan Hoang et al. (2021), Impacts of COVID-19 Pandemic on the Global Energy System and the Shift Progress to Renewable Energy: Opportunities, Challenges, and Policy Implications. Energy Policy, 154: 112322. Published online on 28 April 2021, DOI: 10.1016/j.enpol.2021.112322

47. Dave Turk, George Kamiya (2020), The Impact of the Covid-19 Crisis on Clean Energy Progress, License CC BY 4.0, Report — 11 June 2020.

48. LY Zhang, H Li, WJ Lee, et al. (2021), COVID-19 and Energy: Influence Mechanisms and Research Methodologies. Sustainable Production and Consumption, 27:2134–2152. https://doi.org/10.1016/j.spc.2021.05.010

49. IRENA, ILO (2021), Renewable Energy and Jobs – Annual Review 2021, International Renewable Energy Agency, in collaboration with international labour organization of Abu Dhabi, Geneva. ISBN: 978-92-9260-364-9

50. Asian Development Bank (2021), Covid-19 and Energy Sector Development in Asia and the Pacific Guidance Note, https://www.adb.org/sites/default/files/publication/714916/covid-19-energy-sector-asia-pacific-guidance-note.pdf

51. Arthouros Zervos (2017), Renewables Global Futures Report: Great Debates Towards 100% Renewable Energy, REN21.2017 (Paris: REN21 Secretariat), https://www.ren21.net/wp-content/uploads/2019/06/GFR-Full-Report-2017_webversion_3.pdf

2 An Overview of Global Renewable Energy Resources

Present Scenario, Policies, and Future Prospects

V. Manimegalai[1], V. Rukkumani[2], A. Gayathri[1], P. Pandiyan[3], and V. Mohanapriya[4]
[1]Department of EEE, Sri Krishna College of Technology, Coimbatore, Tamil Nadu, India
[2]Department of EIE, Sri Ramakrishna Engineering College, Coimbatore, Tamil Nadu, India
[3]Department of EEE, KPR Institute of Engineering and Technology, Coimbatore, Tamil Nadu, India
[4]Department of EEE, Bannari Amman Institute of Technology, Erode, Tamil Nadu, India

CONTENTS

DOI: 10.1201/9781003369554-2

2.1 INTRODUCTION

Human evolution has been based on the availability and use of energy. From the use of fire and animal power in the early days to the increasingly widespread use of electricity and cleaner sustainable fuels for various purposes, energy has been a fundamental driving force for human progress. In every country, the need for energy can be seen in a variety of sectors, including the provision of key services such as lighting, cooking, cooling, mobility, heating, and the functioning of machines and telecommunications equipment. Today, the inadequacy of reliable and clean energy supplies is regarded as one of the most considerable barriers to improving human wellbeing worldwide. Fossil fuels are expected to be exhausted due to the substantial growth of the population that has caused a significant change in the global energy demand. Numerous renewable energy resources have been used to combat this concern, such as wind, biomass, solar, geothermal, and tidal.

The energy demand increased from 25% of the global average in 1990 to 60% in 2019, with coal accounting for most of the increase, while carbon dioxide emissions increased to more than 15% of the global average. The forces of urbanization and industrialization have been creating energy demand. Specific renewable energy sources (RES) like solar, wind, biomass, hydro, tidal, and energy will be used to meet the demand for growth. The sudden disruption caused by COVID-19 forced each country to make changes in the energy sector, despite continuous work on the target. A 40-day lockdown will be implemented to slow the spread of the virus, which resulted in an 8% drop in GDP for the year 2020. So, each country faces an inevitable impact on its energy demand.

There are various RES, such as solar, biogas, wind, tidal, and hydro, which can be used as an alternate source of energy to reduce emissions and live an environmentally friendly life. All over the world, electricity demand is the most significant factor to be considered. Even with the help of these alternate sources, the load demand cannot be fulfilled. The problems should be sorted out, and necessary challenges should be taken to overcome these problems. The usage of alternate resources with certain control techniques is used to get the desired output.

Compared to other carbon-based fossil fuels, RES play a significant role in lowering greenhouse gas emissions [1]. Most countries have started to focus on RES because of their numerous benefits. Every country has framed various policies, agencies, and government bodies to increase the exploitation of renewable technologies. In 2010, India was one of the world leaders in installed capacity of RES with 17594 MW [2]. A specific sustainable development code has been initiated in order to achieve a sustainable, cleaner, and environmentally friendly energy production in the future. Global countries also followed the Sustainable Development Goals (SDGs) by making certain targets based on their countries [3]. Figure 2.1 depicts the global installed renewable energy capacity from 2011 to 2020 [4]. Renewable energy capacity installed around the world in 2020 [4] is depicted in Figure 2.2.

The share of global electricity generated using RES is 29% in 2020, whereas it was 27% in 2019. The worldwide electricity power generation capacity using wind energy in 2020 was 733 GW, solar energy was 714 GW, bioenergy was 127 GW, and geothermal energy was 14 GW. Even though there is an impact of COVID-19, China

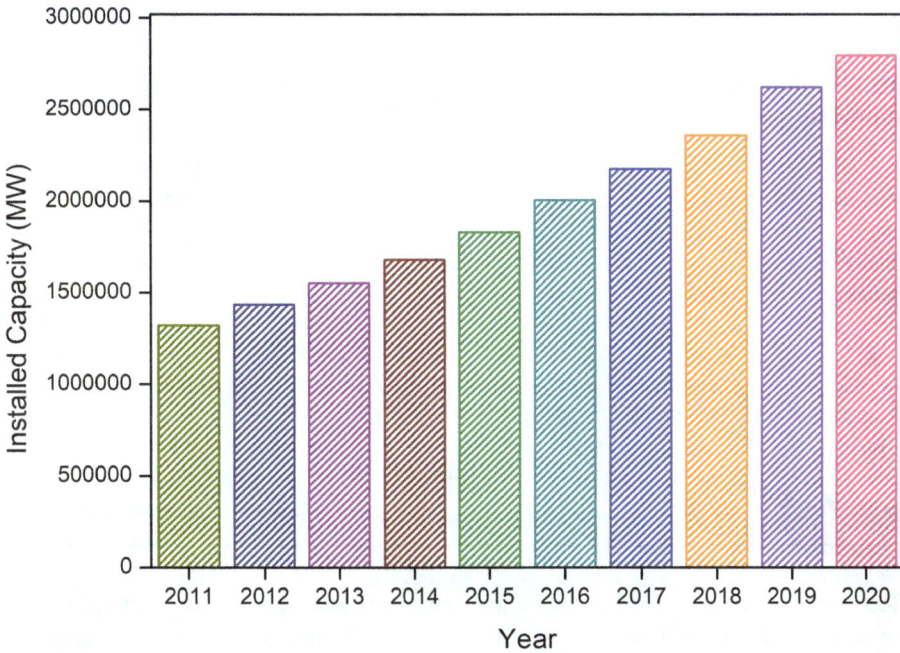

FIGURE 2.1 Global installed renewable energy capacity from 2011 to 2020.

stands first in installing renewable energy like solar, wind, hydro, and bioenergy. Next to China are the United States, Japan, South Africa, Germany, etc. Table 2.1 illustrates the installed capacity of RES in different parts of the world [4].

The structure of the proposed study, and the breakdown of the rest of the chapter, is as follows: Section 2.2 examines different types of renewable energy resources.

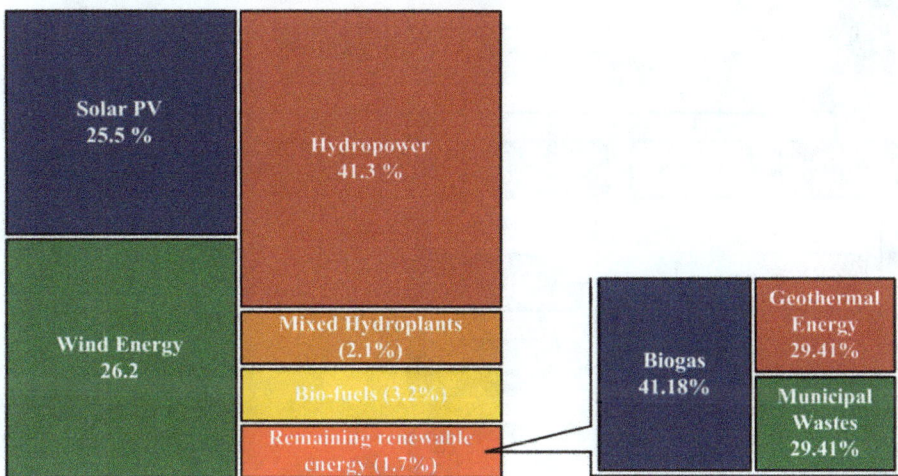

FIGURE 2.2 Global renewable energy sources installed capacity by 2020.

TABLE 2.1

Installed Capacity of Renewable Energy Sources in Different Parts of the World

Country	Solar (MW)	Wind (MW)	Geothermal (MW)	Hydro (MW)	Bioenergy (MW)	Marine (MW)
China	254354.8	281992.67	–	339840	18686.89	4.75
United States	75571.7	117743.8	2586.8	71423.3	11980.1	–
Germany	53783	62184	40	4236	10366.5	–
India	39211.16	38558.6	–	45954.77	–	10532.25
Japan	68665.49	4371	481	22522	–	1826
Sweden	1417.15	9688	–	16380	5298.5	–
South Africa	5989.58	2636	–	747.4	264.69	–
Australia	17344	8603	–	5913	875	1
Italy	21600.35	10839.46	797.19	25204.1	3554.15	0.21
Switzerland	2943.45	87	0.31	4226	21130	–

Section 2.3 contains information on how renewable energy resources can aid in sustainable development. The effect of climate change due to renewable energy resources is discussed in Section 2.4. Section 2.5 provides the policies and incentives followed for implementing renewable energy resources-based electricity generation. The challenges faced in realizing renewable energy generation are given in Section 2.6. Section 2.7 elaborates the benefits of RES. Section 2.8 concludes the chapter with future scope.

2.2 CLASSIFICATION OF RENEWABLE ENERGY SOURCES

RES can be grouped into several categories. Figure 2.3 represents the types of RES [5].

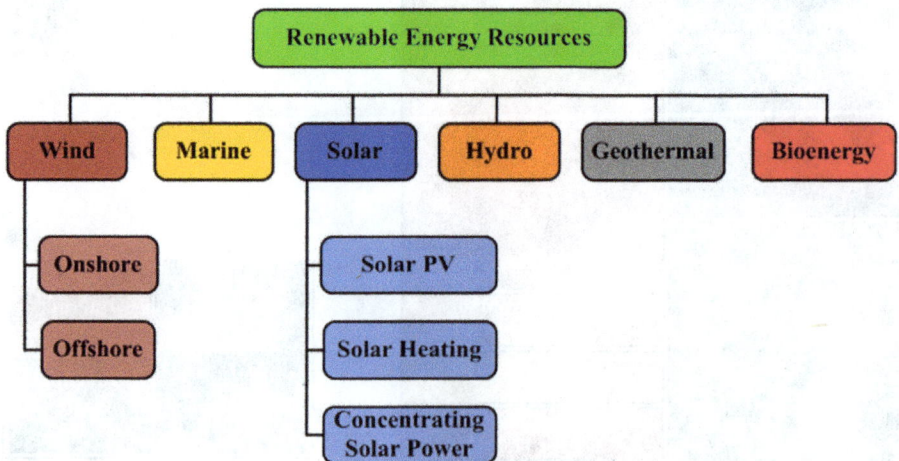

FIGURE 2.3 Types of renewable energy resources.

2.2.1 BIOMASS

"Biomass" refers to a substance that is mainly obtained from plants, trees, and crops. Through the photosynthesis process, they are able to capture and store the sun's energy. Biomass energy is the process of converting biomass into various forms of energy. Biomass energy is directly derived from land, crops, or wastes [6, 7]. Although biomass energy is inexhaustible and long-lasting, it shares many properties with conventional fuels. Biomass is able to be burned directly for energy; it can also be used as a feedstock for the generation of a variety of liquid and gas fuels (biofuels). Biofuels are transportable and storable, allowing for an on-demand cogeneration process. This could be in an energy mix with a high reliance on alternating energy sources. Because of these commonalities, biomass is likely to play a significant part in future energy prospects. As a result, biorefinery and biotransformation technologies are being developed as a novel way to convert biomass feedstock into clean energy sources. In the carbon cycle, biomass energy main features are shown in Figure 2.4 [8]. Thermochemical and biochemical conversion methods can turn biomass feedstock into bioenergy. The process for converting bioenergy into various end products is shown in Figure 2.5 [9].

2.2.2 GEOTHERMAL ENERGY

Natural processes can extract renewable energy from the earth through geothermal energy, which is the most dominant and effective way to use it. This type of energy can supply heat for a residential unit and produce energy through a geothermal

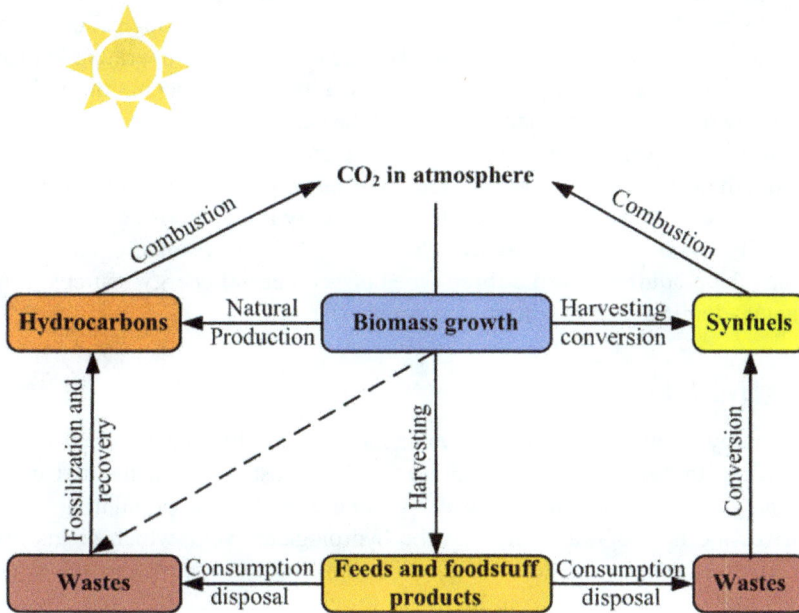

FIGURE 2.4   Biomass energy: Main features.

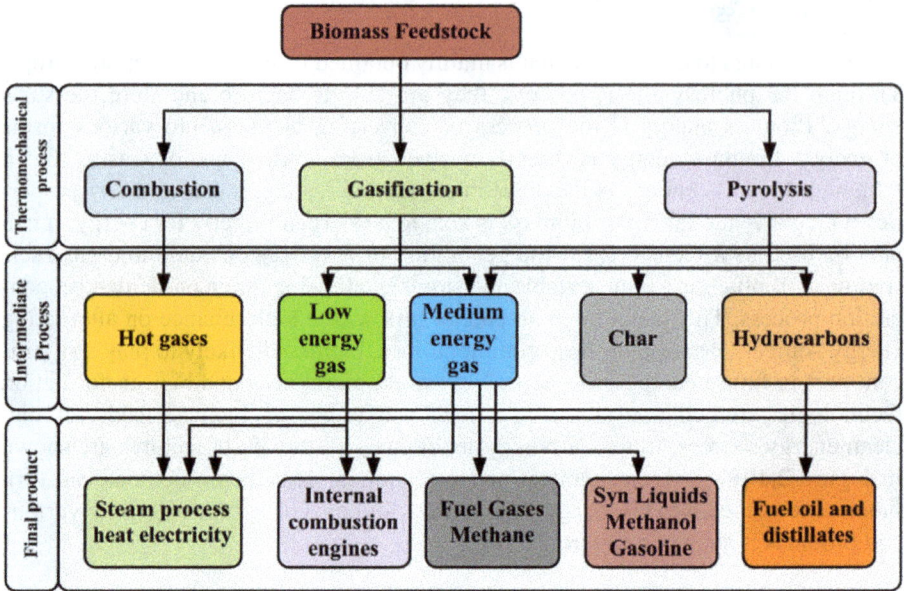

FIGURE 2.5 Processes for converting bioenergy into various end products.

power plant. During the rock formation process, a small amount of thermal energy is trapped in the rock. This causes steam or liquid water to be trapped at a deep geological depth, making geothermal energy possible. With varying reservoir temperatures and depths, geothermal systems can be found in a variety of geological contexts.

Recent volcanic eruptions have been associated with high-temperature hydrothermal systems (more than 180°C) where strong heat production occurs through radioactive decay of isotopes or when aquifers are heated by circulated water from deep fault zones. Such systems can be both intermediate temperatures and low temperatures (less than 100°C). When used under optimal conditions, geothermal fields can provide both electricity and heat. Geothermal energy is a low-cost, dependable, and ecologically beneficial energy source [8]. Hydrothermal systems, conductive systems, and deep aquifers are the three types of geothermal energy sources. Liquid- and vapour-dominated hydrothermal systems exist.

2.2.3 HYDRO ENERGY

Hydro energy is obtained from the water movement. This can generate electrical energy using turbines and water movement. The most common method used for hydro energy is dams, but tidal and wave power are pretty more popular. Solar radiation drives the flow of water through the hydrological cycle, which yields hydropower. Hydropower can be generated by the movement of water in rivers, which is aided by gravity, as it flows from higher to lower elevations. There are three types of hydropower plants, based on operation and water flow. Depending on the hydrology and terrain of the river basin, hydropower plants can range in size from small to big. Hydropower has been around for almost a century and has proven to be reliable. This takes a direct conversion from water flow to electricity, so hydropower is now an exceptionally available power technology with one of the highest conversion processes compared to other sources. It is more durable and less expensive.

2.2.4 SOLAR ENERGY

Solar-based energy generation is the process of capturing the sun's radiant energy to generate hot water or electricity through solar cells and concentrated solar power (CSP) systems. Many technologies have been implemented around the world by demonstrating their technical viability [9].

2.2.4.1 Photovoltaic

This photovoltaic (PV) system converts the sun's energy directly into electricity. One of the most important components of this system is the solar cell, which is made up of a semiconductor material that transforms sun's energy into direct current power. PV systems are made up of PV modules and a variety of other application-specific system components. The solar panels are arranged in flat or otherwise in tree-like structure [10] to extract the energy. Modules combined with a PV system can produce electricity with a capacity of up to 10 MW. Solar PV systems with silicon-based technology are the most common. Thin-film modules, made up of non-silicon materials, have become gradually more popular in recent years. Silicon-based modules have higher efficiency than thin films, but they are less expensive per unit of capacity. PV, which concentrates sunlight over a smaller area, is on the verge of becoming widely available. This type of PV cell can achieve efficiencies of up to 40%. On the other hand, organic solar cells are still in the early stages of development [11].

Solar PV offers two benefits. Module production can be done in big plants, resulting in cost savings. In contrast, PV technology is a very integrated technology. PV has an advantage over concentrated varieties in that it uses both direct and diffuse sunlight to generate energy, allowing it to generate electricity even when the sky is cloudy. This capacity enables efficient implementation in far more global locations than CSP. The types of PV systems are off-grid and grid-connected. A solar PV system operating off-grid represents an exceptional economic opportunity for villagers living in unserved areas. Mini-grid systems are easily installed in rural areas, proving to be a preferable method of electrifying rural areas in recent years. An inverter is used to convert direct current to alternating current in grid-connected systems, and then it feeds the generated power into the power grid. Because the grid acts as a buffer, system costs are lower than with an off-grid installation. These systems that are connected to the grid are divided into two categories: distributed and centralized.

2.2.4.2 Concentrating Solar Power

This system generates energy by concentrating direct rays to heat a gas, solid, or liquid, which is then used in a series of processes to generate electricity. The most prevalent concentrating method of solar energy in large-scale CSP systems is reflection, not refraction. Trough or linear Fresnel systems concentrate light on a line (linear focus), whereas central receiver or dish systems focus light on a point (point focus).

There are many CSP applications, ranging from very small distributed systems with a few kilowatts to very large centralized plants with hundreds of megawatts. Trough systems have become increasingly popular in CSP as the power sector expands [12, 13].

2.2.4.3 Solar Thermal Heating and Cooling

For residential, commercial, and industrial applications, this energy is used to provide heating of water, heating of buildings, etc.

2.2.5 WIND ENERGY

A wind turbine converts wind energy into various forms, mainly electricity and mechanical power. Sails are also used to propel ships. When electricity is generated from wind, it converts the kinetic energy into mechanical energy, then to electrical energy. Therefore, the wind turbine industry is tasked with designing turbines and power plants that are both efficient and cost-effective. According to theory, wind speed cubes increase the kinetic energy by extracting it from the wind. Nevertheless, a turbine extracts available energy (40%–50%) in a fraction of time; it can be designed by focusing on the capture of maximizing energy over a range of wind speeds. All the parameters should be taken into account in order to diminish the cost of wind energy. It is possible to reduce the material usage of wind turbines while increasing their size, reliability, and efficiency. The design of wind turbines must balance these goals.

The development of horizontal and vertical axis onshore wind turbines was investigated from 1970 to 1980. The horizontal axis gradually took over, specifically, the number of blades as well as blade fixed positions. A group of onshore wind turbines is known as a wind power plant or wind farm. The standard range of wind power facilities is 5–300 MW; however, other power sizes are possible. Onshore wind energy technology is more established than offshore wind energy technology, but it also demands a higher level of investment. Several factors motivate offshore wind energy development, including the good quality of wind resources at sea, the possibility of operating larger turbines, and the need for infrastructure with less land-based transmission, which can form larger power plants [14, 15].

2.2.6 MARINE ENERGY

Waves, ocean currents, tidal range, salinity gradients, thermal energy conversion, and currents are the six main sources of marine (ocean) renewable energy. They all have a different origin, require different transformation methods, and offer different benefits. Most technologies based on ocean energy, except tidal barrages, are in the concept or prototype phases. A study found that ocean energy will yield 7400 EJ/year of energy, which is significantly greater than current and expected future energy needs. There have been few evaluations of the technical potential of various ocean energy systems, and such potential will vary depending on future technological advancement [16–18]. As ocean energy is still at a very early stage, less than 5417.4 MW of wave energy, and tidal energy was installed annually between 2010 and 2020 [8].

2.3 RENEWABLE ENERGY AND SUSTAINABLE DEVELOPMENT

RES reloads themselves and can include bioenergy, geothermal energy, hydropower, solar energy, ocean energy (tides and waves), and wind energy; therefore, they do not deplete the earth's resources. The global population growth and the rapidly rising energy demand have resulted in the use of conventional energy sources (like gas, coal, and oil), which has created some issues such as greenhouse gas emissions, fuel price fluctuations, depletion of fossil fuel reserves, and geographical tensions [5]. This results in unsustainable conditions and is a permanent danger to human societies. RES are the best alternative to the growing challenges. The renewables 2019 global status reports that in 2017, renewable energy accounted for 18.1% of global energy production [19]. The reliable supply of energy is imperative in all economies for heating, industrial equipment, transportation, lighting, and other purposes. When RES are used instead of conventional fuels, they most probably reduce gas-based emissions. RES should be supportable in order to obtain emissionless energy flows on our earth. Sustainability requires renewable energy to be infinite, to deliver environmentally friendly products and services, and to be non-harmful to the environment [20].

A reliable supply of energy resources is widely recognized as a necessary but insufficient condition for societal development. Moreover, sustainable development requires the availability of energy resources that are convenient, cost-effective, and able to perform all essential duties without detrimental impacts on society. Conventional fuels and uranium are widely understood as limited ones. However, renewable energy is an energy source obtained from sunlight, wind, and falling water, and, therefore, it is long-term sustainable. Similarly, waste (which can be converted into usable energy forms, such as incineration) and biomass fuels are commonly identified as sustainable energy sources [21]. Environmental considerations are an essential component of long-term development. Activities that cause constant degradation of the environment are not sustainable for several reasons. For example, the cumulative effect exhibited by such activities can be fatal for humans, ecological systems, and more. The use of energy resources is responsible for a major percentage of a society's environmental effects. In this world, if our society starts to use energy resources for sustainable growth, then the environment won't have any negative impact.

In addition, the growing global population necessitates the formulation and successful implementation of sustainable development. Certain important factors can contribute to a society's long-term success. Some of them are public awareness, environmental education and training, promoting RES, etc. [22].

2.4 RENEWABLE ENERGY AND CLIMATE CHANGE

"Climate change" is currently gaining much awareness worldwide, both in scientific and political circles. Even though climate change has been occurring since the beginning of time, in recent years, climate change has had one of the greatest impacts on the environment, leading to risks in the future. Carbon dioxide emissions have grown over the years [23, 24]. It is found that, before 1995, annual averages were at 1.4 ppm, while after 1995, they were around 2.0 ppm [25]. The UNFCCC describes climate change as a change in the universal atmosphere directly or indirectly related to manmade activities and results in natural calamities [26]. The goal of limiting global warming to below 2°C has been a major focus of the international climate debate for more than a decade. Carbon dioxide emissions have been rising rapidly since 1850, when fossil fuels became the primary energy source worldwide. By the end of 2010, data showed that conventional fuel consumption was responsible for the bulk emissions of greenhouse gases (GHG), with absorptions of greater than 390 ppm (39%) above preindustrial levels [27]. The use of renewable technologies reduces environmental effects, produces minimal secondary waste, and is economically and socially viable in the face of present and future demands. Renewable technologies offer a unique chance to reduce GHG and global warming by replacing traditional energy sources (conventional-based fuels) [28].

2.5 POLICIES AND INCENTIVES

Government regulations will continue to influence renewable energy technology (RET) adoption and implementation, particularly in areas other than electricity generation. As renewable energy becomes more affordable and innovative, the policy will be even more crucial. Most countries across the globe have renewable energy support programmes in place by 2020, but they have varying levels of ambition. The year 2020 was a precarious year for tracking advancements in renewable energy goals. Over 165 countries have set goals for expanding renewable energy use in various sectors by year's end. The objective was largely associated with the electricity sector, followed by total final energy consumption, heating and cooling, and transportation. In the most recent statistics, 80 of the 2020 targets were met, while most (134) did not meet the expected demand (ranging from 2017 to 2020). Several countries were on course to meet their goals, while others had not. In addition, more than 30 countries still need to set new 2020 goals for upcoming years since the 2020 countries' objectives ended this year.

Even though the world is in a pandemic of COVID-19, the renewables policy is strongly supported in 2020. Many nations provided specific aid to renewables in recovery strategies and financing packages to cope with the epidemic. However, the fossil fuel industry received significantly more funding overall. The primary aims of policy initiatives for less-developed renewable energy markets and some developing and rising economies are to increase renewable energy capacity and generation to satisfy demand, improve energy security, and increase energy access. Renewable energy policies can target any end-use sector, including buildings, industry, transportation, and power generation. By 2020, at least five nations have announced climate change plans that include support for renewable energy across various industries, although most of these plans target a single sector. Trade policy plays a major role in product development, exchange, renewable energy development, and demand for renewables within individual countries.

COVID-19 crisis prompted governments worldwide to commit USD 12 trillion in financial stimulus. Among these funds, they utilized USD of 732.5 billion in energy-related assistance. However, as of April 2021, renewable energy incentives only represented around 264 billion USD of the total expanse granted by governments around the globe, while fossil fuel projects donated 309 billion. Examples of direct support for coal include the USD 6.75 billion utilized in coal infrastructure support for India and the USD 2.5 billion bailout of Doosan Heavy Industries, a manufacturer of coal plants. A Canadian pipeline received USD 4.4 billion, especially for loans, and oil and gas firms in the United Kingdom received GBP 1.3 billion (USD 1.7 billion) in low-interest loans. Regional development accounts for roughly 30% of the European Union budget of EUR 750 billion (USD 921 billion). The COVID-19 impetus package included energy efficiency and renewable energy generation, retrofitting buildings to save energy, producing renewable heat, developing renewable hydrogen, and electrifying automobiles. Despite supporting coal in their recovery plans, China, India, and the Republic of Korea have all pledged to invest in renewable energy. As part of Colombia's strategy, the USD 4.6 million was slated to finance 27 renewable energy and transmission projects. Government agencies have contributed about USD 95 billion to the power sector in response to COVID-19. Although some governments contributed money for additional renewable power generation, this was mainly to assure the continuity of services and decrease consumers' bill burdens rather than to incentivize renewables. The Israeli recovery plan proposed spending USD 2 billion on developing an additional 2 GW of solar plants. A solar PV home system is scheduled to be installed in 5 million homes in Nigeria as part of the stimulus package. The stimulus package set aside USD 620 million for this project. A total of USD 900 billion was utilized for the US rescue package, including production extensions and investment tax credits for PV and onshore wind power. Low-income households will receive USD 1.7 billion for renewable energy and USD 4 billion for research and development of solar, wind, hydroelectricity, and geothermal energy.

Most energy-related stimulus funding went to developing renewable heat and improving the energy efficiency of existing structures in the building and industrial sectors. Taking a larger goal at least by 2050 to repair the country's entire building stock due to COVID-19, France's cover package includes a USD 8.6 billion to promote building revamps, particularly those supporting renewable energy. Aviation was the greatest stimulus beneficiary in the transportation industry, but only three nations – Austria, France, and Sweden-included "green" requirements for aviation incentives. KRW 2.6 trillion (USD 2.4 billion) was included in the Republic of Korea's recovery package to encourage electric and hydrogen vehicles. The French proposal includes EUR 11 billion (USD 13.5 billion) in funding for electric vehicles and charging station infrastructure. As a part of Germany's economic stimulus package, EUR 5.9 billion (USD 7.3 billion) will be allocated to EV and charging infrastructure subsidies, and EUR 7 billion (USD 8.5 billion) will be allocated to renewable hydrogen that will reduce greenhouse gas emissions. Electrifying public transportation, improving electric vehicle charge stations, and supporting the purchase and charging of low-emission vehicles are all part of Spain's EUR 3.8 billion (USD 4.6 billion) auto industry aid package [29]. Australia's ocean renewable energy potential is the world's largest; by 2050, these energy sources could fill up to 11% of Australia's needs. Solar energy is also a significant potential resource in Australia, with an annual solar radiation release of over 58 million PJ. Wind energy, another established technology, is the fastest growing renewable energy source in Australia because of its large coastline. Australia follows certain policies and planning for each resource in order to attain future sustainable resources [30].

2.6 CHALLENGES AFFECTING RENEWABLE ENERGY SOURCES

Renewable energy may be the primary energy source for countries with low carbon emissions. In order to make use of RES extensively, energy systems should be far away from disruptive changes. The major problem of the energy transition from non-sustainable to RES started in the 21st century. The most significant impediment to using RES is a country's policy and policy instruments, which influence the technological progress and cost. Furthermore, technological advances influence the cost of RETs, leading to market failures and limited adoption. To overcome these issues, a renewable energy strategy must consider the interconnectivity of issues that affect RES to attain sustainability, illustrated in Figure 2.6 [31]. The policy recommendations for mitigating climate change and its effects are as follows:

  i. Sectors and regions can contribute to reducing global warming by investing in renewable energy and encouraging policies that minimize it.
 ii. Climate change mitigation can be greatly aided by reducing the carbon footprint through lifestyle and behaviour patterns adjustments.
iii. Research investigates ideas and technologies that can reduce land use while reducing renewable energy-related accidents and the risk of resource rivalry, such as in bioenergy, where food production competes with energy production.
 iv. By improving international collaboration and support for developing countries in infrastructure expansion and technological upgrades, one can reduce climate change and its effects.

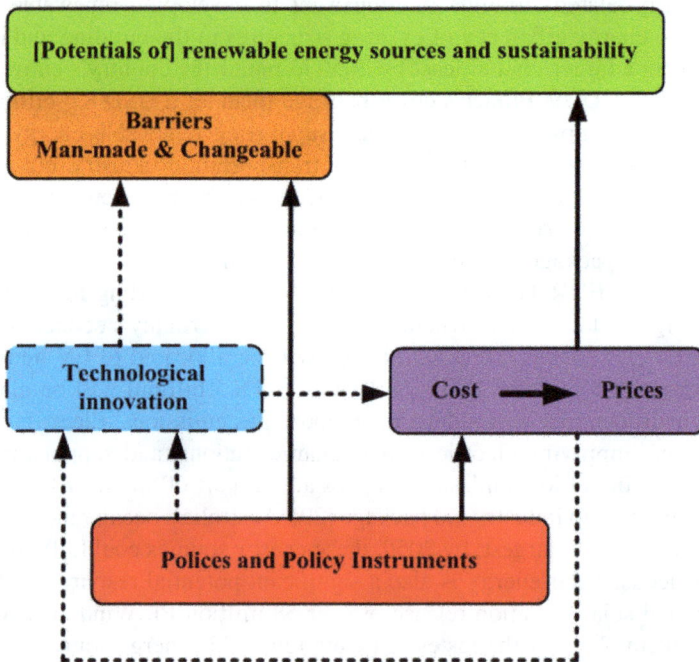

FIGURE 2.6   Factors influencing renewable energy generation.

Research and development centred around sustainable energy resources and systems in the wake of the oil crisis in the early 1970s. At that time, renewable energy was the most feasible form of energy conversion due to price expectations, cost estimates, and ease of implementation. Developing RETs leads to commercially viable products. Advancements in RETs can be developed as profitable and environmentally friendly alternatives to traditional energy generation. It is possible to increase the support of renewable sources in order to decrease the energy demand. Also, this would provide more employment opportunities and also provide benefits to the environment. The government energy sector is aware of this opportunity and supports its renewable energy industry in capitalizing on it. The benefits of RES can be realized in terms of energy, the economy, and the environment. Some of them are as follows:

- Research and development
- Assessment based on technology
- Development of standards
- Transfer of technology

Such initiatives will help persuade potential users to think about the advantages of RETs [22].

2.7 BENEFITS OF RENEWABLE ENERGY SOURCES

The primary sources of energy in rural areas are biomass and kerosene. Promoting RES for rural energy requirements is needed to reduce poverty and improve the use of natural energy resources that reduce oil imports. This shift leads to economic growth towards greater sustainability, as well as environmental and social stability. There is a lack of information about socio-economic issues. The data that are accessible are distributed and lack quality. As a result, it is difficult to determine the whole socio-economic and environmental impact of renewable energy on the country [32]. The different socio-economic benefits are briefly explained in the following sections.

- *Cost reduction in electricity transmission and distribution*
  Bangladesh has around 87,319 villages, but they are not linked to the national grid. Only about 22% of our rural residents are connected to the grid, according to estimates. Electrification by grid extension or supplementary power plants can only reach a fraction of the population in remote locations. As a result of wide distribution and low demand, producing, transmitting, and especially distributing energy is prohibitively expensive. An independent family lighting kit, such as a PV system, may be a viable solution for these situations. And other advantages are that they do not need any grid lines to deliver electricity to each home. Losses and theft of power are avoided when distribution and transmission lines are not present.
- *Foreign currency savings opportunity*
  Around the country, all sectors, like government offices, banks, schools, industries, are either employing conventional methods (candles, kerosene wick lamps, etc.) or running their diesel generator sets in their off-grid

settings and also have their own electricity budgets, which may be readily met with PV applications. In Bangladesh, hard-earned dollars are spent on petrol importation. Replacing diesel generators with RETs will increase the energy mix and save foreign currency at the same time. It is a well-known fact that by diversifying the energy supply, RETs can help to ensure energy security and price stability. Another issue with the traditional grid is line disruptions, which are far less common with RETs and can be prevented if the user is well trained.

- *Creating a better learning environment in rural areas*
  The majority of rural off-grid schools are without electricity. These rural schools can use RETs for various purposes. Modern benefits will attract more students, but they will also help keep talented teachers and staff who are currently hesitant to work in non-electrified locations. Kerosene lanterns and candles supplied insufficient lighting, polluted the environment, and offered a fire threat. The lighting from RETs/PV is more consistent and brighter. Besides improving schoolwork visibility, high-quality illumination increases the ease with which homemakers can carry out valuable duties in the evening with little to no effort. PV systems are also more cost-effective than dry-cell batteries in terms of lighting.
- *Possibility of employment*
  RETs employ up to three times as many people as coal or nuclear power plants. Danish wind energy manufacturers estimate that for each megawatt of the produced wind turbine, there are 17 jobs, and for each megawatt of installed wind energy, there are five jobs. The wind energy business employed over 85,000 people worldwide in 2000, and it is expected to employ up to 1.8 million people by 2020.
- *Rural women's lifestyle development as well as a safe environment*
  The provision of electricity has a number of beneficial implications for women. PV does not solve the critical need for a satisfying energy source for cooking. Domestic cooking consumes a significant amount of lignocellulosic biomass. Women are the only ones involved in this activity. Women spend several hours a day collecting tree biomass, sometimes with the help of children. Even though it is done for free, releasing women from menial tasks empowers them. Women will be able to devote more time to self-improvement activities, such as literacy, nutrition education, family planning, sewing classes, and so on, which, in turn, will result in better child care, gardening, and, therefore, increased revenue for the sector.

2.8 CONCLUSION AND FUTURE SCOPE

Recent decades have seen a growing interest in alternative and cleaner ways of generating power due to global warming, rising energy costs, and the energy crisis. As a result, the United Nations established SDGs to achieve a future of sustainable, cleaner, and environmentally friendly energy production. Several leading nations are at the forefront of renewable energy development through the rapid expansion of renewable energy capacities and targeted policies and deadlines in order to meet their goals. In order to promote suitable renewable energy, each country has determined its energy policies based on its physical and geological landscape. This chapter addressed the prospects and growth of renewable energy worldwide and existing policies that encourage the integration and hybridizing of renewable energy resources, as well as significant barriers that inhibit the increased use of renewable energy.

The future scope of the renewable energy resources is presented as follows. Combined heat and power (CHP) will assume an increasingly important role in all future energy scenarios as people become more aware of the significant negative impacts of conventional energy generation. In addition to being considered a modern generation, renewable technologies can be developed. As energy consumption increases, most developed countries have implemented various strategies to reduce GHG emissions, increase renewable penetration, and facilitate energy transitions. However, investments in renewable technologies and the associated uncertainties about generation are the primary concerns in the energy sector. It is expected that future projects will concentrate on implementing compact RES-CHP systems that allow generation from multiple sources and can accommodate a growing share of variable renewables on a large scale. Aside from being a significant low-cost source of low-emission electricity, solar energy also appears to be a key component of energy strategies for 2030 and 2050 regarding electric mobility, thermal energy production, and off-grid electricity usage.

Regular improvements in solar cell technology and fast growth in PV panels result in increased installed capacity. Hence, it will dominate wind power by 2025, hydropower by 2030, and coal-fired plant capacity by 2040. Over the next few years, solar cell technology will become more efficient and more advanced. The use of solar energy, especially distributed generation, is environmentally beneficial. Although its increased use will generate additional demands for households and the grid, more energy will also be required. In this regard, controlling distributed generation in every household and ensuring utility companies have flexible power systems are the primary challenges. The energy sector will likely devote its attention primarily to developing efficient and flexible regulatory frameworks that will allow and encourage distributed generation based on renewable technologies. Economic dispatch will be resolved more effectively by developing more effective forecast strategies for both production and consumption. In addition, improvements in energy storage will certainly improve the economics of renewable energy. As a result, energy storage integration will be a fascinating topic that is likely to contribute to the success of renewable energy. Geothermal energy will be used more frequently soon because of numerous ongoing research and development projects.

REFERENCES

1. Srivastava, A. R., Khan, M., Khan, F. Y., & Bajpai, S. (2018). Role of renewable energy in Indian economy. In *IOP Conference Series: Materials Science and Engineering* (Vol. 404, No. 1, p. 012046). IOP Publishing.
2. Elavarasan, R. M., Shafiullah, G. M., Padmanaban, S., Kumar, N. M., Annam, A., Vetrichelvan, A. M., ... & Holm-Nielsen, J. B. (2020). A comprehensive review on renewable energy development, challenges, and policies of leading Indian states with an international perspective. *IEEE Access, 8*, 74432–74457.

3. Arora, D. S., Busche, S., Cowlin, S., Engelmeier, T., Jaritz, J., Milbrandt, A., & Wang, S. (2010). *Indian Renewable Energy Status Report: Background Report for DIREC 2010* (No. NREL/TP-6A20-48948). National Renewable Energy Lab. (NREL), Golden, CO.

4. https://www.irena.org/Statistics/View-Data-by-Topic/Capacity-and-generation/Technologies

5. Ellabban, O., Abu-Rub, H., & Blaabjerg, F. (2014). Renewable energy resources: Current status, future prospects and their enabling technology. *Renewable and Sustainable Energy Reviews, 39,* 748–764.

6. Srirangan, K., Akawi, L., Moo-Young, M., & Chou, C. P. (2012). Towards sustainable production of clean energy carriers from biomass resources. *Applied Energy, 100,* 172–186.

7. Sriram, N., & Shahidehpour, M. (2005). Renewable biomass energy. In IEEE Power Engineering Society General Meeting, 2005 (pp. 612–617). IEEE.

8. Hammons, T. J. (2003). Geothermal power generation worldwide. In *2003 IEEE Bologna Power Tech Conference Proceedings,* (Vol. 1, pp. 8). IEEE.

9. Byrne, J., Kurdgelashvili, L., Mathai, M., Kumar, A., Yu, J., Zhang, X., ... & Timilsina, G. (2010). World solar energy review: Technology, markets and policies. *Center for Energy and Environmental Policies Report.*

10. Pandiyan, P., Saravanan, S., Prabaharan, N., Tiwari, R., Chinnadurai, T., Babu, N. R., & Hossain, E. (2021). Implementation of different MPPT techniques in solar PV tree under partial shading conditions. *Sustainability, 13*(13), 7208.

11. Bhuiyan, A. G., Sugita, K., Hashimoto, A., & Yamamoto, A. (2012). InGaN solar cells: Present state of the art and important challenges. *IEEE Journal of Photovoltaics, 2*(3), 276–293.

12. Sioshansi, R., & Denholm, P. (2010). The value of concentrating solar power and thermal energy storage. *IEEE Transactions on Sustainable Energy, 1*(3), 173–183.

13. Machinda, G. T., Chowdhury, S., Arscott, R., Chowdhury, S. P., & Kibaara, S. (2011). Concentrating solar thermal power technologies: a review. In *2011 Annual IEEE India Conference* (pp. 1–6). IEEE.

14. Islam, M. R., Mekhilef, S., & Saidur, R. (2013). Progress and recent trends of wind energy technology. *Renewable and Sustainable Energy Reviews, 21,* 456–468.

15. Eltamaly, A. M. (2013). Design and implementation of wind energy system in Saudi Arabia. *Renewable Energy, 60,* 42–52.

16. Bhuyan, G. S. (2010). World-wide status for harnessing ocean renewable resources. In *IEEE PES General Meeting* (pp. 1–3). IEEE.

17. Aly, H. H., & El-Hawary, M. E. (2011). State of the art for tidal currents electric energy resources. In *2011 24th Canadian Conference on Electrical and Computer Engineering (CCECE)* (pp. 001119–001124). IEEE.

18. Elghali, S. B., Benbouzid, M. E. H., & Charpentier, J. F. (2007). Marine tidal current electric power generation technology: State of the art and current status. In *2007 IEEE International Electric Machines & Drives Conference* (Vol. 2, pp. 1407–1412). IEEE.

19. Murdock, H. E., Gibb, D., Andre, T., Sawin, J. L., Brown, A., Ranalder, L., ... & Brumer, L. (2021). Renewables 2021-Global status report, REN21.

20. Twidell, J., & Weir, T. (2015). *Renewable Energy Resources.* Routledge.

21. Anon. (1989). Energy and the Environment: Policy Overview. Geneva: International Energy Agency (IEA).

22. Dincer, I. (2000). Renewable energy and sustainable development: A crucial review. *Renewable and Sustainable Energy Reviews, 4*(2), 157–175.

23. Asumadu-Sarkodie, S., & Owusu, P. A. (2016). Multivariate co-integration analysis of the Kaya factors in Ghana. *Environmental Science and Pollution Research, 23*(10), 9934–9943.

24. Asumadu-Sarkodie, S., & Owusu, P. A. (2016). The relationship between carbon dioxide and agriculture in Ghana: A comparison of VECM and ARDL model. *Environmental Science and Pollution Research*, *23*(11), 10968–10982.
25. Butler, J. H., & Montzka, S. A. (2016). The NOAA annual greenhouse gas index (AGGI). *NOAA Earth System Research Laboratory, 58.*
26. Fräss-Ehrfeld, C. (2009). *Renewable Energy Sources: A Chance to Combat Climate Change* (Vol. 1). Kluwer Law International BV, United States.
27. Edenhofer, O., Pichs-Madruga, R., Sokona, Y., Seyboth, K., Matschoss, P., Kadner, S., ... & von Stechow, C. (2011). IPCC special report on renewable energy sources and climate change mitigation. *Prepared by Working Group III of the Intergovernmental Panel on Climate Change.* Cambridge University Press, Cambridge, UK.
28. Panwar, N. L., Kaushik, S. C., & Kothari, S. (2011). Role of renewable energy sources in environmental protection: A review. *Renewable and Sustainable Energy Reviews*, *15*(3), 1513–1524.
29. Ranalder, L., Busch, H., Hansen, T., Brommer, M., Couture, T., Gibb, D., ... & Sverrisson, F. (2020). Renewables in Cities 2021 Global Status Report, REN21.
30. Li, H. X., Edwards, D. J., Hosseini, M. R., & Costin, G. P. (2020). A review on renewable energy transition in Australia: An updated depiction. *Journal of Cleaner Production*, *242*, 118475.
31. Owusu, P. A., & Asumadu-Sarkodie, S. (2016). A review of renewable energy sources, sustainability issues and climate change mitigation. *Cogent Engineering*, *3*(1), 1167990.
32. Islam, M. R., Islam, M. R., & Beg, M. R. A. (2008). Renewable energy resources and technologies practice in Bangladesh. *Renewable and Sustainable Energy Reviews*, *12*(2), 299–343.

3 Plastic Waste Conversion
A New Sustainable Energy Model in the Circular Economy Era

Aditya Dharaiya[1,2] and Rushika Patel[2,3]
[1]Department of Geology, Savitribai Phule Pune
University, Pune, Maharashtra, India
[2]Wildlife and Conservation Biology Research
Foundation, Patan, Gujarat, India
[3]Gujarat Biotechnology Research Centre, Department
of Science and Technology, Government of Gujarat,
Gandhinagar, Gujarat, India

CONTENTS

3.1 INTRODUCTION

The first synthetic polymer says plastic was created in the early 20th century to reduce the usage of plants, papers, and make reusable things for several routine life purposes [1]. Then after, several different forms of plastics were generated by petrochemical and various hydrocarbons processing, which resulted in seven different categories of plastic. The details of plastics are shown in Table 3.1. By looking after these details, most of the things from bottles to money are made up of plastic.

DOI: 10.1201/9781003369554-3

TABLE 3.1

Details of Seven Major Plastic Types

Type and Recycling Code	Usage	Biodegradability	References
PET, 1	Short time single-use, packaging, food and beverage bottles, world market: 10.2%	Poor, takes >50 years for biodegradation, 110 µm/year under marine environment after acceleration by UV rays or heat	[2–5]
HDPE, 2	Shampoo and shop bottles, detergents packaging, bucket, bins, toys, fencing and piping materials, world market: 16.1%	Highly resistant, 0–11 µm/year under marine environment, half-life: 58–1200 years	[3, 6, 7]
PVC, 3	Wire coatings, furnishing, packaging, credit cards, world market: 11.8%	Poor, adherence and growth was observed at a minor rate by white-rot fungi and other bacteria	[3, 8, 9]
LDPE, 4	Grocery bags, milk pouch, food packaging, flexible container lids, vehicles, world market: 19.9%	The landfill is effective in the biodegradation process, 1.5% after 150 days in an aqueous system, 1.6–83 µm/year in landfill and marine environment after acceleration by UV rays or heat, half-life 1.4 to >1000 years	[3, 10, 11]
PP, 5	Kitchenware, disposable items, diapers, world market: 21.1%	3–23% weight loss after 50 days in composite, 0.51–7.5 µm/year in landfill and marine environment after acceleration by UV rays or heat, 53–780 years	[3, 12, 13]
PS(E), 6	Packaging form, boxes and trays, world market: 7.8%	Not able to degrade even after 32 years in the soil	[3, 14]
Others, 7 (multilayered plastic, PU, SBR, Epoxy)	CD, DVDs, bottles and storage containers, glasses, hard materials	Very high half-life and non-biodegradable, 20–1400 µm/year in landfill and marine environment after acceleration by UV rays or heat	[3]

Nowadays, new terms are also implemented such as the biopolymer and plastic sector moving forward in the same direction. Biopolymers are polymeric materials made up of biological materials, that is, plants, microbial biomass, or natural organic matters. Based on the sources, plastic is classified into oil-derived or bio-material derived categories. Biopolymers are either biodegradable or non-biodegradable upon release in the environment, whereas some oil-based plastic is biodegradable [15]. Polycaprolactone (PCL) and poly (butylene succinate) (PBS) are oil-based but are biodegradable, whereas polyethylene (PE) and nylon 11 (NY11) are made up of plants and other biomass but not degraded by microorganisms. The poly(hydroxybutyrate) (PHB), poly(lactide) (PLA), and starch blends are produced from biomass and also

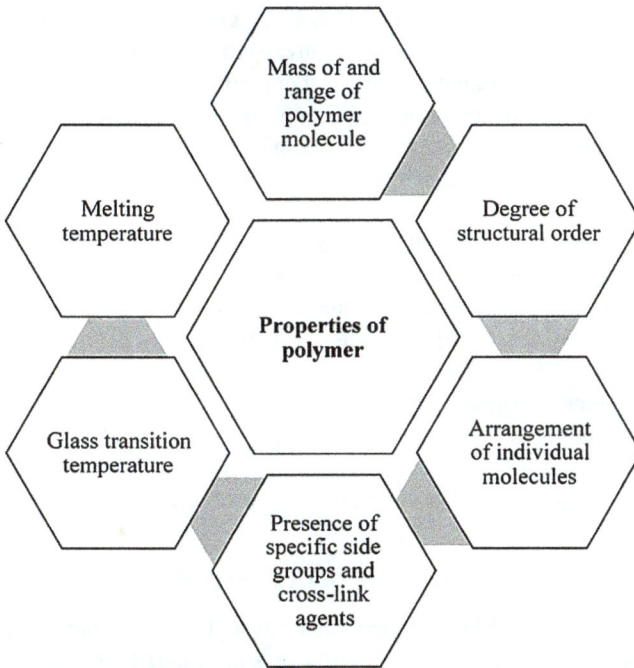

FIGURE 3.1 Polymer properties to be considered for the biodegradation process.

degraded by microbes. In the case of acetyl cellulose (AcC) bioplastic, it may be biodegradable if having low acetylation. Biodegradability also depends on the substitutions ratio and the cross-linked agents used to prepare plastic films. The biodegradability of any polymer depends on the surface condition and structure. The several properties of the polymer are also affecting biodegradation processes, such as hydrophobic properties, chemical structure, molecular weight and distribution, temperature, crystal structure, type, and number of the bond present (Figure 3.1) [16]. Non-biodegradable biopolymers may not serve as an alternative to current plastic materials and may lead to waste generation. Most of the plastic is non-biodegradable or takes a very long duration so it is better to recycle it.

Plastic trash is also thrown in landfill, open, roadside, and water bodies or oceans, endangering biodiversity and marine life's health and safety. Moreover, polychlorinated dibenzo-p-dioxins, a carcinogen, is produced by unregulated plastic incineration. Turning plastic waste into various energy sources or other materials has two advantages. First, the risks posed by plastic trash can be decreased and second, we will be able to extract variety of products having a versatile application, which also plays major role in the circular economy (CE). As a result, our reliance on fossil fuels and other non-renewable sources will be reduced to some extent. The main objective of the chapter is to provide knowledge on the threats of plastic waste to biodiversity, current trends in technologies on how to divert waste and their models. This chapter also gives light on the advantage of plastic as an alternative source in the CE era. Moreover, limitations of the current methods and future aspects are discussed and pointed out.

Plastic waste pollution is a tremendous issue globally nowadays, thus the main objective of this chapter is to focus on conversion of plastic waste into an energy source by means of mechanical, thermochemical, and chemical methods. Along with it, it also focuses on the usage and applications of the byproducts generated by such conversion methods and comparison of byproducts from every Waste to Energy (WtE) conversion methods and models with special emphasis, regarding to the principle of CE.

The organization of this chapter is as follows: Section 3.2 introduces major sources, determination method, types, and major threats of plastic pollutants to biodiversity. Section 3.3 describes the amount of plastic waste generated and collected for various purposes, advanced technology for the segregation of plastic waste, and its types. Section 3.4 focuses on the plastic waste conversion which is sub-sectioned in two parts – focus on plastic to new plastic materials and focus on the various waste-to-energy models. Section 3.5 describes comparison of plastic byproduct and its role in the CE. Section 3.6 includes major limitation of current plastic waste conversion model and Section 3.7 concludes the chapter with future scope.

3.2 IMPORTANCE OF DIVERSION OF PLASTIC WASTE MATERIALS

Plastic waste is entered into the environment through various manufacturing to disposal processes. Majorly two entry points or sources: i) Point and ii) Secondary [17]. The point (primary) sources include the production in industries for different purposes and disposal after first use. This plastic waste undergoes various environmental conditions and microbial attacks. It would be converted into micro- and nano-plastic forms which are also known as a major source of secondary plastic pollution. From the pollutant point of view, plastic waste is divided into three categories: i) macro-plastic having >5 mm size, ii) microplastic which is lower than 5 mm but higher than 100 nm, and iii) nano-plastic having <100 nm size [18]. Based on the medium (terrestrial or aquatic) plastic, waste generation may be determined in three ways: i) plastic solid waste (in kg/percentage contributed to the total municipal solid waste), ii) sediments (weight of particles/debris per meter square per kg sediments), and iii) a number of particles present (per liter water sample). Plastic waste sources, determination method, and type based on size are summarized in Figure 3.2.

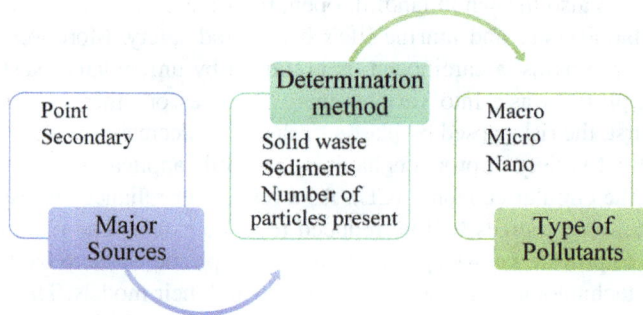

FIGURE 3.2   Plastic waste materials in the environment.

Macro-plastic degrades from the surface and allows limited diffusion of oxygen, water, and light, whereas microplastic is undergoing bulk degradation with high penetration of oxygen, water, and light. Macro-plastic is resistant to microbial attack due to its smoother surface and resistance to surface modifications. Conversion of macro to micro and nano is known as fragmentation and microbial assimilation process. Whenever plastic reaches the environment, major factors affecting the fragmentation and microbial assimilation are pH, oxygen, temperature, sunlight (UV-rays), moisture content, and medium (landfill, freshwater, or marine environment). After degradation and fragmentations, micro and nano-plastics are more dangerous than macro ones both to living organisms and humans [19]. The major causes of macro and nano-plastics are described in Figure 3.3.

There are no parameters to test for plastic waste and its other forms. Before developing new standards or updating current ones, we need to have a better understanding of the situation. Microplastics, which range in size from less than 5 mm, are now regarded as one of the most serious dangers to the global marine ecosystem. They are so little and light that they can simply be folded into themselves. Many fish and water species have misassumption of color plastic and consider it as prey or jellyfish [20]. They consume it, then it accumulates inside the stomach, which causes blockage of the digestive tract, subsequently infection in the stomach or death by starvation. Microplastics can also function as a carrier for other harmful substances, eventually ending up in the bodies of live species that ingest them [21]. Plastics can interact with soil fauna in landfills, altering their health and functioning. Based on research, UN Environment stated, "Earthworms make their burrows differently when microplastics are present in the soil, affecting the earthworm's fitness."

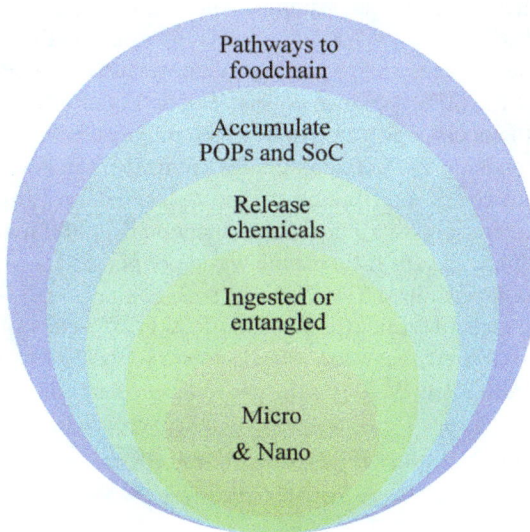

FIGURE 3.3 Threats of micro- and nano-plastic to biodiversity and the environment. Persistent organic pollutants (POPs): PCB, DDT, dyes; other substances of concern (SoC).

Macro-plastic pollutants deprive light in an aqueous system. In wild, birds and animals are attracted by the smell of food-contaminated plastic due to hunger and they eat plastic. Herbivores are the major source of plastic entry points into the terrestrial food chain. This increases bio-magnification and leads to cancer, neurological diseases, or the death of several animals. Further impacts of ingestion are divided into lethal (severe disease or death) and sub-lethal (increase the chemical burden and reduce ideal health conditions) [21]. Macro-plastics may carry microbial strains, viruses, and protists that function as disease vectors [22].

When we try to find out how much plastic waste is produced and disposed of, we may come across different numbers. This is because we don't have accurate information on how much plastic is being used and thrown away by each person. Plastic manufacturing to waste disposal cycle is complicated and there is no standard method for the detection and monitoring of plastic waste. Every country and state to state plastic waste management systems vary. It is difficult to tell the difference between plastic particles that have been maintained in aquatic systems and those that have been retained in terrestrial ecosystems. The determination of actual leakage points and interaction among both systems are next to impossible. It is very important to collect and process plastic waste materials from various sources to protect the environment and conserve biodiversity.

3.3 COLLECTION AND SEGREGATION OF WASTE

Central Pollution Control Board (CPCB) and State Pollution Control Board (SPCB) monitor the amount of solid waste including plastic collected and proceed every year in India. India generates 4059.18 tons of plastic waste every day, contributing 6.42% of total solid waste. Sixty major cities generate remarkable plastic waste. Delhi, Kolkata, Chennai, Mumbai, Kanpur, Ahmedabad, and Surat are major plastic contributor cities [23]. Forty percent of plastic waste remained uncollected and ended up in the river, ocean, and roadsides. Whatever waste enters into the aquatic environment out of the 80% is plastic debris. The debris of microplastic in a single water bottle from the ocean was found to be in the range of 0–10,000 [24]. Over 90% of the plastic that ends up in the ocean comes from rivers in Asia and China, with the Ganga and Indus rivers contributing the most plastic debris. Ingestion, entanglement, and habitat alteration have a negative impact on over 800 marine species [25]. Although, India generates very little plastic waste per person (11 kg) as compared to other developed countries (30–110 kg/person) such as the United States, the United Kingdom, Canada, and Australia, the waste collection efficiency of India was found to be high (80%). However, very low plastic waste (only 28.4%) was treated. The plastic waste generated was 35,680 thousand tons in the United States (for 2018). The amount of plastic waste (in terms of percentage) recycled, processed for energy recovery, and landfilled is represented in Figure 3.4 [26]. Post-consumer plastic waste (5094.1 million pounds) was collected from the United States for recycling purposes in 2019.

The collection of waste materials is a crucial and first step for the recycling and energy recovery processes. Recently, most of the industrial plastic waste materials are diverted to a different sector. However, there are no specific guidelines for

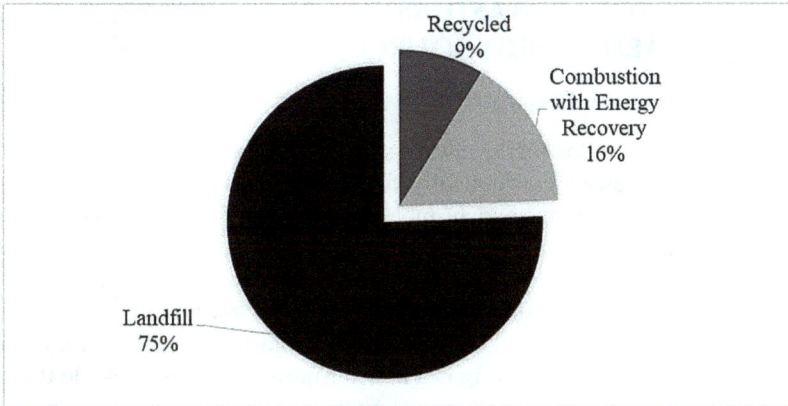

FIGURE 3.4   Plastic waste processing in the United States.

the disposal or collection of household plastic waste in many countries. Ultimately, plastic waste was separated from municipal solid waste by government bodies. A specialized collection of commingled dry recyclables (packaging and other home products mostly consisting of plastics, metals, paper, cardboard, and occasionally glass) now offers an appealing alternative for maximizing trash recovery. The collection method is majorly divided into two tools: home separation and post-collection separation. Both tools include several strategies: curbside, drop, buy-back, deposit, or refund programs. The post-collection separation tool was found to be more feasible for waste recovery and energy production [27]. In India door-to-door and Kabadiwala traditions are also followed for the collection of household plastic waste. Such waste was segregated and transferred for energy recovery.

The segregation method also determines the composition of plastic waste, suitability for downstream, and pre-treatment process. Moreover, the segregation process plays a major role in the overall performance and quality of byproducts. Advancements in technology are very useful in the segregation of plastic waste materials. Gray Level Co-occurrence Matrix (GLCM) was developed to separate plastic waste materials from mixed waste. The pre-defined values of various paper and plastic materials were initially installed and stored in the program which was further compared by device and used to identify plastic waste material. After successful identification and comparison, the signals were transferred to the microcontroller with the help of a Universal Synchronous/Asynchronous Receiver/Transmitter (USART). The vision-controlled robotic arm (prototype) was designed with GLCM and USART tool having the ability to collect plastic waste materials. Such a technique does not require manpower and also avoids physical contact to waste material [28]. Another prototype (ARDUINO CONTROLLER UNO) was developed for the separation of plastic waste based on the size and color (particularly red color). Three sensors were implemented: IR, Proximity, and Color for the sorting of bigger size, metal object, and red color materials, respectively. Three DC stepper motors were also attached for moving plastic materials to the next sensor after rejecting other components. Such methodology helped in the development of automated waste segregation units [29].

3.4 CONVERSION OF WASTE AS A NEW SUSTAINABLE ENERGY SOURCE

3.4.1 WASTE TO NEW PLASTIC

Plastic recycling now follows a downward spiral rather than an infinite loop. Plastics are generally reprocessed mechanically for new plastics, which involves sorting, cleaning, shredding, melting, and re-molding. Plastic's quality deteriorates every time it is recycled in this manner. When the plastic is heated, the polymer chains are partially broken down, decreasing its tensile strength and viscosity, making it more difficult to produce. Most plastics can only be recycled a few times before it degrades to the point of being useless, and new, lower-grade plastic is usually unfit for use in food packaging. Chemical recycling is a new business that aims to tackle this problem by dissolving plastic into its chemical components, which may then be used to create fuels or reincarnate new polymers. In this section, the main focus is on the most advanced technology developed by Mura Technology, in the United Kingdom. Not only chemical recycling, but all types of plastic recycling have been hampered by financial constraints. It is just more expensive to collect, sift, and recycle packaging than it is to produce new packaging when considering a CE. Mura Technology has started the world's first commercial-scale system capable of recycling any type of plastic. The industrial plant can process different types and forms of plastics such as composite plastic, plastic in all states of decay, multilayered plastic contaminated by food or other trash. The major steps involved in the process and advantage of technology are presented in Figure 3.5 (https://muratechnology.com/technology/). INEOS is also the largest industry involved in plastic recycling and CE using advanced technology (https://www.ineos.com/sustainability/circular-economy/recycling/). In India, The Shakti Plastic Industries have developed a large-scale plant for

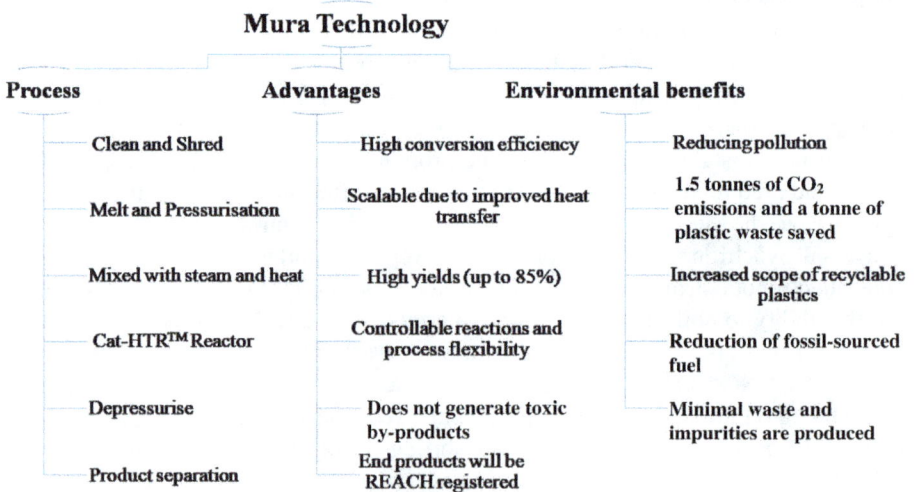

Mura Technology

Process	Advantages	Environmental benefits
Clean and Shred	High conversion efficiency	Reducing pollution
Melt and Pressurisation	Scalable due to improved heat transfer	1.5 tonnes of CO_2 emissions and a tonne of plastic waste saved
Mixed with steam and heat	High yields (up to 85%)	Increased scope of recyclable plastics
Cat-HTR™ Reactor	Controllable reactions and process flexibility	Reduction of fossil-sourced fuel
Depressurise	Does not generate toxic by-products	Minimal waste and impurities are produced
Product separation	End products will be REACH registered	

FIGURE 3.5   Overview of Mura Technology for plastic recycling.

the recycling of plastic. The Shakti Plastic Industries have sponsored many door-to-door and Kabadiwala for collection of waste from the various cities of India. They also play a crucial role in the awareness and training session on plastic waste management and recycling. They have been recognized as a pioneer in the field of plastic recycling in India (https://www.shaktiplasticinds.com/#section-waste-management-specialist). Various WtE models are described in the next section.

3.4.2 WASTE TO ENERGY

There are a variety of well-known WtE conversion methods that may be used to realize the potential of waste as a source of energy. Simple methods for disposing of dry waste to more complicated technology capable of handling enormous volumes of industrial waste are among them [30]. WtE may be defined as the transformation of something that someone does not want into something that someone else requires or desires. WtE may be easily understood by categorizing the feedstock as an end-product rather than the process itself. On the basis of outcomes, the utility of the processes or technologies may be compared. The rate of waste generation is influenced by socio-economic growth, industrialization, and climatic change. In general, the higher the economic success and the larger the percentage of urban people, the more solid waste is generated. The reduction of solid waste volume and bulk is a critical issue, especially in the scarcity of disposal sites in many areas of the world [31]. When WtE is mentioned, the first thing that comes to mind is a large incinerator burning solid MSW. Biogas may be made from the organic portion of municipal solid waste, which is primarily food; refuse-derived fuels (RDF) can be made from flammable materials; and inert materials can be repurposed as fuel. For a comparable time, the utilization of liquid wastes to create biogas has helped agricultural processing industries overcome the difficulties of low-quality electrical and heating energy [32]. Under certain conditions, after the biogas process, the solids were collected and utilized as a low-grade fertilizer. Enerkem's announcement of a large solid municipal waste-to-chemicals-and-fuels facility in Rotterdam exemplifies technology and business model innovation. The capital expenses of WtE projects range from $1000 to $1 billion. Health, urban landscapes, transportation impacts, energy availability, agricultural production, and energy security are just a few of the co-benefits included in WtE initiatives. The following sections offer a basic overview of several aspects of the WtE process, including feedstock types, conversion methods, and outputs.

3.4.2.1 Feedstock for WtE

The energy extracted from waste materials during harvesting and processing may be turned to heat, which can then be converted to electricity and process steam. Waste may be harmful to the environment if it is not properly managed. Even as modern agriculture and forestry operations have developed and become industrialized to meet growing demand, using byproducts directly as fuel remains a cost-effective choice in many instances.

The following are the many forms of waste that may be utilized as feedstock for various energy conversion technologies:

3.4.2.1.1   Food

These are edible crops that were developed and grown to provide food for humans and animals, but they may also be used as a source of energy. Food crops used for fuel include sugarcane, corn, wheat, sugar beets, sweet potatoes, sorghum, soya, and palm oil [33].

3.4.2.1.2   Agricultural Residues

Agricultural leftovers include biogases, rice husk, straw, stem, leaves, shell, stover, peel, and pulp, among other crop wastes. Crop residues are generated in large amounts every year across the world, yet they are mostly underutilized. Some crops generate a variety of trash. Rice produces straws as well as husks, which account for 25% of the rice's weight. These leftovers are currently mixed back into the soil, burned, left to rot, or grazed by cattle. These wastes might be transformed into liquid fuels or thermochemically utilized to create electricity and heat. Open burning pollution will be reduced as a result of this. Several agricultural and biomass research, on the other hand, have indicated that removing and utilizing a portion of crop residue for energy generation, resulting in massive volumes of low-cost materials, may be acceptable. Animal wastes are included in the categorization of agricultural residues. Manure from cow farms and pig farms may be gathered in large quantities and utilized as a bioenergy source. Chicken feces also contain a significant amount of nitrogen, which makes it ideal for energy production. On the other hand, stronger environmental rules regarding odor and water pollution open the opportunities for sustainable development of WtE conversion. Anaerobic digestion is the most common technique of converting animal WtE. The process creates biogas, which may be used as a cooking fuel, a fuel for internal combustion engines, or as a source of electricity while a gas engine is running [30].

3.4.2.1.3   Forestry Residues and Wood Wastes

Plantation thinning, logging road clearance, tree trimming, pulp and lumber stem wood extraction, and residues are all sources of woody materials. Forest leftovers are the byproducts of thinning in young stands or cutting in older stands for wood or pulp, which offer tops and branches that can be used for biomass energy. Additional biomass can be found in stands that have been destroyed by insects or fire. This discarded material is frequently ignored and left to decay on the site. It may, however, be collected and used to create hot gases for steam production in a biomass gasifier [34].

3.4.2.1.4   Animal Wastes

Animal and poultry manure being the most popular sources are utilized to create biomass energy in a number of ways. Previously, such garbage was collected and sold as fertilizer or simply dumped on farms, but environmental regulations regarding odor and water pollution have been strengthened. So, waste management is required, which adds to the motivation for WtE conversion. Anaerobic digestion is the most appealing and useful technique of converting these waste materials to usable form, and food industry and slaughterhouse wastes are also viable anaerobic digestion feedstock [35]. This generates biogas, which may be used to power internal combustion engines, generate electricity from small gas turbines, cook with directly, and heat space and water.

3.4.2.1.5   Sugar Industry Wastes

The majority of sugarcane mills use bagasse to generate power for their own use, although certain mills are able to export a significant quantity of electricity to the grid. The sugarcane business produces a large amount of bagasse each year. It has the potential to be a major source of biomass energy, as well as a boiler feedstock for producing steam for the process of heat and electricity generation [36].

3.4.2.1.6   Solid Wastes

Food waste, municipal solid wastes (MSW), wastewater treatment sludge, and recovered solids from sewage are all examples of solid waste. Millions of tons of household wastes are collected each year, with the most majority being deposited in landfills. Putrescibles, paper, and plastic comprise the biomass resource in MSW, which accounts for around 80% of all MSW collected. MSW may be converted to energy by either direct burning or spontaneous anaerobic digestion in the landfill. The gas generated by natural decomposition of MSW is collected, cleansed, and purified before being fed into internal combustion engines or gas turbines at these plants to create heat and power. MSW in developing countries consist mostly of organic components with high moisture content, as well as inert waste fractions such as sand or ash. Sorting municipal waste at the source enables for the creation of value and energy extraction. The organic component of MSW can be anaerobically stabilized in a high-rate digester to produce biogas for electricity or steam production [37].

3.4.2.1.7   Sewage

Households and industry are the most common sources of sewage. When residential and industrial sewage, as well as runoff from roads and other paved surfaces, is processed in a wastewater treatment plant, sewage sludge is formed, which is a mixture of water, inorganic, and organic components. Sewage sludge is a comparable biomass resource to other animal wastes. Sewage may be converted into biogas using the anaerobic digestion process. The leftover sewage sludge can be thermally treated and pyrolyzed to create additional energy.

3.4.2.1.8   Black Liquor

Using Upflow Anaerobic Sludge Blanket digestion technology, black liquor may be wisely used for biogas generation. The effluent emitted by this industry is extremely diverse, since it contains compounds from wood or other raw materials, processed chemicals, and compounds created during processing. The pulp and paper industry is one of the most polluting industries, using large amounts of energy and water in a variety of unit operations [38].

3.4.2.1.9   Energy Crops

Energy crops are frequently grown as a low-cost, low-maintenance crop to be used in the manufacture of biofuels like bioethanol or to be burned for their energy content to generate electricity or heat [39]. Woody crops like jatropha, willows, and poplars, as well as grasses like elephant grass, Napier grass, miscanthus, and switchgrass, are examples of energy crops.

3.4.2.1.10 Industrial Wastes

In general, these are wastes produced by numerous businesses for the production of biomass energy. The food sector generates a large quantity of wastes and byproducts that may be utilized to generate biomass energy. Biomass wastes are produced in both solid and liquid form in the meat and confectionary sectors. Solid waste includes fruit and vegetable peelings and scraps, food that does not meet quality control standards, pulp and fiber from sugar and starch extraction, filter sludges, and coffee grounds. Liquid wastes include waste produced by washing meat, fruits, and vegetables, cleaning and processing activities, and winemaking, among other things. The airline industry generates a lot of wastes from meals, which is mostly plastic and food leftovers. The waste aspect of the cruise ship industry and huge malls is comparable. Another type of industrial waste is palm oil mill effluent (POME). It is an acidic, thick brownish liquid that comes out of a palm oil mill's sterilizing, clarifying, and separation operations [32]. For every ton of fresh fruit bunch of oil palm processed, 0.65 m^3 of POME is generated. In a traditional palm oil mill, a biomass power plant driven by palm kernel shell and mesocarp fiber generates energy and processes steam. This function can be replaced with a biogas plant that uses POME as an input, allowing more shell and fiber to be used as boiler fuel. This reduces the amount of fuel oil that must be purchased, allowing the mill to make a profit [40].

3.4.2.2 WtE Conversion Methods

The available WtE conversion strategies are summarized in this section. The compendium includes 18 case studies of WtE initiatives from throughout the world, using the feedstock described in the preceding section as a starting point.

3.4.2.2.1 Thermal

3.4.2.2.1.1 Direct Combustion   Direct combustion is the earliest biomass conversion technique, particularly for generating heat and steam as well as, in the presence of oxygen, it burns biomass. Through direct fire, a biomass combustion plant can create steam, electricity, or both (combined heat and power [CHP]). The combustion methods used to convert renewable biomass fuels to heat and electricity are similar to those used to convert fossil fuels to heat and power. The biomass fuel is burned in a boiler to create high-pressure steam that passes through a set of turbine blades, rotating the turbine. The turbine is linked to an electric generator, which generates power. District heating and cooling systems can also benefit from steam. Figure 3.6 shows the schematic process of direct combustion. Co-firing is the process of combining fossil fuels such as coal or natural gas with biomass as a feedstock. Co-firing with biomass might be a viable option for meeting tight emissions restrictions. Heat is a byproduct of electricity generation; as a result, all power plants generate heat, which is generally vented into the atmosphere via cooling towers or discharged into adjacent bodies of water. The waste heat from the CHP process is collected and used in district heating. Cogeneration transforms roughly 85% of the potential energy in biomass into usable energy [41]. When compared to facilities that solely generate electricity, the CHP plant is very resource-efficient, producing more

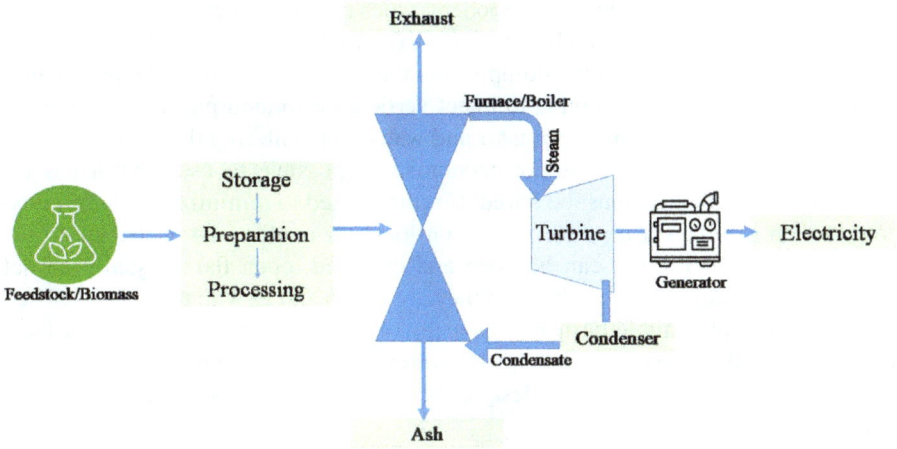

FIGURE 3.6 Direct combustion/steam turbine process.

energy per unit of biomass used. Most biomass-fired facilities are placed in locations with a consistent supply of biomass, such as sugarcane mills, rice mills, and paper mills, for operational efficiency.

3.4.2.2.2 Mechanical and Thermal (Combined)

3.4.2.2.2.1 Mechanical Biological Treatment Mechanical biological treatment (MBT) combines mechanical (e.g., sorting, shredding, milling, separating, or screening) and biological (drying, composting, or anaerobic digestion) components to produce solid recovered fuel or RDF and redirect organic materials for fertilizer and energy [42]. This fuel may be processed further into pellets or briquettes and used as a feedstock in energy plants to replace fossil fuels. MBT has a number of favorable environmental effects, including increased landfill efficacy owing to enhanced leachate and landfill gas output and quality. From mixed waste streams, MBT can also recover a higher percentage of recyclables. MBT is made up of four separate treatment procedures that result in four different sorts of outputs:

  i. RDF has a high calorific value due to the high paper and plastic content.
 ii. Stabilized organic waste is produced by the biological treatment of the organic portion of the waste.
iii. Ferrous and non-ferrous metals used for potential recycling.
 iv. Inert wastes–scraps residues are disposed of in landfills.

3.4.2.2.2.2 Landfill Gas Capture Landfill gas, which is composed of 35–55% methane, is produced during the operation of an engineered or sanitary landfill by the anaerobic decomposition of organic waste in the landfill body. A landfill gas recovery plant, which includes an extraction system and a flaring system, is constructed to catch the methane produced [43].

Extraction system: Different components, such as vertical perforated pipes, horizontal perforated pipes, and ditches, are used to extract gas from landfills. Membrane is occasionally used to cover the dump where the gas is collected. The most popular method of active gas collection is to inject vertical perforated pipes into the waste mass to collect gas while preventing air and water from entering the system.

Flaring system: When it is not economically possible to use landfill gas for energy purposes, the gas must be flared. Flaring is used to minimize methane emissions, which can have a negative impact on local air quality and contribute to the greenhouse effect. Flaring can be open and enclosed, open flares usually do not fulfill emission requirements, although they are cheap and easy to use. The enclosed system consists of a single burner or an array of burners contained in a refractory-lined cylindrical enclosure, also helps to eliminate smells as well as the risk of fire and explosion. As a result of this design, the fire burns more evenly and emits less pollutants.

3.4.2.2.3 Thermochemical

Biomass has several limits when compared to fossil fuels, making it difficult to employ on a wide scale. Poor bulk density, high moisture content, and low calorific value of raw biomass all have an influence on logistics and ultimate energy efficiency. Large amounts of biomass are required because of its low energy density, making storage, transportation, and handling logistically problematic. Biomass has a high moisture content, which decreases the efficiency of the process and raises the cost of fuel generation. The uneven forms of raw biomass are a problem, especially during feeding and co-firing or gasification systems. Because it contains more oxygen than carbon and hydrogen, it is less suitable for thermochemical conversion. Raw biomass must be treated to make it acceptable for energy uses in order to overcome these obstacles.

3.4.2.2.3.1 Torrefaction Torrefaction is a thermal pre-treatment procedure for raw biomass that changes the physical and chemical composition. Torrefaction is the process of heating biomass to temperatures ranging from 200°C to 400°C without the use of air. Moisture evaporates and low-calorific components or volatiles contained in the biomass are pushed out when heated to the specified temperatures. The hemicellulose in the biomass decomposes during this process, converting it from a fibrous low-grade fuel to a product with good fuel properties. This method decreases the biomass weight by around 20–30% while only losing 10–15% of the energy. Torrefaction transforms biomass into a coal-like material that burns more efficiently than the original biomass and it is more brittle, making it easier to process and using less energy. Once torrefied, raw biomass becomes a high-grade biofuel that may be used to replace coal in electricity generation and as a feedstock for gasification operations.

3.4.2.2.3.2 Gasification With the aid of air or steam at 800–1000°C, gasification converts the carbon in organic waste into a synthetic gas (syngas) mostly composed of carbon monoxide and hydrogen. After that, syngas may be burnt to generate heat energy. Gasification is achieved by regulating the absence of extremely low quantities of oxygen during partial oxidation. The following zones occur within a biomass gasifier when producing gas from biomass: (i) drying, (ii) pyrolysis, (iii) combustion, and (iv) reduction.

i. The removal of surface water by filtration, evaporation, or a mix of both is part of the drying process. Evaporation is generally done with waste. Pyrolysis is simply a charring process. To make more flammable gas, the char is reacted with steam or burnt in a little amount of air or oxygen.

ii. Once the temperature hits about 240°C during the pyrolysis stage, biomass quickly decomposes with heat. Biomass decomposes into solids, liquids, and gases. The solid component is often referred to as charcoal, while the mixture of gas and liquid discharged is referred to as tars. When big molecules like tar are exposed to heat, they decompose into lighter gases, this is known as cracking.

iii. Because tar gases produce sticky tar that clogs an engine's valves, the procedure is critical in the generation of clean gas that is suitable with an internal combustion engine. Cracking is also necessary for adequate combustion to take place. Only when combustible gases are completely combined with oxygen can complete combustion occur.

iv. The next step is to reduce or remove oxygen from waste products at a high temperature, resulting in flammable gases. All of the heat used in drying, pyrolysis, and reduction originates from combustion, either directly or indirectly through heat exchange activities in a gasifier. During combustion, tar gases or char from pyrolysis can be used as fuel.

There are five types of gasifier used to achieve the above processes, which are Downdraft, Updraft, Cross draft, Fluidized bed, and Plasma Gasifier.

3.4.2.2.3.3 Pyrolysis Pyrolysis is a thermochemical reaction that takes place in the absence of oxygen at temperatures ranging from 400°C to 600°C. The organic material does not burn because there is no oxygen present, but chemical components like cellulose, hemicellulose, and lignin degrade producing flammable gases, and charcoal pyrolysis generates three products depending on parameters such as temperature, pressure, and heating rate, which are solid, liquid, and gas in the form of bio-oil, syngas, and biochar. Pyrolysis has a number of advantages as it is a simple technique that can handle a wide range of feedstock. Biochar, bio-oil, and syngas, among other pyrolysis products, have the potential to lessen the country's reliance on imported energy supplies by utilizing locally accessible resources. Pyrolysis may be done on a small scale, allowing it to be used even in remote places where biomass is accessible, lowering transportation and handling expenses. Pyrolysis does have greater related expenses to some level, and its profitability is mostly determined by the price of biomass. Bio-oil is likewise of low quality, and it usually requires further refinement before it can be used in fossil-fuel-powered applications.

3.4.2.2.3.4 Liquefaction Hydrothermal liquefaction is a thermochemical method for converting organic matter into liquid bio-crude and co-products. Liquefaction takes place at a moderate temperature between 300°C and 400°C with the addition of a reducing agent, generally hydrogen or carbon monoxide. Because biomass is wet, it is treated via hydrothermal processing, which includes heating aqueous slurries. Temperature and heating rate, solvent, pressure, feedstock content, residence time, and catalysts all impact the conversion of biomass to bio-oil. While any biomass may be turned to bio-oil by hydrothermal liquefaction, the yield and quality of bio-oil are determined by the organic components in the feedstock, that is, cellulose, hemicellulose, protein, and lignin. The conversion process is heavily influenced by temperature as temperatures above the ideal induce more char to develop, which leads to more gas generation, but temperatures below the ideal diminish depolymerization and bio-oil yields. The process of hydrothermal liquefaction is quick. Residence durations are measured in minutes and are influenced by a variety of factors such as temperature, feedstock, and solvent ratio. Water functions as a catalyst in the process, although additional catalysts can be employed in the reaction vessel to improve conversion. The use of hydrothermal liquefaction to create biofuels has the benefit of producing no net carbon emissions.

3.4.2.2.4   *Biochemical*

3.4.2.2.4.1   Fermentation   Fermentation is a type of anaerobic digestion that breaks down glucose in organic materials. Sugar is transformed to alcohol or acid by a sequence of chemical reactions. The biomass material is inoculated with yeast or bacteria, which feed on the sugar and generate ethanol and carbon dioxide. Corn and sugarcane are the most popular agricultural wastes cultivated for industrial ethanol production; nevertheless, more innovative methods using lignocellulosic waste materials as feedstock are being developed. To be utilized as a transportation fuel, bioethanol must go through a distillation process to reach the requisite purity. The leftovers from the fermentation process can be utilized as animal feed, and biogases can be used as boiler fuel. India, the People's Republic of China (PRC), and Thailand are among the world's top ethanol producers, with corn, wheat, cassava, and molasses serving as the primary feedstock.

3.4.2.2.4.2   Anaerobic Digestion   Anaerobic digestion is a method of producing biogas by digesting organic waste in an oxygen-free atmosphere. Fermentation is one of the steps in anaerobic digestion, and biogas facilities employ either wet or dry fermentation. A liquid biomass slurry is maintained within the reactor allowing anaerobic digestion to continue over several weeks in wet fermentation, which is more frequent. It may be necessary to pre-treat organic inputs and/or add fluids to make them acceptable for wet fermentation. In addition, in order for microorganisms to come into touch with the organic matter, the feedstock in the reactor may need to be mechanically cycled or heated to provide a comfortable environment for the bacteria.

3.4.2.3   Outputs

3.4.2.3.1   *Energy Outputs*

3.4.2.3.1.1   Heat   Biomass may generate heat for space heating, hot water, and process heating/steam generation. The most popular way of producing heat from biomass is direct combustion, which uses a variety of feedstock such as agricultural wastes, MSW, wood, and forest leftovers, among others. Biomass heating systems come in a variety of sizes, ranging from kilowatts (kW) to megawatts (MW). District heating systems and process heat applications for businesses that produce biomass, such as sawmills, rice mills, sugar mills, alcohol plants, furniture manufacture, and agricultural drying sites, are the main markets for biomass heating. Because many sectors demand heat throughout the year, using biomass for heating can result in significant fuel cost reductions.

3.4.2.3.1.2 Power A biomass-based power system is a cost-effective way to generate energy without using fossil fuels. Biomass may be used to generate power in a variety of ways. Direct combustion, gasification, anaerobic digestion, and pyrolysis are examples of these processes. As stated in the preceding section, different feedstock kinds can also be utilized for heating reasons. The power generation capacity ranges from kilowatts to megawatts.

3.4.2.3.1.3 Transportation Fuels and Additives Biofuels, together with improvements in fuel efficiency and electrification of the light vehicle fleet, are considered as one of the most realistic solutions for decreasing carbon emissions in the transportation industry. Because electric vehicles and fuel cells are not viable for heavy-duty vehicles, maritime vessels, and airplanes, biofuels will play an increasingly important role in reducing CO_2 emissions. Biomass offers a variety of possibilities for producing gasoline and diesel fuel replacements. Some, like the production of ethanol as a gasoline substitute or processed vegetable oils (biodiesel) as a diesel fuel substitute, are well-known; others, like the gasification of biomass to produce hydrogen for fuel cell vehicles or synthetic hydrocarbons for conventional vehicles, are more speculative.

3.4.2.3.1.4 Bio-oil Bio-oil is a byproduct of pyrolysis, particularly fast pyrolysis, which entails the rapid thermal breakdown of carbon-based materials at moderate to high temperatures. Bio-oil can be utilized as a low-grade diesel fuel and as a chemical feedstock. When compared to fossil fuels, using bio-oil has certain environmental benefits.

3.4.2.3.1.5 Biochar Biochar is a type of charcoal made from the pyrolysis and torrefaction of biomass and used in soil. Because biochar is naturally alkaline, it lowers the overall acidity of the soil. Due to its high porosity, it also helps enhance agricultural production by increasing water and nutrient retention. It is extremely effective in absorbing CO_2 from the atmosphere. It may be utilized as a source of energy, but it is best employed as a high-surface-area scaffold for microorganisms to fix nutrients into soils.

3.4.2.3.1.6 Syngas Synthetic gas, often known as syngas, is created through the processes of gasification and pyrolysis. Syngas is a gas made up of carbon monoxide, methane, and hydrogen that may be used to power turbines and generate energy. It may potentially be used to replace natural gas or turned into a biofuel.

3.4.2.3.1.7 Biogas Biogas is a gas combination that generally contains 50–75% methane, 25–45% carbon dioxide, and trace amounts of other gases. In an airtight holding tank or reactor, a biogas plant allows the anaerobic digestion of organic inputs such as municipal and industrial waste, agricultural and agro-industrial byproducts. In small-scale systems, the biogas is burnt to create heat, which is then turned into electrical energy or utilized for other purposes, such as cooking.

3.4.2.3.1.8 *Compressed Natural Gas*

Compressed natural gas (CNG) is a gaseous form of methane kept in high-pressure tanks and used as an alternative to automobile fuels. CNG cars are widely used in India, Indonesia, Pakistan, the People's Republic of China, Thailand, and Uzbekistan. Biogas from anaerobic digestion of agro-industrial waste or landfill gas, which has been improved for this purpose by removing contaminants and is also known as biomethane, is one source of CNG.

3.4.2.3.2 *Non-energy Outputs*

3.4.2.3.2.1 *Ash*

The burning of biomass produces ash, which accounts for roughly 5–15% of all biomass treated. The inclusion of heavy metals and other inorganic compounds, which are produced as a result of the thermochemical processes that occur when biomass is combusted, limits the use of ash. Depending on the feedstock, ash may be utilized in a variety of applications. Bottom ash can be utilized as an agricultural fertilizer, a building material additive, or a neutralizing and liming agent in organic and homogenous feedstock. The permissible concentration of heavy metals and persistent organic contaminants (such as dioxins) in municipal waste fly ash has been shown through testing and has been utilized as a road base or building materials. The fly ash must be disposed of at a hazardous waste landfill if the testing results are unacceptable. Significant amounts of silica may be present in fly ash from homogeneous processes. The application of this fly ash necessitates testing, which will decide its usage.

3.4.2.3.2.2 *Alcohol*

Bioethanol is the alcohol generated when biomass is fermented. Sugar-rich crops such as sugarcane, potato, beetroot, and corn are used as feedstock in the fermentation process. Bioethanol is made from the sugars in these crops, which are fermented by yeast strains. The use of ethanol as an alternative fuel is one of the options for reducing greenhouse gas emissions; nevertheless, the high cost of manufacturing is currently the biggest barrier.

3.4.2.3.2.3 *Bio-fertilizer*

The leftover slurry from animal manure-based biogas generation has been shown to provide a more significant short-term soil conditioning impact when applied as fertilizer, since nitrogen and other nutrients are more readily accessible, as opposed to undigested manure. Sludge has also been proven to contribute to high rates of stable humus formation, earthworm activity, and nitrogen loss mitigation. While most vegetable crops and many varieties of fruit appear to benefit from sludge fertilization, the fertilizing impact varies by plant and is reliant on climate and soil type. During the manufacturing process, all bio-fertilizers must be checked for contamination.

3.4.2.3.2.4 *Digestate*

Digestate is made up of indigestible matter and dead microorganisms that remain after the anaerobic digestion process. It is a nutrient-dense material that may be used as a fertilizer but is not compost, despite its comparable properties. Digest ate can be applied directly to the soil, but separating the liquid and fiber components, which have different nutritional distributions, is another alternative. The liquor is easy to apply over crops, and the separated fiber can be utilized as a soil conditioner.

3.5 COMPARISON OF BYPRODUCTS IN TERMS OF THE CIRCULAR ECONOMY

The CE idea has acquired the attention of both researchers and industrial experts. Total 114 different definitions of CE were identified and systematically analyzed by 17 different dimensions. Nonetheless, critics guarantee that it implies various things to various individuals. The research findings on CE definition revealed that CE is a combination of reduce, reuse, and recycle activities. Economic success from waste materials is considered the primary goal of the CE, followed by environmental quality, social equality, and future generations [44]. Comparison of waste to byproducts is represented in Table 3.2. The feasibility of thermoplastics recycling is influenced

TABLE 3.2

Comparison of Waste to Byproduct with Respect to the Circular Economy

Sr No.	Amount of Plastic Material	Converted into or Byproduct/Product	Revenue	References
1	1 kg plastic waste	750 ml of automotive grade gasoline	–	[45]
2	1 ton plastic waste	1 km road construction	18.9 lakh INR (saving 6.3 lakh as compared to plain and regular method)	[46]
3	156 kg plastic waste	Toilet	7556 INR (saving 19,444 INR as compared to cement wall structure)	[46]
4	Feeding weight 1110 kg/day	Net oil yield: 666 kg	–	[47]
5	Mixed plastic: Used PE (205 kg) and PP (410 kg)	Yield (kg): 484 hydrocarbon oil; 88 solid residues; 43 off-gas	–	[47]
6	20,000 tons of plastic waste	13,000 tons of fuel	Net profit in first year: 22,545,654,327	[48]
7	Plastic waste: 26,000 tons per annum	Output (tons per annum) Pyrolysis: 16,000 Catalytic depolymerization: 18,000 Gasification w/F-T: 10,000 Gasification w/Bio: 9,000 Gasification w/MTG: 16,000	Operating income £7,814,000 £6,445,000 £5,637,000 £5,379,000 £6,446,000	[49]

Note: '–' represent data not available in the reference.

by two major economic considerations. These are the costs of recycled polymer versus virgin polymer, as well as the cost of recycling as compared to other suitable disposal options.

3.6 LIMITATIONS OF CURRENT MODEL

Plastics are highly designed products that are made in sophisticated and tightly regulated manufacturing operations to meet precise standards that are linked to a set of physical characteristics. Material diversity and the costs of identifying and segregating waste plastics into recognizable grade ranges are the two most major roadblocks in the recycling process. Single-use plastics are the most straightforward to recycle, as long as they can be collected and reprocessed at a reasonable cost [50]. However, due to low numbers and low weight of such materials, it may include high collection and plant implementation cost. Processing of mixed plastic waste to fuel conversion may lead to low yield and low caloric byproduct. Another major issue is separation of byproducts or various oil generated during the process. Apart from that, black oil was also generated after completion of a process which is highly toxic in nature.

3.7 CONCLUSION AND FUTURE SCOPE

The current rate of economic expansion is unsustainable due to the depletion of fossil fuel sources such as crude oil, natural gas, and coal. As a result, many renewable energy sources have been used; however, certain others, such as plastic waste, have yet to be completely established as a full-scale economic business. The short-term activities to be followed immediately include reducing the use of plastics, reusing them multiple times, collecting plastics separately, and recycling as much as feasible. Ideal technology is also required to deal with the sub-national variation of plastic waste materials for upcycling. Biological cycling of plastic waste into biopolymer is still unexplored and need more research. Long-term plans to combat polymeric wastes in the future will require the promotion of research on degradable polymers and a campaign to modify human habitual behavior on plastics usage via adequate teaching from schools to universities. Virgin and most of the single-use plastic are produced from fossil fuels. Only 2% of plastic is produced from old or recycled plastic materials as recycling reduces the quality of products. There is a high need for moving WtE conversion. WtE is a collection of transformational processes that may be utilized on a project-by-project basis to produce energy, reduce waste, improve environmental results, and establish flourishing secondary markets for reused, repurposed, and upcycled materials. To create more CE, the technology may be employed in both smart cities and thriving rural economies. This chapter has demonstrated a variety of technologies that have substantial benefits. The technical presentation of materials and the lack to clarify an economic and developmental justification for their acceptance have hampered the industry's progress. Funding and operational difficulties have plagued the small number of new businesses. As a result, there is a notion that the sector is tough. The road to a completed project with positive cash flows will be shortened thanks to innovation in risk allocation and finance. However, for infrastructure with a life duration of 20–40 years, caution should be used in capacity planning and assumptions. Lack of significant knowledge and awareness regarding the availability of recycled plastics, their quality, and their appropriateness for certain applications can also be a deterrent to using it. Joint efforts of NGO, Government and citizen science approach may help to move faster in the field of plastic as a source of energy in the CE era.

REFERENCES

1. C. Lott *et al.*, "Half-life of biodegradable plastics in the marine environment depends on material, habitat, and climate zone," *Front. Mar. Sci.*, vol. 8, no. May, pp. 1–19, 2021, doi: 10.3389/fmars.2021.662074.
2. K. Hiraga, I. Taniguchi, S. Yoshida, Y. Kimura, and K. Oda, "Biodegradation of waste PET," *EMBO Rep.*, vol. 20, no. 11, pp. 1–5, 2019, doi: 10.15252/embr.201949365.
3. J. Oliveira *et al.*, "Marine environmental plastic pollution: Mitigation by microorganism degradation and recycling valorization," *Front. Mar. Sci.*, vol. 7, no. December, 2020, doi: 10.3389/fmars.2020.567126.
4. I. Taniguchi, S. Yoshida, K. Hiraga, K. Miyamoto, Y. Kimura, and K. Oda, "Biodegradation of PET: Current status and application aspects," *ACS Catal.*, vol. 9, no. 5, pp. 4089–4105, 2019, doi: 10.1021/acscatal.8b05171.
5. H. K. Webb, J. Arnott, R. J. Crawford, and E. P. Ivanova, "Plastic degradation and its environmental implications with special reference to poly(ethylene terephthalate)," *Polymers (Basel).*, vol. 5, no. 1, pp. 1–18, 2013, doi: 10.3390/polym5010001.
6. S. Awasthi, P. Srivastava, P. Singh, D. Tiwary, and P. K. Mishra, "Biodegradation of thermally treated high-density polyethylene (HDPE) by *Klebsiella pneumoniae* CH001," *3 Biotech*, vol. 7, no. 5, pp. 1–10, 2017, doi: 10.1007/s13205-017-0959-3.
7. A. Chamas *et al.*, "Degradation rates of plastics in the environment," *ACS Sustain. Chem. Eng.*, vol. 8, no. 9, pp. 3494–3511, 2020, doi: 10.1021/acssuschemeng.9b06635.
8. H. Fakhrul, A. Fazal, R. Farooq, R. Sohaib, G. Abdul, and S. Muhammad, "Assessment of biodegradability of PVC containing cellulose by white rot fungus," *Malays. J. Microbiol.*, vol. 10, no. 2, 2014, doi: 10.21161/mjm.55113.
9. B. Y. Peng *et al.*, "Biodegradation of polyvinyl chloride (PVC) in *Tenebrio molitor* (Coleoptera: Ttenebrionidae) larvae," *Environ. Int.*, vol. 145, no. August, p. 106106, 2020, doi: 10.1016/j.envint.2020.106106.
10. B. Cichy *et al.*, "Investigating the degradability of HDPE, LDPE, PE-BIO, and PE-OXO films under UV-B radiation," *J. Spectrosc.*, vol. 2015, pp. 965–970, 2015.
11. B. M. Kyaw, R. Champakalakshmi, M. K. Sakharkar, C. S. Lim, and K. R. Sakharkar, "Biodegradation of low density polythene (LDPE) by pseudomonas species," *Indian J. Microbiol.*, vol. 52, no. 3, pp. 411–419, 2012, doi: 10.1007/s12088-012-0250-6.
12. J. Arutchelvi, M. Sudhakar, A. Arkatkar, M. Doble, S. Bhaduri, and P. V. Uppara, "Biodegradation of polyethylene and polypropylene," *Indian J. Biotechnol*, vol. 7, no. 1, pp. 9–22, 2008.
13. P. Luthra, K. K. Vimal, V. Goel, R. Singh, and G. S. Kapur, "Biodegradation studies of polypropylene/natural fiber composites," *SN Appl. Sci.*, vol. 2, no. 3, pp. 1–13, 2020, doi: 10.1007/s42452-020-2287-1.
14. B. T. Ho, T. K. Roberts, and S. Lucas, "An overview on biodegradation of polystyrene and modified polystyrene: The microbial approach," *Crit. Rev. Biotechnol.*, vol. 38, no. 2, pp. 308–320, 2018, doi: 10.1080/07388551.2017.1355293.
15. L. P. Wackett, "Bio-based and biodegradable plastics: An annotated selection of World Wide Web sites relevant to the topics in microbial biotechnology," *Microb. Biotechnol.*, vol. 12, no. 6, pp. 1492–1493, 2019, doi: 10.1111/1751-7915.13502.
16. Y. Tokiwa, B. P. Calabia, C. U. Ugwu, and S. Aiba, "Biodegradability of plastics," *Int. J. Mol. Sci.*, vol. 10, no. 9, pp. 3722–3742, 2009, doi: 10.3390/ijms10093722.

17. A. C. Vegter *et al.*, "Global research priorities to mitigate plastic pollution impacts on marine wildlife," *Endanger. Species Res.*, vol. 25, no. 3, pp. 225–247, 2014, doi: 10.3354/esr00623.

18. N. Laskar, and U. Kumar, "Plastics and microplastics: A threat to environment," *Environ. Technol. Innov.*, vol. 14, p. 100352, 2019, doi: 10.1016/j.eti.2019.100352.

19. L. Peng, D. Fu, H. Qi, C. Q. Lan, H. Yu, and C. Ge, "Micro- and nano-plastics in marine environment: Source, distribution and threats—A review," *Sci. Total Environ.*, vol. 698, p. 134254, 2020, doi: 10.1016/j.scitotenv.2019.134254.

20. C. Wilcox, E. Van Sebille, B. D. Hardesty, and J. A. Estes, "Threat of plastic pollution to seabirds is global, pervasive, and increasing," *Proc. Natl. Acad. Sci. U. S. A.*, vol. 112, no. 38, pp. 11899–11904, 2015, doi: 10.1073/pnas.1502108112.

21. P. S. Puskic, J. L. Lavers, and A. L. Bond, "A critical review of harm associated with plastic ingestion on vertebrates," *Sci. Total Environ.*, vol. 743, p. 140666, 2020, doi: 10.1016/j.scitotenv.2020.140666.

22. T. S. M. Amelia, W. M. A. W. M. Khalik, M. C. Ong, Y. T. Shao, H. J. Pan, and K. Bhubalan, "Marine microplastics as vectors of major ocean pollutants and its hazards to the marine ecosystem and humans," *Prog. Earth Planet. Sci.*, vol. 8, no. 1, 2021, doi: 10.1186/s40645-020-00405-4.

23. CPCB, *Assessment and Characterisation of Plastic Waste Generation in 60 Major Cities*, 2015.

24. S. A. Mason, V. G. Welch, and J. Neratko, "Synthetic polymer contamination in bottled water," *Front. Chem.*, vol. 6, no. September, 2018, doi: 10.3389/fchem.2018.00407.

25. S. Harding, *Marine Debris: Understanding, Preventing and Mitigating the Significant Adverse Impacts on Marine and Coastal Biodiversity*, No. 83. Quebec, Canada: Secretariat of the Convention on Biological Diversity, 2016.

26. US EPA, *Plastics: Material-Specific Data*, 2018. https://www.epa.gov/facts-and-figures-about-materials-waste-and-recycling/plastics-material-specific-data.

27. E. Dijkgraaf, and R. Gradus, "Post-collection separation of plastic waste: Better for the environment and lower collection costs?" *Environ. Resour. Econ.*, vol. 77, no. 1, pp. 127–142, 2020, doi: 10.1007/s10640-020-00457-6.

28. R. P. Subin, S. Jeyanthi, and S. Rajesh, "An approach for real time plastic waste segregation," *Appl. Mech. Mater.*, vol. 787, pp. 138–141, 2015, doi: 10.4028/www.scientific.net/amm.787.138.

29. R. Sureshkumar, S. J. Suji Prasad, T. Annas, L. Aruna, and D. G. Iniya, "Automatic segregation of plastic waste for recycling industries," *Int. J. Sci. Technol. Res*, vol. 9, no. 2, pp. 424–427, 2020.

30. M. U. Muhammad, L. Abubakar, A. Musa, and A. S. Kamba, "Utilization of wastes as an alternative energy source for sustainable development: A review," *ChemSearch J*, vol. 4, no. 1, pp. 57–61, 2013.

31. D. Hoornweg, L. Thomas, and U. D. Sector, *What a Waste : Solid Waste Management in Asia*, 1999. [Online]. Available: https://agris.fao.org/agris-search/search.do?recordID=US2012400855.

32. A. S. Rahayu *et al.*, *Handbook POME-to-Biogas Project Development in Indonesia*, 2nd ed. The United States Agency for International Development (USAID), 2015.

33. A. Paschalidou, M. Tsatiris, and K. Kitikidou, "Energy crops for biofuel production or for food? – SWOT analysis (case study: Greece)," *Renew. Energy*, vol. 93, pp. 636–647, 2016, doi: 10.1016/j.renene.2016.03.040.

34. R. Ruan *et al.*, "Chapter 12 – Gasification and pyrolysis of waste," in *Current Developments in Biotechnology and Bioengineering*, R. Kataki, A. Pandey, S. K. Khanal, and D. Pant, Eds. Elsevier, Amsterdam, Netherlands, 2020, pp. 263–297.

35. M. Svanberg, C. Finnsgård, J. Flodén, and J. Lundgren, "Analyzing animal waste-to-energy supply chains: The case of horse manure," *Renew. Energy*, vol. 129, pp. 830–837, 2018, doi: 10.1016/j.renene.2017.04.002.
36. N. Mohan, S. Awasthi, and A. Agarwal, "Sugar industry-sustainable source of bio-energy/renewable fuel," *IJERT*, vol. 10, no. 01, pp. 543–549, 2021.
37. A. S. Nugroho, M. Rahmad, F. N. Chamim, and Hidayah, "Plastic waste as an alternative energy," *AIP Conf. Proc.*, vol. 1977, pp. 1–6, 2018, doi: 10.1063/1.5043022.
38. X. Chen *et al.*, "Recycling of dilute deacetylation black liquor to enable efficient recovery and reuse of spent chemicals and biomass pretreatment waste," *Front. Energy Res.*, vol. 6, no. June, pp. 1–11, 2018, doi: 10.3389/fenrg.2018.00051.
39. K. Launder, *Energy Crops and Their Potential Development in Michigan*, Lansing, MI 48909, 2002. [Online]. Available: http://michiganbiomass.com/
40. M. J. Chin, P. E. Poh, B. T. Tey, E. S. Chan, and K. L. Chin, "Biogas from palm oil mill effluent (POME): Opportunities and challenges from Malaysia's perspective," *Renew. Sustain. Energy Rev.*, vol. 26, pp. 717–726, 2013, doi: 10.1016/j.rser.2013.06.008.
41. S. Zhang, *Thermal Conversion*, 2015. https://allaboutbiofuels.wixsite.com/biofuels.
42. F. Fei, Z. Wen, S. Huang, and D. De Clercq, "Mechanical biological treatment of municipal solid waste: Energy efficiency, environmental impact and economic feasibility analysis," *J. Clean. Prod.*, vol. 178, pp. 731–739, 2018, doi: 10.1016/j.jclepro.2018.01.060.
43. H. Terraza, and H. Willumsen, *Guidance Note on Landfill Gas Capture and Utilization*, 2009. [Online]. Available: https://publications.iadb.org/publications/english/document/Guidance-Note-on-Landfill-Gas-Capture-and-Utilization.pdf.
44. J. Kirchherr, D. Reike, and M. Hekkert, "Conceptualizing the circular economy: An analysis of 114 definitions," *Resour. Conserv. Recycl.*, vol. 127, no. April, pp. 221–232, 2017, doi: 10.1016/j.resconrec.2017.09.005.
45. R. R. N. S. Bhattacharya, K. Chandrasekhar, M. V. Deepthi, P. Roy, and A. Khan, *Fact Sheet on Plastic Waste in India*, Delhi, Inida, 2018. [Online]. Available: http://www.teriin.org/sites/default/files/files/factsheet.pdf.
46. S. Sharma, and S. Mallubhotla, *Plastic Waste Management Waste Issues, Solutions & Case Studies*, 2nd ed., no. March. Ministry of Housing and Urban Affairs, Government of India, 2019.
47. United Nations Environment Programme, *"Converting Waste Plastics into a Resource Compendium of Technologies,"* Osaka/Shiga, Japan, 2009.
48. I. Fahim, O. Mohsen, and D. Elkayaly, "Production of fuel from plastic waste: A feasible business," *Polymers (Basel).*, vol. 13, no. 6, pp. 1–9, 2021, doi: 10.3390/polym13060915.
49. S. Haig, L. Morrish, R. Morton, U. Onwuamaegbu, P. Speller, and W. Simon, *Plastics to Oil Products*, 2015. [Online]. Available: www.zerowastescotland.org.uk.
50. J. Hopewell, R. Dvorak, and E. Kosior, "Plastics recycling: Challenges and opportunities," *Philos. Trans. R. Soc. B Biol. Sci.*, vol. 364, no. 1526, pp. 2115–2126, 2009, doi: 10.1098/rstb.2008.0311.

4 Layout Planning of a Small-Scale Manufacturing Industry Using an Integrated Approach Based on AHP-PSI

Parveen Sharma[1], Anil B. Ghubade[1],
Dipesh Popli[2], and Tazim Ahmed[3]
[1]School of Mechanical Engineering, Lovely
Professional University, Phagwara, India
[2]Mechanical Engineering Department, GD
Goenka University, Gurgaon, India
[3]Industrial and Production Engineering, Jessore
University of Science and Technology, Bangladesh

CONTENTS

4.1 INTRODUCTION

The layout of the facilities on the shop floor is the most critical and important part of the design when setting up an industry. It deals with organizing the shop floor facilities such as machines, appliances, departments, etc., in an appropriate manner to get the high production from them. Without proper facility layout, many adversities can be arising, such as longer queue, higher inventory work-in-process, and overload of equipment for material handling [1, 2]. The quality of a good facility layout on the shop floor has efficient utilization of energy, less inventory of work-in-process, easy moving of vehicles on the site, etc. There are numerous factors that affect the facilities layout. The two categories of factors are: qualitative and quantitative. The factors which are linked to quality and cannot measure directly in any mathematical form are considered as qualitative factors such as space occupation and utilization, flexibilities in shop floor, person issues, maintenance possibilities, etc. Conversely, if factors that could be measured directly and can express in mathematic ways are considered as quantitative factors. Some quantitative factors are product output, overall time in the system, overall distance traveled, etc. During the facility layout design, all these factors cannot be considered simultaneously. Therefore, their analysis becomes important during layout planning and design.

The layout analysis plays an incredibly significant part in the layout design stage. There are numerous strategies and methodologies used in layout planning. The multi-criteria decision making (MCDM) techniques have been implemented by researchers' practitioners to solve layout-based difficulties. Numerous methodologies are available in the literature, which fall into the class MCDM, such as SAW, TOPSIS, GTMA, AHP, and VIKOR [3–10]. Kuo et al. [11] studied and implemented the gray relational analysis model in his work for solving multiple attribute decision-making layout problems. In 2003, Yang and Kuo [12] studied and adopted an AHP/DEA based technique for the problem of layout design. In 2013, Hadi-Vencheh and Mohamadghasemi [13] projected an integrated approach based on AHP–NLP for solving the layout plan problem. Chaudhary and Uprety [14] implemented AHP in telecom industry. Maniya and Bhatt [15] also implemented preference selection index (PSI) approach for an industry for selection of FMS. Sharma and Singhal [16] performed a comparative analysis for selection of layout by implementing PSI.

If a design has a poor layout to a composition, the new view is going to be turned off and either consciously or subconsciously, they will reject and move away from the design. Graphic design is a visual communication, and utilizing quality layout design will communicate the right message to any viewer. So, mastering layout and composition design is essentially setting yourself up for making quality product in less time on the shop floor.

Reviewing a design normally needs a place to rest on something of interest as it means to reduce the material handling, otherwise we have to look at other design and quickly move to something else. In the designed layout a crucial point has a benefit, it is also a good idea to communicate the most crucial information it doesn't need to be an image or graphic because typography can be used as a focal point, if done correctly so yeah build a focal point increase the scale and give it some room to breathe.

A study conducted by Sharma and Singhal [17] described the utilization of MCDM methodology to solve problem of layout design. Analytic hierarchical process (AHP) was also used by Sharma and Singhal for comparative study of the procedural approaches to the layout problem [18, 19]. Yang and Hung [20] worked on MCDM to handle the problem of layout design. Partovi and Burton [21] described AHP for facility layout. AHP is a technique that is used in a wide range of research applications [7, 22–25]. This approach is capable of systematically integrating tangible and intangible variables and offers a straightforward organized approach to the issue of layout design.

To achieve the goal of present research, a PSI-based AHP approach has been used for the study of the alternatives established because of some chosen qualitative factors in the Indian context for the automotive parts industry.

The objective of the chapter is to propose an integrated approach based on AHP-PSI, to solve the problem of layout of small-scale manufacturing industry.

4.1.1 ORGANIZATION OF THE CHAPTER

The chapter is organized as follows: Section 4.2 highlights the proposed methodology. Real time example is elaborated in Section 4.3, followed by results and discussion in Section 4.4. Section 4.5 concludes the chapter with future scope.

4.2 PROPOSED METHODOLOGY

An AHP-PSI-based integrated approach has been implemented in this proposed work as follows:

4.2.1 ANALYTIC HIERARCHICAL PROCESS METHODOLOGY

In 1990, Saaty [7] developed the AHP, a multi-attribute decision-making (MADM) technique, Figure 4.1 shows basic steps of analytic hierarchical process. It is a great

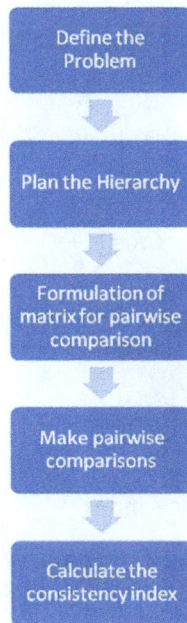

FIGURE 4.1   Basic steps of analytic hierarchical process considered for proposed work.

TABLE 4.1

Relative Importance of Factors

Numerical Rating	Verbal Judgments of Preferences
1	Equally preferred
2	Equally to moderately
3	Moderately preferred
4	Moderately to strongly
5	Strongly preferred
6	Strongly to very strongly
7	Very strongly preferred
8	Very strongly to extremely
9	Extremely preferred

tool for dealing with difficult decision-making situations. The implementation of Saaty's [7] AHP is enlisted as follows:

Step 1: Problem definition and goal identification.
Step 2: Plan the hierarchy from the top–bottom approach.
Step 3: For each factor and option, a matrix is created for pairwise comparison. Table 4.1 shows the pairwise comparison scale.
Step 4: The consistency of pairwise comparisons is resolute by utilizing the eigenvalue (λ_{max}), which is used to construct the consistency index (*CI*) as shown in Equation (4.1):

$$CI = \left(\frac{\lambda_{max} - n}{n - 1} \right) \qquad (4.1)$$

The consistency ratio (*CR*) is used to assess consistency, and *n* is denoted for matrix size.

$$CR = \left(\frac{CI}{RI} \right) \qquad (4.2)$$

where *RI* is average random consistency (Table 4.2). If *CR* is less than 0.1, the decision will be revisited and revised; if it is more than 0.1, the choice will be reconsidered and amended.

TABLE 4.2

The Average Random Consistency Index

Matrix size	1	2	3	4	5	6	7	8	9	10
Random consistency value	0	0	0.58	0.9	1.12	1.24	1.32	1.41	1.45	1.49

4.2.2 PREFERENCE SELECTION INDEX APPROACH

Maniya and Bhatt in 2010 [26] introduced the PSI approach for decision-making. This technique has a range of advantages compared to other related methods, such as the elimination of estimation of relative weight of the characteristics involved in the problem. Due to this consistency of the PSI system, it is ideal for problem layout of facilities in order to rank alternatives [15]. Figure 4.2 shows the steps of PSI implementation. The following steps are there to implement the present study:

Step 1: The first step in the PSI process is to construct the decision matrix as illustrated in Equation (4.3).

Let $A = (A_i$ for $I = 1,2,3, \dots n)$ be a set of layout alternatives.
$C = (C_j$ for $j = 1,2,3, \dots, m)$ be a set of criteria.
X_{ij} = performance of alternative A_i with respect to criteria C_j.

$$
\begin{array}{c}
\begin{array}{ccccccccc}
 & C_1 & C_2 & C_3 & \cdots & \cdots & C_j & \cdots & \cdots & C_m
\end{array} \\
\begin{array}{c}
A_1 \\
A_2 \\
A_3 \\
\vdots \\
\vdots \\
A_i \\
\vdots \\
\vdots \\
A_n
\end{array}
\left[
\begin{array}{ccccccccc}
X_{11} & X_{12} & X_{13} & .. & .. & X_{1j} & .. & .. & X_{1m} \\
X_{21} & X_{22} & X_{23} & .. & .. & X_{2j} & .. & .. & X_{2m} \\
X_{31} & X_{32} & X_{33} & .. & .. & X_{3j} & .. & .. & X_{3m} \\
\vdots & \vdots & \vdots & :: & :: & \vdots & :: & :: & \vdots \\
\vdots & \vdots & \vdots & :: & :: & \vdots & :: & :: & \vdots \\
X_{i1} & X_{i2} & X_{i3} & .. & .. & X_{ij} & .. & .. & X_{im} \\
\vdots & \vdots & \vdots & :: & :: & \vdots & :: & :: & \vdots \\
\vdots & \vdots & \vdots & :: & :: & \vdots & :: & :: & \vdots \\
X_{n1} & X_{n2} & X_{n3} & .. & .. & X_{nj} & .. & .. & X_{nm}
\end{array}
\right]
\end{array} \quad (4.3)
$$

Step 2: Formulate a hierarchical matrix of judgment. Each design criterion for each facility has different units, which requires normalization. Normalized matrix appears in Equation (4.6). Standardization is achieved as shown below in the current system:

If the selection criteria for the design of the facility are greater in expectation, the better:

$$
N_{ij} = \frac{X_{ij}}{X_{j,\max}}; \forall i,j \quad (4.4)
$$

where $X_{j,\max} = \max_{j}\{X_{ij}\}; \forall i,j.$

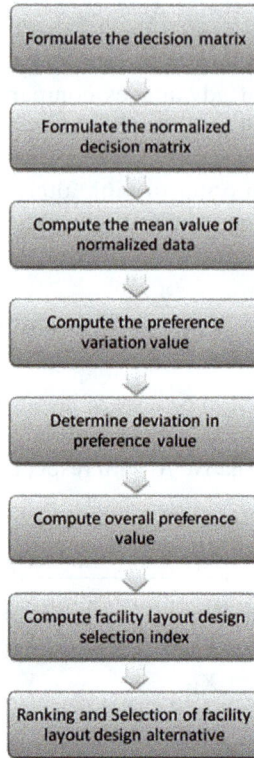

FIGURE 4.2 Preference selection index steps adopted for considered problem.

If the expectation of the design attributes or criteria of the facility is smaller:

$$N_{ij} = \frac{X_{j,\min}}{X_{ij}}; \forall i, j \qquad (4.5)$$

where $X_{j,\min} = \min_{j}\{X_{ij}\}; \forall i, j.$

	C_1	C_2	C_3	...	...	C_j	...	...	C_m
A_1	N_{11}	N_{12}	N_{13}	..	..	N_{1j}	..	..	N_{1m}
A_2	N_{21}	N_{22}	N_{23}	..	..	N_{2j}	..	..	N_{2m}
A_3	N_{31}	N_{32}	N_{33}	..	..	N_{3j}	..	..	N_{3m}
$\vdots$	$\vdots$	$\vdots$	$\vdots$	::	::	$\vdots$	::	::	$\vdots$
$\vdots$	$\vdots$	$\vdots$	$\vdots$	::	::	$\vdots$	::	::	$\vdots$
A_i	N_{i1}	N_{i2}	N_{i3}	..	..	N_{ij}	..	..	N_{im}
$\vdots$	$\vdots$	$\vdots$	$\vdots$	::	::	$\vdots$	::	::	$\vdots$
$\vdots$	$\vdots$	$\vdots$	$\vdots$	::	::	$\vdots$	::	::	$\vdots$
A_n	N_{n1}	N_{n2}	N_{n3}	..	..	N_{nj}	..	..	N_{nm}

$$(4.6)$$

Step 3: In the next step, find the mean value of the normalized data as following:

$$\overline{N_j} = \frac{1}{n}\sum_{i=1}^{n} N_{ij}, \forall i, j \tag{4.7}$$

Step 4: Compute the preference variation value (Π_j) using Equation (4.8):

$$\prod_j = \sum_{i=1}^{n} [N_{ij} - \overline{N_j}]^2 \tag{4.8}$$

Step 5: Find out the deviation (Φ_j) in preference value as given by the equation:

$$\varphi_j = \left| 1 - \prod_j \right| \tag{4.9}$$

Step 6: In this step, overall preference value (Ψ_j) needs to be found out:

$$\psi_j = \frac{\varphi_j}{\sum_{j=1}^{m} \varphi_j} \tag{4.10}$$

Step 7: Calculate the facility layout design selection index (Ω) using Equation (4.11):

$$\Omega = \begin{bmatrix} \Omega_1 \\ \Omega_2 \\ \Omega_3 \\ \vdots \\ \Omega_i \\ \vdots \\ \Omega_m \end{bmatrix} = \begin{bmatrix} (N_{11}\times\psi_1)+(N_{12}\times\psi_2)+(N_{13}\times\psi_3)+\cdots+(N_{1j}\times\psi_j)+\cdots+(N_{1n}\times\psi_n) \\ (N_{21}\times\psi_1)+(N_{22}\times\psi_2)+(N_{23}\times\psi_3)+\cdots+(N_{2j}\times\psi_j)+\cdots+(N_{2n}\times\psi_n) \\ (N_{31}\times\psi_1)+(N_{32}\times\psi_2)+(N_{33}\times\psi_3)+\cdots+(N_{3j}\times\psi_j)+\cdots+(N_{3n}\times\psi_n) \\ \vdots \\ (N_{i1}\times\psi_1)+(N_{i2}\times\psi_2)+(N_{i3}\times\psi_3)+\cdots+(N_{ij}\times\psi_j)+\cdots+(N_{in}\times\psi_n) \\ \vdots \\ (N_{m1}\times\psi_1)+(N_{m2}\times\psi_2)+(N_{m3}\times\psi_3)+\cdots+(N_{mj}\times\psi_j)+\cdots+(N_{mn}\times\psi_n) \end{bmatrix} \tag{4.11}$$

Step 8: after calculating index value (Ω), based on this value, rank the layout alternative.

4.3 AN EXEMPLARY APPLICATION

In this section, the methodology implemented to an automobile parts manufacturing industry has been depicted.

4.3.1 PROBLEM FORMULATION

A manufacturing unit of automotive parts, near Majholi village (Himachal Pradesh, India), was selected to implement the proposed process. This company supplies

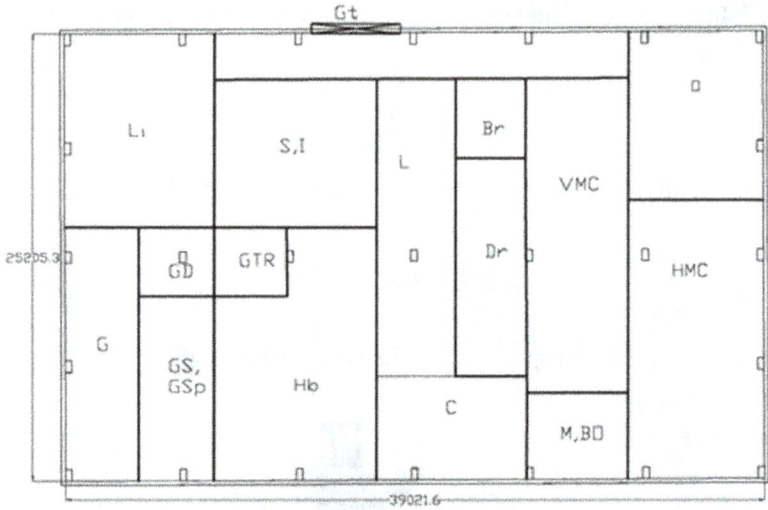

FIGURE 4.3   Layout 1 for selected industry.

vehicle parts to various Indian tractor industries. Several problems were identified after consulting with the company's management and production personnel. Some of the major problems were lack of flexibility in the production system, poor utilization of floor area, and inability to meet the production target. There were 15 divisions with 40 different types of machines and the floor space available was 39×25.2 m². There were five alternative designs for the facility layout of the industry among which the only one had to be chosen to mitigate the problems. These layout designs have been denoted as *L1, L2, L3, L4, and L5* (Figures 4.3–4.7). The management

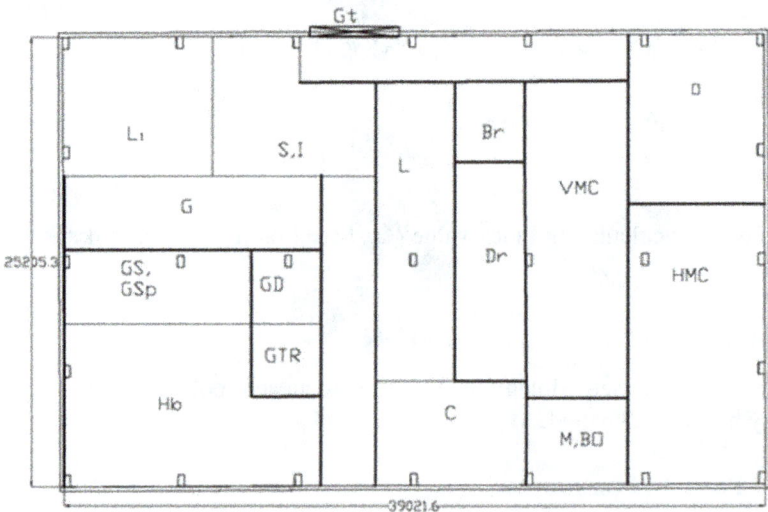

FIGURE 4.4   Layout 2 for selected industry.

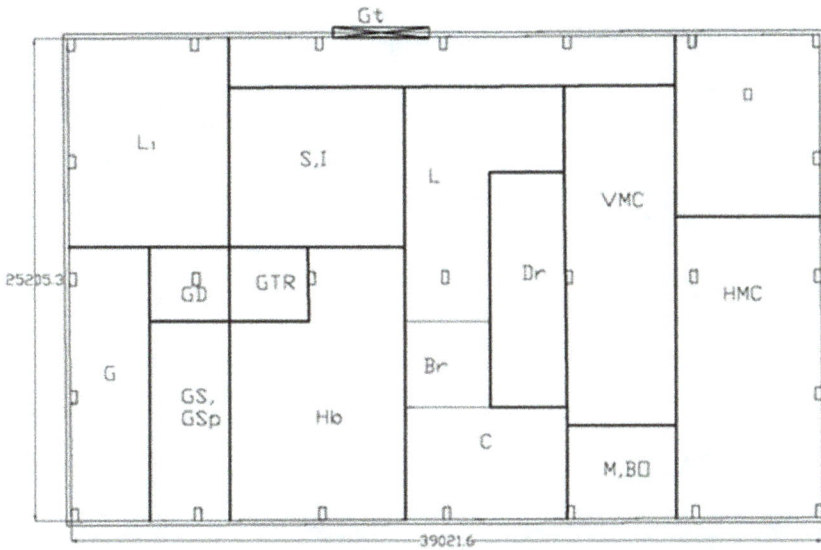

FIGURE 4.5   Layout 3 for selected industry.

team of the company had selected some criteria to evaluate these design alternatives. Those evaluation criteria were: (i) human factors, (ii) routing flexibility, and (iii) production area utilization.

Human factors include easiness of monitoring of the shop floor and its control, as well as the most critical protection for workers. The management issue was choosing

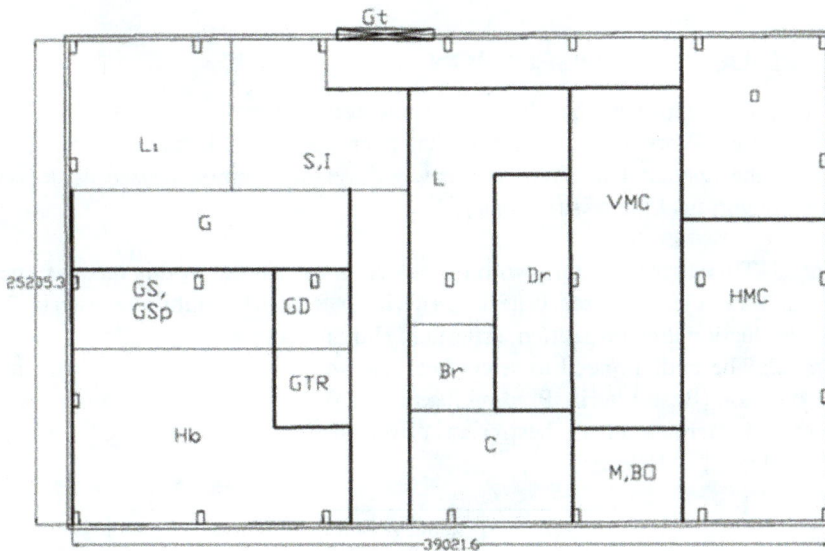

FIGURE 4.6   Layout 4 for selected industry.

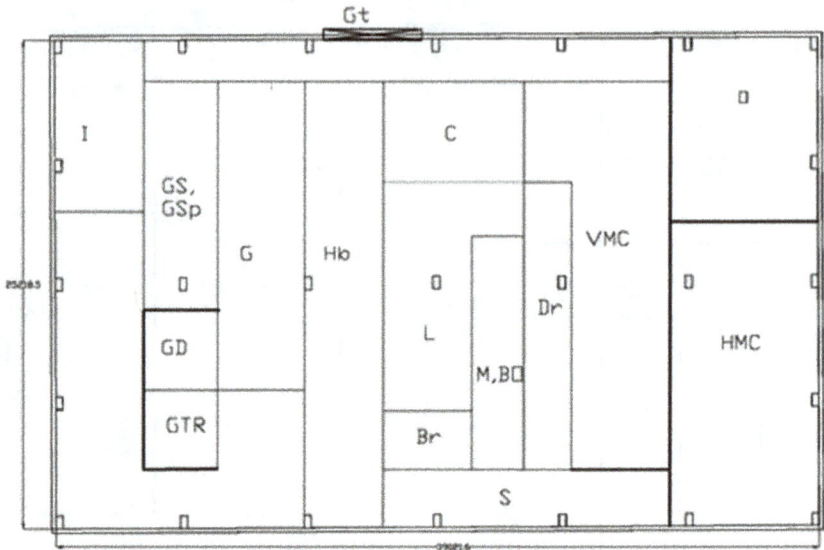

FIGURE 4.7 Layout 5 for selected industry.

the most appropriate and efficient layout design of the facility with respect to these defined parameters. Routing flexibility is the industry's ability to conduct manufacture in the occurrence of any obstacle on the shop floor on any path. It is also illustrated by the capacity of the tool used to move between various departments using several routes. Usage of the production area is the element relating to use of the floor space; it includes a repair room and space for potential expansion.

4.4 RESULTS AND DISCUSSION

The proposed approach has been implemented and various results have been obtained, Best Worst Method (BWM) has been also implemented for cross-verification of the results. The BWM is a newly developed approach used as decision-making in various kind of situations [27].

Step 1: This approach starts with the selection of criteria; in this method, the authors selected three criteria (criteria 1: Routing flexibility; criteria 2: Production area utilization; criteria 3: Human factors).

Step 2: The authors need to select best and worst criteria to be used for the decision (Best criteria: Routing flexibility; Worst criteria: Human factor).

Step 3: Preference of the best criteria over all the other criteria needs to be evaluated as follows:

Best to others	Routing flexibility	Production area utilization	Human factors
Routing flexibility	1	3	7

Step 4: Preference of each of the other criteria over the worst criteria needs to be determined as follows:

Others to the worst	Human factors
Routing flexibility	6
Production area utilization	3
Human factors	1

Step 5: Need to find the optimal weights is given as follows:

Weights	Routing flexibility	Production area utilization	Human factors
	0.66	0.24	0.1

Step 6: Need to find the reliability of the findings; for the given example, the determined value is given as Ksi* = 0.06 (as the value is near to zero, the solution is more reliable).

After the AHP has been implemented to get the ranking, in this step, the AHP method has been applied to select the most effective facility layout design. In the next step, all the layout design alternatives were evaluated against each of these criteria. Evaluation of alternatives against routing flexibility is represented in Table 4.3. In this case, Layout 5 has ranked first with the height relative weight for routing flexibility. Layouts 1 and 2 have got the same weights valued 0.177 and have ranked second together. Therefore, it can be said that Layout 5 has the highest priority when the only considerable parameter is routing flexibility.

Using the same procedure, pairwise assessment matrix of layout design alternatives for production area utilization has been constructed. Table 4.4 shows the assessment matrix and relative weights of the alternatives in case of production

---

TABLE 4.3

Pairwise Assessment Matrix of Design Alternatives for Routing Flexibility

	L1	L2	L3	L4	L5	Relative Weight	Local Rank
L1	1	2	1	2	1/3	0.177	2
L2	1/2	1	1/2	1	1/4	0.096	3
L3	1	2	1	2	1/3	0.177	2
L4	1/2	1	1/2	1	1/4	0.096	3
L5	3	4	3	4	1	0.452	1

$CI = 0.01$

TABLE 4.4

Pairwise Assessment Matrix of Design Alternatives for Production Area Utilization

	L1	L2	L3	L4	L5	Relative Weight	Local Rank
L1	1	4	1	4	6	0.368	1
L2	1/4	1	1/4	1	3	0.108	2
L3	1	4	1	4	6	0.368	1
L4	1/4	1	1/4	1	3	0.108	2
L5	1/6	1/3	1/6	1/3	1	0.048	3

$CI = 0.01$

area utilization. Here, Layouts 1 and 2 have got the highest relative weights and both have been ranked first for production area utilization. Layout 5 has got the least weight and ranked last.

Figure. 4.8 shows the relative weight and local rank of design alternatives with respect to human factors. Layout 3 has got the maximum weight and ranked first. Layout 5 has got the least weight value of 0.054.

Finally, the overall priority order of the layout designs was obtained by combining all of these. Table 4.5 shows the global weights and global ranks of the alternatives.

Figure 4.9 shows that Layout 5 has got the highest global weight and has ranked first. Although this alternative has got the least weight in the case of production area utilization and human factors, it has ranked first due to relatively more weight

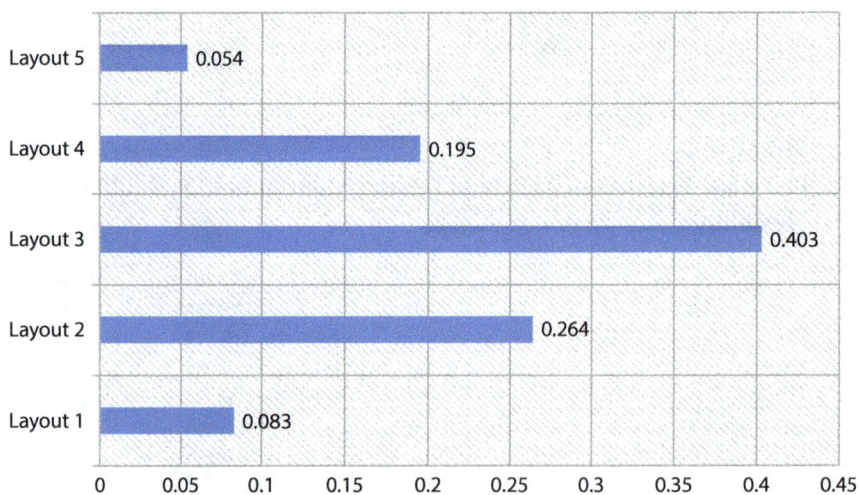

FIGURE 4.8 Relative weights of layout design alternatives for human factors.

TABLE 4.5

Overall Rank of the Facility Layout Design Alternatives

	Routing Flexibility (0.660)	Production Area Utilization (0.240)	Human Factors (0.100)	Global Weight	Global Rank
L1	0.177	0.368	0.083	0.201	3
L2	0.096	0.108	0.264	0.120	4
L3	0.177	0.368	0.403	0.241	2
L4	0.096	0.108	0.195	0.111	5
L5	0.452	0.048	0.054	0.326	1

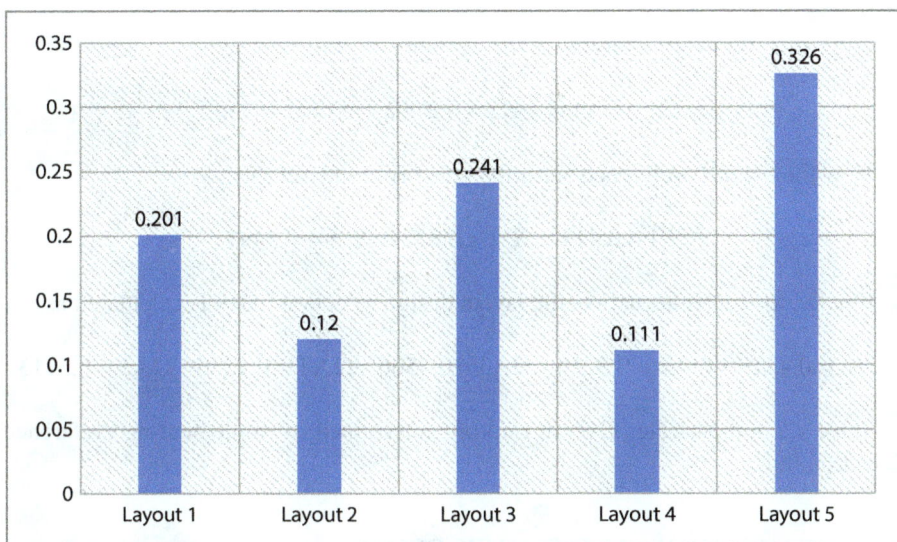

FIGURE 4.9   Global weights and global rank of the layout alternatives.

of routing flexibility than other criteria. Layout 3 has got 0.241 which is the second highest among these five alternatives. Layout 4 has valued the least which is 0.111 and ranked last.

Finally, the assignment order dependent on the preference index shows the first option between the others. The mean values of each layout alternative are calculated after normalization:

$$\bar{N}1 = 0.3842 \quad \bar{N}2 = 0.5430 \quad \bar{N}3 = 0.4011$$

Now for every alternative, preferable variable value is determined as:

$$\Pi 1 = 0.5156 \quad \Pi 2 = 0.7213 \quad \Pi 3 = 0.5907$$

TABLE 4.6

Overall Ranking Using Preference Selection Index Methodology

	Preference Selection Index Value	
L1	0.4219	4
L2	0.3216	5
L3	0.4566	2
L4	0.4280	3
L5	0.5113	1

After then for every alternative, the deviation is computed as:

$$\varnothing 1 = 0.4844 \quad \varnothing 2 = 0.2787 \quad \varnothing 3 = 0.4093$$

Now the overall preference variation value is found out as:

$$\Psi 1 = 0.4132 \quad \Psi 2 = 0.2377 \quad \Psi 3 = 0.3491$$

Finally, for each alternative, the layout design selection index is calculated as:

$$\Omega A1 = 0.4219 \quad \Omega A2 = 0.3216 \quad \Omega A3 = 0.4566 \quad \Omega A4 = 0.4280 \quad \Omega A5 = 0.5113$$

Table 4.6 demonstrates Overall ranking determined by implementing Preference Selection Index.

4.5 CONCLUSION AND FUTURE SCOPE

In the current study, the PSI-AHP approach is implemented to analyze the layout. The efficiency of the designed layout is directly influenced by quality factors. Three qualitative factors were analyzed: routing flexibility, efficiency of the production area, and human issue. The steps of AHP have been implemented, values of λ_{max}, CI, RI, and CR have been calculated for each selected factor. Relative weight has been calculated for each criterion by implementing BWM approach. From the results, it was found that routing flexibility criteria demonstrate the best relative weight. For production area utilization, Layouts 1 and 3 showed higher weights than others. On the other hand, for human issues, alternative 3 showed the best value of relative weight. Finally, Layout 5 has got the highest value after considering all the evaluation criteria and ranked first among the five alternatives. Other MCDM approaches can be used to structure the study of the choices accessible in future work. It is also obvious to look at the created alternatives for quantitative aspects and rate them based on the combination of qualitative and quantitative criteria.

REFERENCES

1. Vats, G., & Vaish, R. (2014). Selection of optimal sintering temperature of $K_{0.5}Na_{0.5}NbO_3$ ceramics for electromechanical applications. Journal of Asian Ceramic Societies, 2(1), 5–10.
2. Şahin, R., & Türkbey, O. (2009). A simulated annealing algorithm to find approximate Pareto optimal solutions for the multi-objective facility layout problem. The International Journal of Advanced Manufacturing Technology, 41(9-10), 1003–1018.
3. Rao, R. V. (2006). A material selection model using graph theory and matrix approach. Materials Science and Engineering: A, 431(1), 248–255.
4. Rao, R. V. (2008). A decision making methodology for material selection using an improved compromise ranking method. Materials & Design, 29(10), 1949–1954.
5. Rao, R. V. (2013). A novel subjective and objective integrated multiple attribute decision making method. In Decision Making in Manufacturing Environment Using Graph Theory and Fuzzy Multiple Attribute Decision Making Methods (pp. 137–157). Springer London.
6. Psarras, J., Capros, P., & Samouilidis, J. E. (1990). Multicriteria analysis using a large-scale energy supply LP model. European Journal of Operational Research, 44(3), 383–394.
7. Saaty, T. L. (1990). How to make a decision: The analytic hierarchy process. European Journal of Operational Research, 48(1), 9–26.
8. Hambali, A., Sapuan, S. M., Ismail, N., & Nukman, Y. (2009). Application of analytical hierarchy process in the design concept selection of automotive composite bumper beam during the conceptual design stage. Scientific Research and Essay, 4(4), 198–211.
9. Milan, L., Kin, B., Verlinde, S., & Macharis, C. (2015). Multi-actor multi-criteria analysis for sustainable city distribution: A new assessment framework. International Journal of Multicriteria Decision Making, 5(4), 334–354.
10. Opricovic, S., & Tzeng, G. H. (2007). Extended VIKOR method in comparison with outranking methods. European Journal of Operational Research, 178(2), 514–529.
11. Kuo, Y., Yang, T., & Huang, G. W. (2008). The use of grey relational analysis in solving multiple attribute decision-making problems. Computers & Industrial Engineering, 55(1), 80–93.
12. Yang, T., & Kuo, C. (2003). A hierarchical AHP/DEA methodology for the facilities layout design problem. European Journal of Operational Research, 147(1), 128–136.
13. Hadi-Vencheh, A., & Mohamadghasemi, A. (2013). An integrated AHP–NLP methodology for facility layout design. Journal of Manufacturing Systems, 32(1), 40–45.
14. Chaudhary, A., & Uprety, I. (2015). Evaluation and precedence of factors affects telecom service quality by AHP and fuzzy analysis. International Journal of Advanced Operations Management, 7(1), 22–40.
15. Maniya, K. D., & Bhatt, M. G. (2011). An alternative multiple attribute decision making methodology for solving optimal facility layout design selection problems. Computers & Industrial Engineering, 61(3), 542–549.
16. Sharma, P., & Singhal, S. (2016). A review of objectives and solution approaches for facility layout problems. International Journal of Industrial and Systems Engineering, 24(4), 469–489.
17. Sharma, P., & Singhal, S. (2016). A review on implementation of meta-heuristic approaches for layout problems in dynamic business environment. International Journal of Multivariate Data Analysis, 1(1), 6–27.
18. Sharma, P., & Singhal, S. (2016). Comparative analysis of procedural approaches for facility layout design using AHP approach. International Journal of Manufacturing Technology and Management, 30(5), 279–288.

19. Sharma, P., & Singhal, S. (2016). Design and evaluation of layout alternatives to enhance the performance of industry. Opsearch, 53(4), 741–760.
20. Yang, T., & Hung, C. C. (2007). Multiple-attribute decision making methods for plant layout design problem. Robotics and Computer-Integrated Manufacturing, 23(1), 126–137.
21. Partovi, F. Y., & Burton, J. (1992). An analytical hierarchy approach to facility layout. Computers & Industrial Engineering, 22(4), 447–457.
22. Abadi, F., Sahebi, I., Arab, A., Alavi, A., & Karachi, H. (2018). Application of best-worst method in evaluation of medical tourism development strategy. Decision Science Letters, 7(1), 77–86.
23. Al-Harbi, K. M. A. S. (2001). Application of the AHP in project management. International Journal of Project Management, 19(1), 19–27.
24. Bayazit, O. (2005). Use of AHP in decision-making for flexible manufacturing systems. Journal of Manufacturing Technology Management, 16(7), 808–819.
25. Deng, H., Yeh, C. H., & Willis, R. J. (2000). Inter-company comparison using modified TOPSIS with objective weights. Computers & Operations Research, 27(10), 963–973.
26. Maniya, K., & Bhatt, M. G. (2010). A selection of material using a novel type decision-making method: Preference selection index method. Materials & Design, 31(4), 1785–1789.
27. Jajodia, S., Minis, I., Harhalakis, G., & Proth, J. M. (1992). CLASS: Computerized layout solutions using simulated annealing. The International Journal of Production Research, 30(1), 95–108.

5 ATC Enhancement due to Charging/Discharging of BESS in Wind Power Integrated Systems

Aishvarya Narain
Department of Electrical Engineering,
United College of Engineering and Research,
Prayagraj, Uttar Pradesh, India

CONTENTS

DOI: 10.1201/9781003369554-5

5.1 INTRODUCTION

The electric power industry has undergone rapid and irreversible changes since the 1980s. Traditionally, large utilities have total control in a vertically integrated system. These utilities control production, transmission, and distribution in a region. According to the rules of the monopoly market, a single utility can produce, transmit, and distribute electricity in an area. Utilities have to operate in accordance with government plans, rules, and regulations. At the same time, they had to be assured of reasonable returns. In most parts of the world, this has led to inefficiency and sluggish attitudes in the industry. They focused more on profit rather than technological innovation and customer satisfaction. These monopolistic strategies were necessary for the expansion, standardization, and development of power industries. The power sector was restructured in many countries to operate under the control of state and federal governments with the main objective of increasing the production and distribution of electricity. Several countries such as Chile, Great Britain, Spain, Argentina, New Zealand, Norway, Sweden, and the United States have undergone restructuring to encourage privatization of the power sector, more choices to the consumers, and competition among producers [1].

The purpose of the power industry restructuring is to eliminate monopoly in the production and trading sectors, thereby, introducing competition at various levels wherever possible. Generating companies may have contracts for the supply of generated power to electricity dealers/distributors or wholesale consumers. They can also sell electricity in a pool in which electricity producers and customers also participate. Thus, electricity trading will be freed from traditional rules and become competitive. The restructuring of the existing rules and regulations to provide competitive electricity market is known as deregulation. According to Willis and Phillipson [2], *"Deregulation is a restructuring of the rules and economic incentives that governing authority sets up to control and drive the electric power industry"*.

Initially, there was no competition in generation, transmission, and distribution as all three are considered as a single entity, but in the deregulated system, all the activities are different. The contract of the generation and distribution company determines the power scheduling and load dispatching. Therefore, power flow in deregulated system is different from the traditionally existing environment. All the companies will try to get maximum profit and consumers will try to get cheaper source power even at the cost of transmission limit.

This can lead to overloading of transmission lines and cause congestion in the system and can exceed voltage, thermal, and stability limits, which in-turn can reduce system security.

The restructuring of the electricity supply industry brought competition among generation companies (GENCOs), distribution companies (DISCOMs), and buyers. Buyers have the option to choose between the available generation to the most afford-able generation. This desire of buyers to get the cheapest electricity may result in a network congestion. Restructuring means breaking down the original structure and reforming them into the new structure to achieve better efficiency, performance, and reliability. Deregulation is the restructuring of rules and regulations of the government policy and creating competition in the market [3, 4]. The objective of deregulation is to provide fair and equal opportunities to the power companies to compete. The competition will help in technological development, improving efficiency, and customer satisfaction. Low cost, good service, cheap and reliable electricity are other benefits of deregulation.

To fulfil the purpose of restructuring, buyers should get power at a low cost with good power quality and reliability [5]. The competitive market encourages players to choose new technology to compete. Developing countries are not able to fit it completely due to increasing demand so investors focused on restructuring the electricity market [6]. A group of participants such as producers, consumers, prosumers, and combinations are known as aggregators in an electrical power system. Pantoš [7] proposed the utilization of these aggregators, generators, and loads for market-based congestion management. The power transfer distribution factor, topological load distribution factor, and topological generation distribution factor are calculated to obtain the relation between these quantities and line power flow. The two-level optimization problem is solved by linear programming for congestion management (CM).

Price-based zone detection using locational marginal price, k-means clustering, and queen/rook spatially constraint clustering had been discussed by Poyrazoglu [8]. A case study on high voltage transmission in the Turkish power system has also been conducted to determine the zones in a competitive market.

Free and fair electricity trade is done in the open-access market. Hence, there is a need for an operator (system operator (SO)/grid operator) to manage the security and reliability of the system. The SO manages the transmission system transactions ensuring open access to all the market entities. To transact power through a common transmission line, it is necessary to know the status of the available power transfer capacity of the system. Additional power can be transferred to the system without exceeding the available transfer capability (ATC) limits. Restructuring of the electric power system leads to many new issues and challenges, while on the other hand, many existing players have redefined the scope of their activity to gain the advantages. One of the major issues is congestion in the system. The other trending issues are integration of renewable energy sources (RES) with battery energy storage system (BESS) into the system. This chapter considers these technical issues for contingency analysis. This chapter uses ATC-based CM with proposed BESS size and charging criteria.

Therefore, the main purpose of the chapter is

- To calculate average of TCDF (ATCDF) using transmission congestion distribution factor (TCDF) values for different congested lines;
- To create the congested zones for integration of wind power generation (WPG) using ATCDF values;
- To calculate wind farm (WF) power and integrate BESS using ATCDF value to make the system more reliable;
- Three cases have been used for different BESS size with WPG;
- And, to obtain enhanced ATC values by proposed charging/discharging of BESS method.

In this chapter, the survey on RES, wind energy scenarios, ESS, its types and challenges in WF power calculation, its integration with BESS is done. The chapter also discusses the ATC-based CM method and its types. The novel approach has been proposed to install the RES and BESS for ATC enhancement under line outage contingency conditions.

The chapter is organized as follows: Section 5.2 discusses CM methods. Section 5.3 describes the ATC-based CM method and its types. In Section 5.4, a brief introduction about RES is shown. Section 5.5 discusses the wind energy scenarios in the world and Section 5.6 reviews on wind energy scenario that describes the challenges in power output calculation of WF. The survey on ESS, its types and challenges and opportunities for setting up RES with BESS has been done in Sections 5.7 and 5.8, respectively. The proposed methodology is discussed in Section 5.9. Section 5.10 discusses results. And, Section 5.11 concludes the chapter with future scope.

5.2 CONGESTION MANAGEMENT

The restructuring of the power industry brings several benefits to the sellers and buyers. Competition in the electricity market improves efficiency and provides consumers with the option of getting cheap electricity from different sellers. It also poses many challenges to the SOs. The congestion in the system is one of the major issues. The commercial cause of congestion is the freedom of buyers to demand any amount of cheap power from the supplier; however, line outage and overloading are the technical reasons for congestion.

It is desirable to transmit power from one node of the system to another node without violating the operating constraints. The voltage, thermal, and stability limits are considered as transmission constraints. The system is said to be congested when the transfer of power is not possible due to violation of any constraints. This is undesirable in the system and needs to be eliminated as soon as possible to maintain the system's security and stability. Congestion affects the electric power system, that is it causes an increase/decrease in generation, changes in the revenue of generating companies, and an increase in tariff.

The techniques used to alleviate congestion in the transmission lines are called CM methods [9]. There are several methods discussed in the literature, such as sensitivity factor-based method, auction-based, price-based, generator redispatch, optimum power flow (OPF), ATC-based, generation rescheduling, ready to pay, load curtailment, flexible AC transmissions systems devices based, and demand response.

5.3   ATC-BASED CONGESTION MANAGEMENT

The value of ATC in a transmission line gives the amount of power that can be transferred within the range of the transmission line limits. These values are posted at open access same time information system (OASIS) for commercial uses. Anyone requiring a transaction must visit the OASIS website, then decide whether a transaction is feasible based on ATC information. There are two types of ATC as follows:

5.3.1   STATIC ATC

The ATC calculation provides the information of network available power transfer capability for further transactions without violating the system security constraints. These constraints are bus voltage limit, line thermal limit, and steady-state stability limit also known as steady-state constraints. When these constraints are considered for ATC calculation, it is known as static ATC.

5.3.2   DYNAMIC ATC

The transient stability limit [10–13], signal (small/large) stability limit [14], and voltage stability limit [15] are considered as dynamic constraints. When static as well as dynamic constraints are considered for ATC calculation, then it is known as dynamic ATC. In dynamic ATC, transient state of system is mainly studied.

5.4   RENEWABLE ENERGY SOURCES: THE FUTURE OF POWER INDUSTRY

Renewable energy generation has been increasing over the years to meet consumer electricity demand. All forms of renewable energy production around the world are used in conjunction with conventional generation sources. There are many challenges in generating electricity from RES. Several policies have been put in place to encourage renewable energy production, such as lowering renewable energy contracts price, tax holidays, incentives, facilitating sale and purchase, etc. The integration of RES concerning the flexibility of the power system had been reviewed by Impram et al. [16] for smooth power generation during an unexpected situation, balance generation according to demand, and maintain consistency in uncertain conditions. The RES integration affects the transient, frequency, and small-signal stability of the power system. The RES is used to meet most of the demand in modern power systems due to the large production period and priority.

Sandhu and Thakur [17] reviewed the technical and non-technical issues, challenges, utilization, causes, and impact of RES. A survey on energy transformation

Estimated Potential of Renewable Power (MV) as of March 2021

Wind Power, 695509, 46.66%

Solar, 748990, 50.24%

Other, 46228, 3.10%

Small Hydro Power, 21134, 1.42%

Bio Mass Power, 17538, 1.18%

Cogeneration Bagasse, 5000, 0.34%

Waste to Energy, 2556, 0.17%

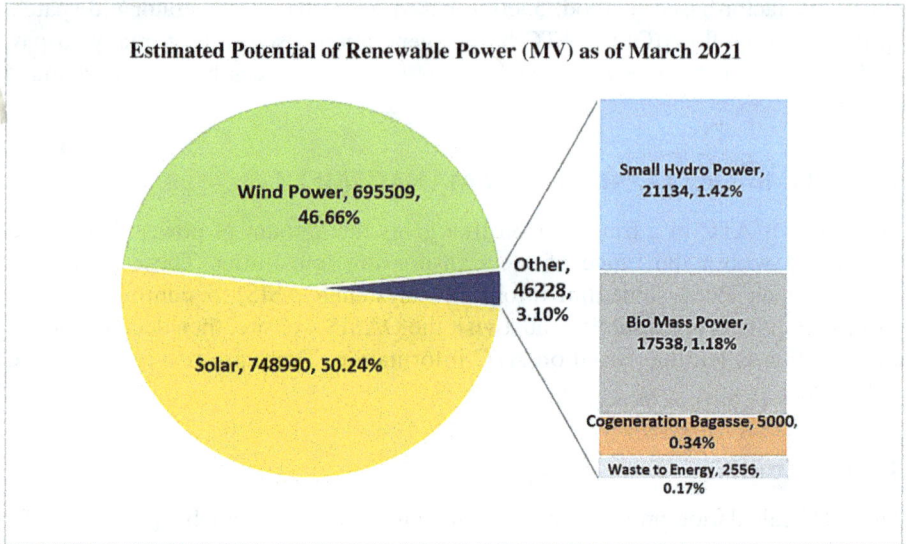

FIGURE 5.1   Estimated renewable power potential by source in India [20].

from fossil fuels to renewable energy was presented by Kalair et al. [18]. The cost of conventional energy sources is increasing, and the cost of RES is decreasing due to falling fabrication costs. Although RES with energy storage system (ESS) cannot completely replace conventional energy sources, it can still make the maximum contribution to power generation in the future.

The United States and China are in first and second place in the renewable energy country attractiveness index (RECAI) released by Ernst & Young Global Ltd (EY) on 19-05-2021. India has moved to the third position due to its exceptional performance on the solar photovoltaic front. As per the report, India's solar sector is likely to grow substantially after the COVID-19 pandemic [19]. India has huge potential to generate power from renewable resources. The total renewable power generation capacity is estimated at around 1490727 MW as of 31-03-2021. This includes solar power potential of 50.24%, wind power potential of 46.66% at 120 m hub height, small-hydro potential of 1.42%, biomass power potential of 1.18%, biogases-based cogeneration in sugar mills of 0.34% and 0.17% from waste as given in Figure 5.1 [20]. According to the ministry of new and renewable energy, the total installed renewable power in India as of 31-03-2021 is 94,433.79 MW [21]. This is far away from estimated data that provide the opportunity to generate electricity from renewable resources (mostly wind and solar) in the future.

5.5   WIND ENERGY

WPG is one of the major sources of electricity generation which is increasing day by day in the world. Figure 5.2 shows the year-wise cumulative installed capacity

Global Wind Power Cumulative Capacity (MW)

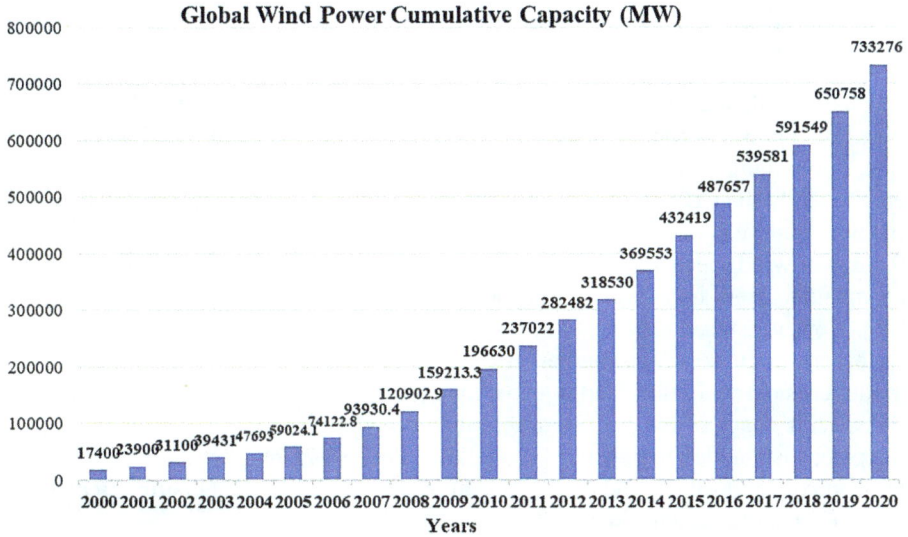

FIGURE 5.2   Cumulative global wind installed capacity from 2000 to 2020 [23].

of wind power globally. According to the report, the total wind installed capacity worldwide is 733,276 MW and is expected to grow rapidly in the coming years.

WPG capacity in India has also increased in recent years as shown in Figure 5.3. India's total wind installed capacity as of 31-03-2021 was 39,242.05 MW [21, 22]. This huge installed capacity makes India the fourth among the countries with the largest installed wind power capacity in the world. The impact of WPG on ATC values during the contingency condition has been discussed in this chapter.

Wind power installed capacity of India

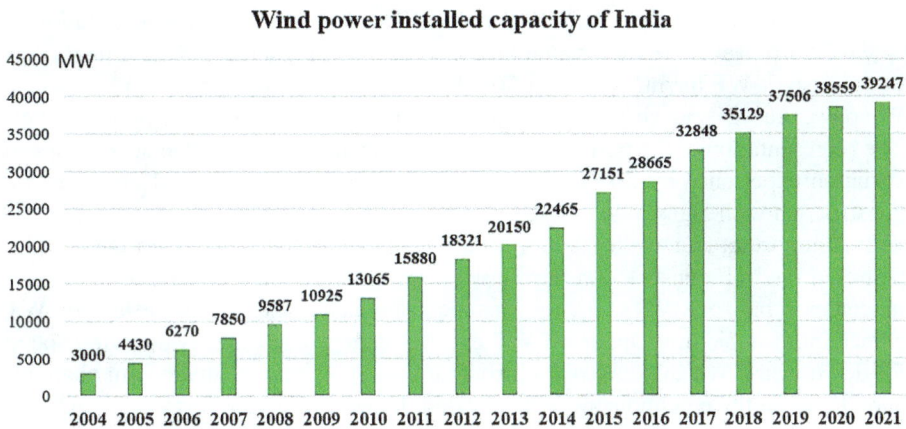

FIGURE 5.3   Wind power installed capacity of India from 2004 to 2021 [20,24].

5.6 CHALLENGES IN WIND FARM POWER OUTPUT CALCULATION

The main challenge in WPG is the correct estimation of power output. The wind turbine (WT) power output depends on several factors such as wind speed, wind direction, hub height, location and weather conditions, precise power curve estimation, etc. To study the electric power supplied by the WT or a WF based on wind speed, the power curve is used. Manufacturers obtain this curve by taking readings at every 0.5 m/s of wind speed and corresponding power generated [25]. The graph provided by manufacturers is helpful to find output power at a given speed and hub height [26]. However, when dealing with big data, the graph is not sufficient for analysis of the power curve so the mathematical expression is required. The mathematical model uses pairs of point readings to frame piecewise continuous functions. Due to the specific shape of the power curve, several models are proposed to analyse it. In the piecewise model, mostly used models are linear [27], quadratic [28], cubic [29], least-square, and spline model [30]. However, they show a lack of continuity of slope and error near rated wind speed.

A review of the deterministic WT power curve was discussed by Villanueva and Feijóo [31]. The factors involved in the WT power curve are based on models, the number of parameters, ease of use, and accuracy. The parameter logistic functions (PLE) are named according to parameters used like 6-parameters (6PLE), 5-parameters (5PLE), etc. It is suggested to choose the methods based on the objective taken, that is minimizing the number of parameters, ease of use, relationship with WT parameters, and priority of the accuracy. It is recommended that the 5PLE method should be used for large parameters, and 3PLE method is used when the minimum number of parameters are needed for consideration. Villanueva and Feijóo [32] presented a comparison of different PLE (6PLE, 5PLE, 4PLE, 3PLE) models for the WT power curve. The results illustrate that the 6PLE and 5PLE give similar results but 6 PLE was not considered as it is difficult to find six parameters of the turbine. 5PLE gives better results than 4PLE but as parameters are less in 4PLE, 5PLE is discarded. 3PLE is better than 4PLE in terms of results and parameter consideration. It is recommended to use either 5PLE or 3PLE model for power curve estimation.

The WT power curve representation contains many mathematical operations, thus in the case of WF having a lot of WTs, the power curve cannot be easily obtained. WF power curve is not simply the summation of all the individual WT power curves. The representation of the power curve in the WF is more complex due to wind speed variation experience by different WTs. Another problem is the land area for WF construction with many WTs.

It is important to consider the minimum area (due to the economy of land) while planning the WF construction. However, minimizing the area leads to a decrease in distance between WTs. As the distance decreases, the shadow effect of WT comes into the picture. Upstream WT extracts energy and reduces the wind speed for downstream WT and causing a turbulent flow. This mutual shadowing phenomenon between the neighbouring WT is called the wake effect or array effect. Due to this, the downstream WTs would not run at full capacity and produce less power. Consequently, the total power output of WF reduces. There are numerous methods for improving WF power output calculation in literature, that is global sensitivity analysis methods [33], cooperative control method for power output optimization [34], active power setpoint method under the influence by varying the meteorological and operational parameters [35], thrust coefficient control strategy [36], probabilistic estimation model of power curve [37], the impact of mountain waves on WPG [38], new multiple wake model for prediction of turbulence intensity [39] for power maximization, etc.

González-Longatt *et al.* [40] discussed the static and dynamic behaviour of a WT. Multiple wake effects, wind direction, and wind speed delay on downstream WT were also discussed. The efficiency of WF depends on the geometry of WT based on wind direction considering the wake effects.

An artificial neural network based approach for WT power output prediction was proposed by Bilal et al. [41]. It is observed that wind output power depends on the site location and climate data (temperature, solar radiation, and humidity). The optimization scheduling model considering wake effect and wind direction for maintenance of offshore WF was discussed by Ge et al. [42]. The objective function was considered to maximize power generation and minimize the maintenance cost. It was observed that wind direction and maintenance both have a significant influence on the wake effect on the downstream WT.

The CM in the presence of a wind energy source at a proper location was discussed by Suganthi et al. [43]. An improved differential evolution algorithm was used to solve the nonlinear optimization problem for generator rescheduling and installing a WF. The use of WPG to relieve the overloaded line for CM was discussed by Farzana and Mahadevan [44]. Minimizing congestion cost and power loss had been taken as the main objective functions in the literature. Abdolahi et al. [45] discussed the probabilistic model for CM of active distribution networks with the integration of RES. The results were compared with the risk-based strategy, and it was suggested that SOs choose the best method for decision-making according to requirements. Gope *et al.* [46] discussed the Moth flame optimization algorithm for rescheduling the conventional generator with penetration of WF. The bus sensitivity factor was used to find the optimal location of a WF. In WF, WTs were arranged in rows and columns which affect the total output power generation. Therefore, in the present work, a new mathematical equation is used to calculate the WF power considering the effect of wind speed, wind direction, and wake effect to calculate more realistic power output.

The power output of RES is not controlled by the grid SO, therefore, the energy storage devices are used to smooth the power and make it dispatchable. With proper placement of ESS, the operator can manage the dynamic behaviour of electricity very effectively. It also helps the operator to control the power transfer to the consumers and provides cost-effective management.

5.7 ENERGY STORAGE SYSTEM

In an electric power system, the generation of electricity can be assumed to be almost certain for a short period of time but the demand cannot be assumed fix. To operate the power system securely, economically and stably, energy storage devices are used

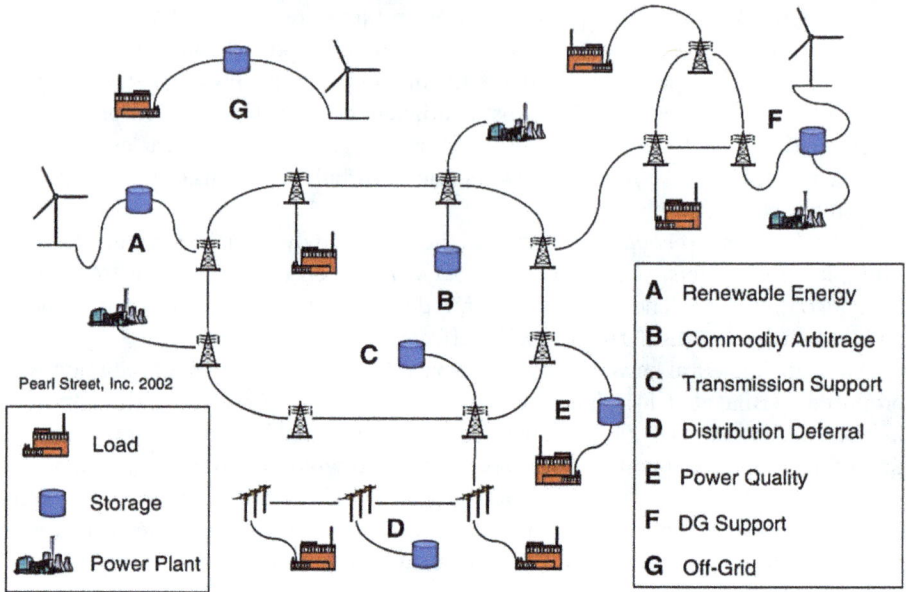

FIGURE 5.4 Application of energy storage system into grid [48].

to supply the amount of power during the peak demand of the system. A comprehensive review on optimal size, location, cost, issues, and challenges in the integration of ESS into modern power scenarios was discussed by Pattanaik et al. [47].

ESS has a wide field of application into the grid as it can be utilized in generation, transmission, and distribution side of the system [48]. ESS on the generation side store the bulk electricity generated in the night. The energy is used during the peak load period time. Stored energy is also used in area control by preventing the unbalance power transfer between one node and another. One main application of ESS in the generation side is to maintain frequency during irregular grid conditions. The main application of ESS in transmission and distribution side is to maintain voltage at every node of the system, power levelling in case of integration with RES, stability, etc. Figure 5.4 illustrates the application of ESS integration into the grid system.

The classification of energy storage used in electrical power systems was discussed by Rosen and Fayegh [49]. Energy storage was classified into two main categories according to the purpose of energy storage. The technology in which energy was stored as thermal energy and released in the same form is known as thermal energy storage technology, whereas when energy is stored in the form of thermal energy and released in electrical form is known as electrical energy storage.

5.7.1 ELECTRICAL ENERGY STORAGE

Energy storage plays an important role as the world moves towards renewable energy generation to achieve low carbon emissions. The integration of ESS into the existing electrical market and its balancing mechanism was analysed by Vahidinasab and Habibi [50]. A comprehensive survey was conducted on ESS formulas, status, problems, issues, and challenges in deployment.

The intermittent nature of RES is the major challenge that makes them unreliable for steady electricity generation. ESS stores energy when the power generated is excess and re-uses it when there is a shortage of power. This technology of storing and reusing electricity makes renewable energy generation stable and reliable. In the present power scenario, the integration of ESS should be considered as complementary services rather than competing services. Technologies used to convert electricity into an easily storable form are mechanical energy storage, chemical energy storage, superconducting magnetic energy storage, cryogenic energy storage, and electrochemical energy storage [51].

5.7.2 MECHANICAL ENERGY STORAGE

An electromechanical system is used to convert electrical energy into a storage form. For example, pump hydro, fly wheel, compressed air, gravity, and liquid piston.

5.7.3 CHEMICAL ENERGY STORAGE

In this technology, electrical energy is used to produce a chemical compound that stores electricity and can be reused when required. Hydrogen, methane, hydrocarbons, and methanol are mostly used for energy storage. The electrolysis of water produces hydrogen while the carbon compound produces other chemical compound with hydrogen.

5.7.4 ELECTROCHEMICAL ENERGY STORAGE

Electrochemical energy storage is used when fast charging/discharging of energy is required due to randomly changing demand. Battery is used to store electricity by chemical reactions. This is an older technology and is frequently used due to the availability of various sizes and capacities. Many types of batteries are used in the system, such as sodium sulphur battery, sodium nickel chloride battery, iron chromium battery, zinc air battery, zinc bromide battery, vanadium redox battery, lead acid battery, lithium-ion battery, flow battery, nickel cadmium battery, etc.

The second type of electrochemical energy storage is the supercapacitor. The supercapacitor stores energy in a series capacitor which is placed between the two electrodes. This makes supercapacitors more compact, have higher density, and have larger capacitance than conventional capacitors. The response time of supercapacitors ranges from ten to hundreds of milliseconds.

5.7.5 SUPERCONDUCTING MAGNETIC ENERGY STORAGE

In this technology, a cryogenically cooled method uses a magnetic field to store energy. This storage has three main parts: cryogenically cooled refrigeration, power conditioning system, and superconducting coil/magnet. During the charging condition, the current in the coil increases and vice versa. Although it has high efficiency and a long life span, the major drawback is its cost.

5.7.6 Cryogenic Energy Storage

In this storage technique, the energy is stored in the form of liquid air. This technology is used by Highview Power Storage Company, UK [52]. The first step is to compress the air and then liquefy it to store it in a tank at atmospheric pressure. Liquid air is pumped out and, after several intermediate steps, is expanded to generate electricity. Its volume is 1/7000 times less than that of gaseous air so it requires very less storage space. The advantage of using this storage is that it is cheap, long-lasting, and did not use toxic substances. The drawback is its efficiency which varies from about 40% to 70%.

5.7.7 Thermal Energy Storage

It consists of energy storage devices that store electricity in the form of thermal energy. Three methods are used in thermal ESSs which are discussed as follows.

5.7.7.1 Sensible Heat Storage

In this form of storage, the temperature of the material changes. There is no change in the phase of the material required. The materials used in this storage application are molten salt, cost ceramics, and concrete. Molten salts are used in solar thermal applications.

5.7.7.2 Latent Heat Storage

In this form of storage, when the material stores latent heat, the material undergoes a phase change (usually from the solid phase to the liquid phase). They are also known as phase change materials (PCM). There are two main categories of PCM, that is organic PCM and inorganic PCM. The materials used in this storage are paraffin wax, esters, glycols, fatty acids, and salts hydrate.

5.7.7.3 Thermochemical Heat Storage

In this form of energy storage, heat is stored in the endothermic process and released by the exothermic process. During charging, thermal energy is used to separate the chemical reactant into products in an endothermic reaction. Products are stored separately and used when energy is needed. During the discharge stage, the stored products are mixed to form the initial reactant in an exothermic reaction. The heat released during the reaction is used as an energy source. The materials used in this storage type are hydroxides, oxides, metal hydrides, carbonates, organic systems, and ammonia systems.

The classification of energy storage according to the power supply side, power grid side, and power distribution side is shown in Figure 5.5. Further ESSs are also classified according to their technical and economic aspects.

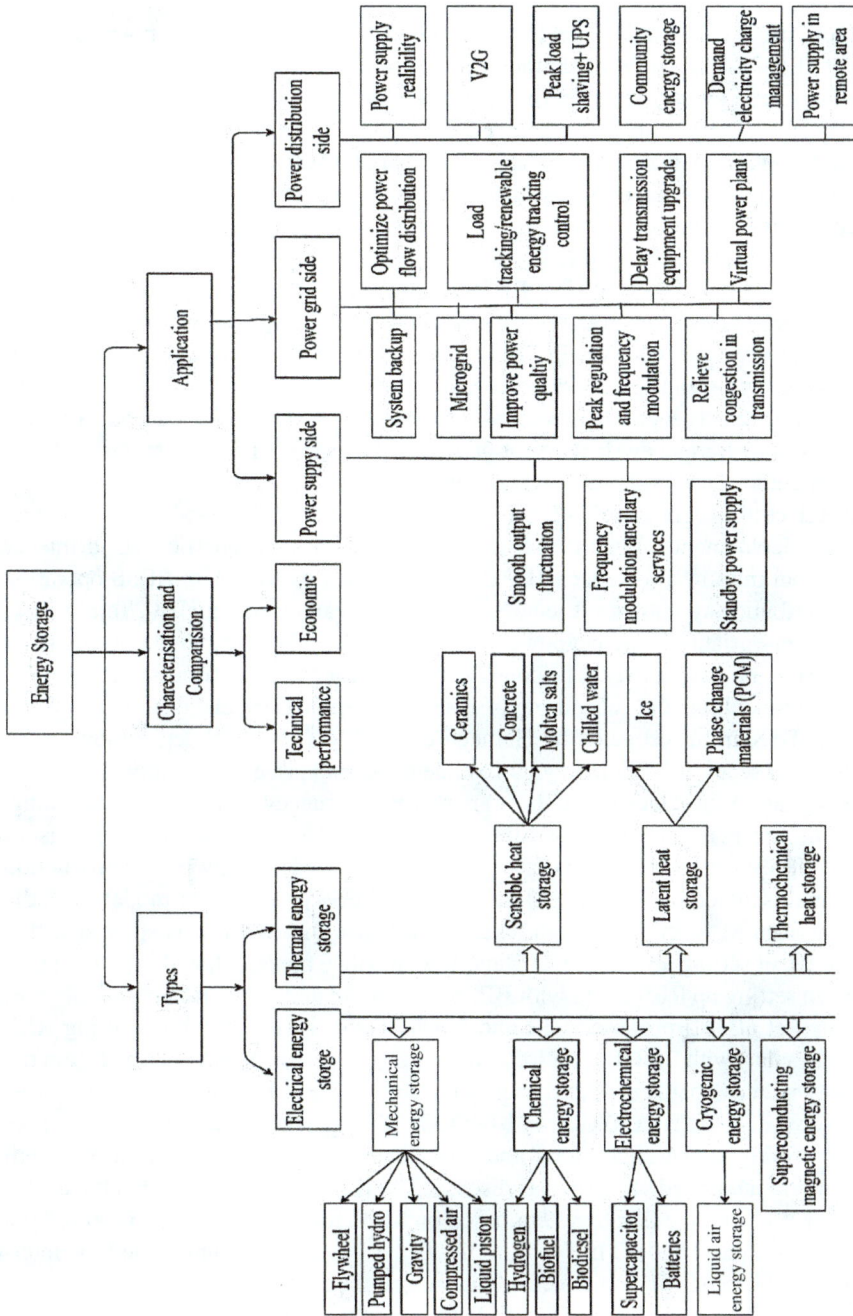

FIGURE 5.5 Classification of energy storage used in electrical power systems.

5.8   CHALLENGES AND OPPORTUNITIES
IN SETTING UP RES WITH BESS

The literature review on the ESS for wind power stationary application was done by Zhao et al. [53]. The different characteristic parameters are the specific power, used technology, energy and power ratings, specific energy, and capital cost of different ESS systems. Yang et al. [54] discussed ESS selection, estimation criteria, modelling, and solution methods. Active power control of a WF with BESS using supervisory MPC was discussed by Wang et al. [55]. Weighted ampere-hour (WAh) is taken as a cost function considering twisted angel variation and index.

A comprehensive review of BESS with the integration of the RES application was discussed by Datta et al. [56]. The BESS was classified based on active areas, reactive areas, and active–reactive power areas. The BESS applications were reviewed based on battery type, control algorithm, connection, application purpose, charging/discharging, and standby period in RES-integrated power systems. The BESS was used to regulate active power as well as reactive power to reduce the impact of RES on voltage and frequency. It was not advisable to install BESS for reactive power management because of its installation cost.

The effect of integration of BESS in support to RES was discussed by Santos et al. [57]. The deployment problem of BESS is formulated as stochastic Mixed-Integer Linear Programming Problem (MILP). A model based on MILP for BESS operation in the distribution system was discussed by Mehrjerdi and Hemmati [58]. Total hourly active power variation, reactive power variation, and voltage magnitude during peak hours are the main objectives. The consideration of reactive power offers extra benefits during the modelling and scheduling of BESS. The limitations and criteria for placement of BESS are investigated by Hameed et al. [59]. The BESS deployment in the Indian power sector is hindered by its cost, demand, and price uncertainty, absence of regulatory authorities, lack of ancillary market for frequency regulation and ramping. Jindal and Shrimali [60] addressed these issues and challenges on the BESS installation with respect to India's coal power versus renewable energy power production target and discussed its feasibility. The RES plus BESS procurement models in India based on agency, function, auction, and role are also discussed. It is suggested that irrespective of setting up a new coal plant, Indian policy makers should explore opportunities in setting up RES along with BESS from profit and commercial point of view.

Gumpu et al. [61] proposed CM and reliability improvement method using RES (wind and solar) with ESS installation. The optimal size of ESS was obtained from the reliability indices, that is expected energy not supplied and expected interruption cost values for GENCO, TRANSCO, and DISCOs. Proper charging/discharging of ESS is used to mitigate the uncertainty and congestion problem. Modelling of WF with BESS in grid integrated scenario was discussed by Raji and Gwabavu [62]. Dynamical control of WF with the BESS system based on MPC was proposed which showed the better performance of MPC in WF with the BESS system. It also mitigated the intermittency problem of WT and effectively increased the stability of the system.

The optimal operation of BESS with WPG using the wind forecast uncertainty in a day-ahead market was developed by Nguyen and Le [63]. The objective was to minimize the generating cost and operational cost of BESS in the AC OPF model. Pulazza et al. [64] proposed the battery-based energy storage transportation (BEST) approach to solve congestion issues. The idea was to transport BEST by train, car, bus, and truck to deploy at the optimal position with variable renewable energy. The objective was to minimize operational cost (generation fuel cost, transportation cost, and load shedding penalty) and investment cost of the storage system.

Mohamed *et al.* [65] proposed CM method using the optimal location, charging/discharging, and size of BESS in medium voltage distribution network. The objective function was considered as minimizing installation cost by selecting the optimal size of the battery. A day-ahead of battery operation planning used by the French transmission SO (RTE) for CM was discussed by Straub et al. [66]. The result showed that the battery was used for CM initially, and after three years of its operation, it can be used for arbitrage purposes.

The high penetration of WF may also cause congestion in the system. The WF power curtailment for CM with BESS was discussed by Straub et al. [67]. A hybrid multi-objective particle swarm optimization approach to minimize the cost and improve the voltage profile was proposed by Wen et al. [68]. The optimal location of BESS under consideration of uncertainties in WPG improves the voltage profile and reduces the operating cost. Del Rosso and Eckroad [69] investigated the ESS to enhance transfer capability limits in a thermal limited transmission line. The probabilistic ATC assessment by integrating WPG based on the canonical low rank approximation method was discussed by Sun et al. [70]. However, this literature did not discussed ATC enhancement.

The ATC values are continuously displayed on the OASIS that gives information about possible transaction. Thus, it is required to calculate ATC very fast and accurate to help ISO. AC power transfer distribution factor (ACPTDF) method are used to calculate ATC in this chapter. When the active power flows in the transmission line exceed its limits (thermal limit, voltage, and/or stability), it is called congestion of lines. ATC helps to provide the available power transfer information in the system to flow extra power from that corridor.

As the WF power injected into the grid is not constant due to the stochastic nature of wind speed, the system requires some additional backup to supply power to end users without interruption. BESS is used in the system to balance the power generation from the wind energy system. When the system has high load demand, the battery will discharge to support the grid, and if there is low load condition, the battery will charge. Proper charging and discharging of battery make the system stable and reliable. Technical challenges, that is selection of critical zones, optimum location for WPG with different BESS size, ATC enhancement using proper charging/discharging of battery in wind battery system, and CM are considered in this chapter.

5.9 PROPOSED METHODOLOGY

Line outage causes the congestion in the transmission lines. Many methods have been used to manage congestion in the lines. In the present work, ATC-based CM approach is used. ATC enhancement in wind battery system is proposed in this chapter.

5.9.1 WIND FARM POWER CALCULATION

The power is calculated using the mathematical equation as discussed by Narain et al. [71]. It is assumed that the wind direction is not changing during the simulation period.

Full shadowing case is considered for the WF power calculation. The mathematical equation to calculate WF power is given by Equation (5.1).

$$P_{WF} = \begin{cases} \displaystyle\sum_{i'=1}^{l} \frac{m * P_r}{1+e^{4s\left(v_{ip}-\tilde{V}_{wake}(x)_{i'}\right)/P_r}} & 0° < \theta < 20° \\[3mm] \displaystyle\sum_{i'=1}^{l} \frac{(m+n+1-2i') * P_r}{1+e^{4s\left(v_{ip}-\tilde{V}_{wake}(x)_{i'}\right)/P_r}} & 20° < \theta < 70° \\[3mm] \displaystyle\sum_{i'=1}^{l} \frac{n * P_r}{1+e^{4s\left(v_{ip}-\tilde{V}_{wake}(x)_{i'}\right)/P_r}} & 70° < \theta < 90° \end{cases} \tag{5.1}$$

where m = number of rows; n = number of columns; l = number of rows or column either of which one has lesser values in case of unequal number of row and column; $V_{wake}(x)_{i'}$ = speed of wake velocity for $i'th$ WT in shadow group; P_{WF} = total power of WF considers wind direction and wake effect.

When row and column will be different in the WF, for $0° < \theta < 20°$ only, row (m) will be considered, for $20° < \theta < 90°$, the row and column ($m + n$) both will be considered, and for $70° < \theta < 90°$, column (n) will be considered in Equation (5.1). In this chapter, row and column are equal and it is assumed θ is not changing during the simulation. For full shadowing case, the angle varies between either $0° < \theta < 20°$ or $70° < \theta < 90°$.

5.9.2   ATC CALCULATION

ATC is calculated using the ACPTDF method given by Equations (5.2) and (5.3) [72]. The bilateral transaction between seller bus 6 and buyer bus 28 is considered for the ATC calculation. That shows line 6–28 has minimum ATC value among all the lines and are known as limiting lines.

$$T_{ab,cd} = \begin{cases} \dfrac{\left(P_{ab}^{max} - P_{ab}^0\right)}{ACPTDF_{ab,cd}}; & ACPTDF_{ab,cd} > 0 \\[3mm] \propto (Infinite); & ACPTDF_{ab,cd} = 0 \\[3mm] \dfrac{\left(-P_{ab}^{max} - P_{ab}^0\right)}{ACPTDF_{ab,cd}}; & ACPTDF_{ab,cd} < 0 \end{cases} \tag{5.2}$$

$$ATC_{cd} = \min\{T_{ab,cd}\}, \ ab \in N_L \tag{5.3}$$

where ACPTDF = power transfer distribution factor between bus a and b due to the bilateral transaction-k between c and d; P_{ab}^{max} thermal limit of the line between buses a and b; P_{ab}^0 = base case power flow between buses a and b; ATC_{cd} = available transfer capability in MW for line violating the limits; and N_L = total number of lines.

TABLE 5.1

Buses in Superimposed Zones for Line Outage 4–6

	Zone 1	Zone 2	Zone 3	Zone 4
Buses	1, 2, 5, 6, 7, 8, 28	3, 4, 9, 11, 25, 26, 27, 29, 30	10, 17, 19, 20, 21, 22, 24	12, 13, 14, 15, 16, 18, 23

5.9.3 ATCDF Value for Optimal Location of WF and BESS

ATCDF value is obtained taking the average value of TCDF for congested line. The mathematical equation to calculate ATC is given in Equation (5.4) [73].

$$ATCDF = \frac{TCDF_1 + TCDF_2 + \dots\dots TCDF_N}{N} \qquad (5.4)$$

where $TCDF_1$ = TCDF value for one congested line; N = number of the congested line due to line outage. Narain et al. [73] suggested the optimal location for generating source integration as given in Table 5.1.

5.9.4 Charging and Discharging Criteria for BESS According to Load Variation

Figure 5.6 shows the variation of load in per unit for 24 hours [74]. The BESS will charge when there is low load condition, that is when it is below base line and discharge when the load of the system will be high. The criteria for charging and discharging of BESS considering load profile are made using the concept as discussed by Mehr et al. [75]. The BESS will charge from time 22:00 to time 07:00 and start to

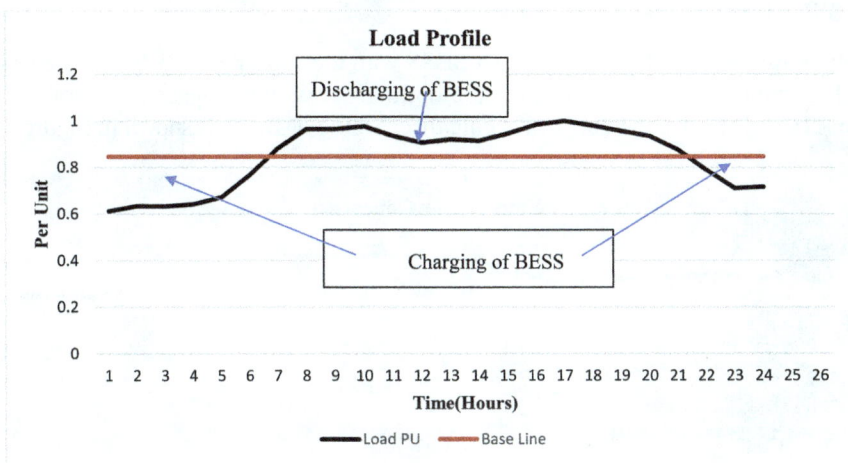

FIGURE 5.6 Charging/discharging criteria of battery energy storage system under load variation.

discharge at time 07:00 to time 22.00 as indicated in Figure 5.6. It is assumed that the BESS has sufficient capacity to deliver power to the system. If the BESS will charge and there is no requirement of discharge, it will be kept in standby mode for future requirements.

5.10   RESULTS AND DISCUSSION

There are three cases that have been considered for the simulation. The three cases are given as follows:

- Case 1: Line outage 4–6 to create congestion, and WF with one BESS is installed.
- Case 2: Line outage 4–6 to create congestion, and WF with two BESSs (each one has half of the capacity of main BESS) is installed.
- Case 3: Line outage 4–6 to create congestion, and WF with three BESSs (each one has one-third of the capacity of main BESS) is installed.

5.10.1   WIND FARM POWER

WF of 50 MW capacity having 25 WT in row and column symmetrical layout is considered. The power generated from the WF for 24 hours is given in Figure 5.7.

5.10.2   CASE 1

The BESS of the capacity of 10 MVA and energy 30 MWh has been selected. It has enough capacity to supply power with proper charging/discharging manner. The ATC value for Case 1 is shown in Figure 5.8. The ATC values enhance in case of high load condition as BESS supply the power to the grid.

5.10.3   CASE 2

In this case, instead of one BESS, two BESS with the capacity of 15 MWh is considered for simulation. It is installed at the optimal location as suggested in Table 5.2. The ATC value variation due to charging/discharging of BESS is shown in Figure 5.9.

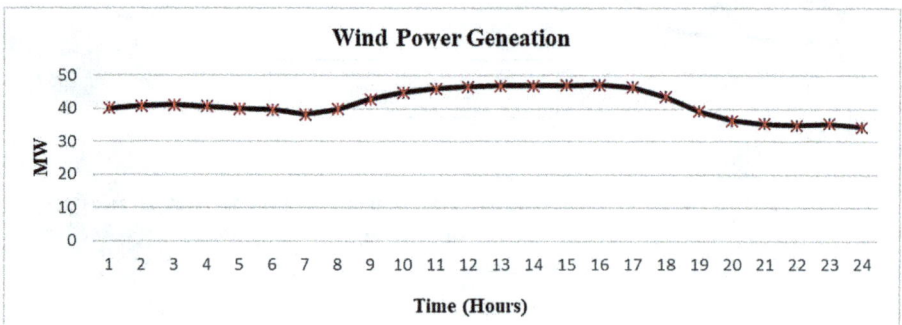

FIGURE 5.7   Wind farm power output for 24 hours.

ATC (MW)

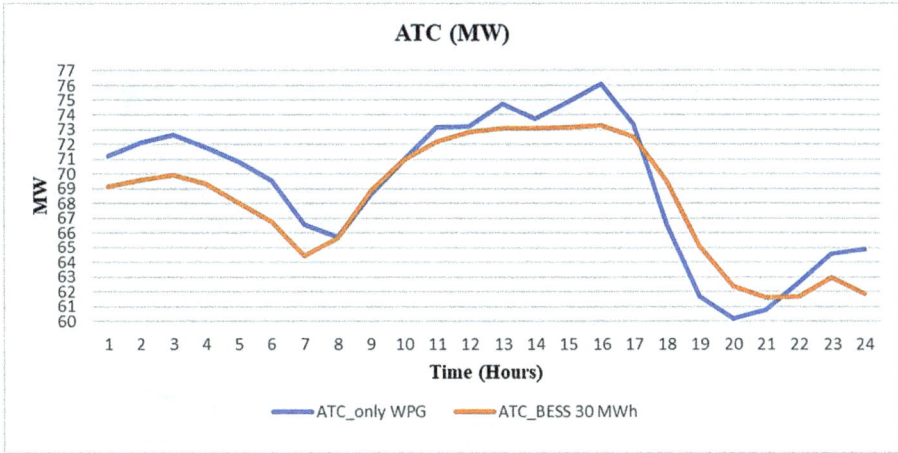

FIGURE 5.8 Available transfer capability variation in limiting line for 24 hours in Case 1.

TABLE 5.2

Bus Location for WF and BESS Installation for Maximum ATC Enhancement

Bus No.	Case 1		Case 2		Case 3	
	WF	BESS	WF	BESS	WF	BESS
	28	30	28	30, 29	28	30, 29, 27

ATC (MW)

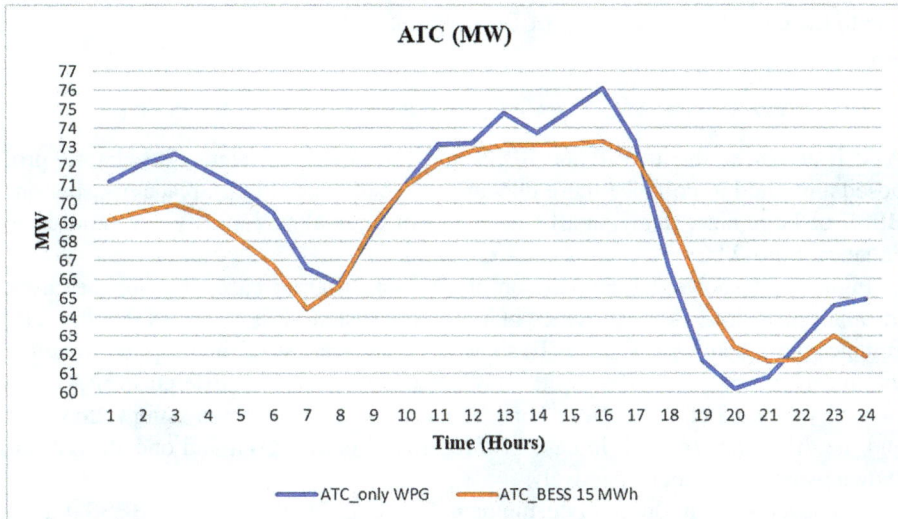

FIGURE 5.9 Available transfer capability variation in limiting line for 24 hours in Case 2.

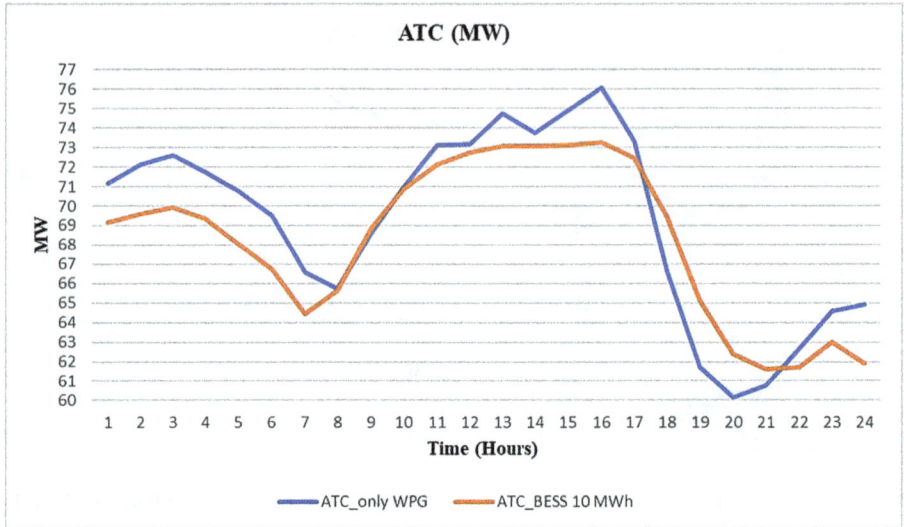

FIGURE 5.10   Available transfer capability variation in limiting line for 24 hours in Case 3.

5.10.4   CASE 3

In this case, instead of one BESS, three BESS with the capacity of 10 MWh is considered for simulation. It is installed at the optimal location as suggested in Table 5.2. The BESS is installed at bus 29, bus 28, and bus 27, respectively. Based on the hierarchical order, it is placed in the zone. All three BESS supplied the power to the grid to enhance the ATC. The enhancement during peak load conditions due to the charging/discharging of BESS is shown in Figure 5.10.

5.10.5   ANALYSIS OF ATC ENHANCEMENT USING DIFFERENT BESS SIZES

ATC is enhanced due to WF and BESS integration in the system. The chapter proposed a method to enhance more during peak load condition. In place of using one BESS of large size, many small size BESS of equivalent capacity can be used to enhance more ATC.

Figure 5.11 shows the comparison of ATC in different cases. For all the cases, the analysis is done in the presence of WF in the system. It is observed that the ATC enhancement is more in Case 3. To show the effectiveness of the proposed method, the difference of ATC in kilowatt (KW) is also shown. The difference of ATC in case of two BESS of 15 MWh and one BESS of 30 MWh is shown using yellow line, and the difference of ATC in case of three BESS of 10 MWh and one 30 MWh is shown using blue line, respectively.

It is observed that although during peak load conditions small-size BESS did not increase ATC value more as compared to single large-size BESS, but during off-peak load, the ATC enhanced more due to small-size BESS.

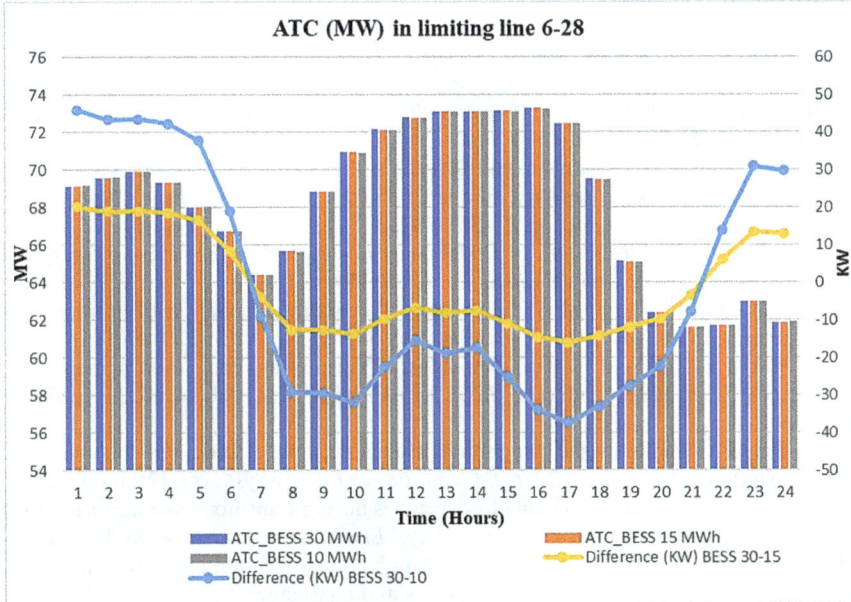

FIGURE 5.11   Comparison of available transfer capability for different cases.

5.11   CONCLUSION AND FUTURE SCOPE

In a deregulated environment, many conventional and RES are available. The ESSs are transferable and can be deployed to manage congestion by charging and discharging criteria. During low load conditions, it charges, and during peak load, it discharges to manage congestion in the system. The placement of BESS plays a very important role in the case of WPG. It is suggested that instead of deploying large capacity BESS at one place, small capacity BESS should be deployed at many places for ATC enhancement. The novel concept of ATC enhancement using charging/discharging of BESS with WF is present in this chapter. The optimal location of WF and BESS is obtained using the proposed equations. The result shows significant improvement in the ATC when BESS is properly charged and discharged based on the load variation. The comparison of different BESS size cases for ATC is also highlighted in the chapter. In place of one BESS, if more than one BESS of small size is integrated into the system, the ATC improves more than one BESS case in off-peak load conditions. That helps SOs to transact more power during low load conditions.

   In this chapter, main focus was only on one technical issue, that is CM to enhance the ATC, whereas other technical issues like power loss, voltage profile, cost, and BESS state of charge/depth of discharge can be included for future research work. The optimal size of BESS can also be obtained by using optimization techniques, and comparative analysis can be done. To create congestion, only one-line outage is considered for the simulation, multiple line outages contingency condition can be considered. Load variation is taken only for 24 hours; however, monthly and yearly load variations can be used for further analysis.

REFERENCES

1. Y. R. Sood, N. P. Padhy, and H. O. Gupta, "Deregulation of power sector – A bibliographical survey," *Int. J. Glob. Energy Issues*, vol. 11, no. 1–4, pp. 195–202, 2019.
2. H. L. Willis, and L. Philipson, *Understanding electric utilities and de-regulation*, vol. 27. CRC Press, 2018.
3. J. W. M. Cheng, F. D. Galiana, and D. T. Mcgillis, "Studies of bilateral contracts with respect to steady-state security in a deregulated environment," *Proc. 20th Int. Conf. Power Ind. Comput. Appl.*, pp. 31–36, 1997.
4. M. I. Alomoush, and S. M. Shahidehpour, *Restructured electrical power systems: Operation: Trading, and volatility.* vol. 1, CRC Press, 2017.
5. M. I. Alomoush, and S. M. Shahidehpour, "Generalized model for fixed transmission rights auction," *Electr. Power Syst. Res.*, vol. 54, no. 3, pp. 207–220, 2000.
6. H. Singh, S. Hao, and A. Papalexopoulos, "Transmission congestion management in competitive electricity markets," *IEEE Trans. Power Syst.*, vol. 14, no. 3, pp. 877–883, 1999.
7. M. Pantoš, "Market-based congestion management in electric power systems with exploitation of aggregators," *Int. J. Electr. Power Energy Syst.*, vol. 121, pp. 1–10, 2020.
8. G. Poyrazoglu, "Determination of price zones during transition from uniform to zonal electricity market: A case study for Turkey," *Energies*, vol. 14, no. 4, pp. 1–13, 2021.
9. A. Narain, S. K. Srivastava, and S. N. Singh, "Congestion management approaches in restructured power system: Key issues and challenges," *Electr. J.*, vol. 33, no. 3, pp. 1–8, 2020.
10. X. Zhang, Y. H. Song, Q. Lu, and S. Mei, "Dynamic available transfer capability (ATC) evaluation by dynamic constrained optimization," *IEEE Power Eng. Rev.*, vol. 19, no. 2, pp. 1240–1242, 2004.
11. M. Eidiani, and M. H. M. Shanechi, "FAD-ATC: A new method for computing dynamic ATC," *Int. J. Electr. Power Energy Syst.*, vol. 28, no. 2, pp. 109–118, 2006.
12. I. A. Hiskens, M. A. Pai, and P. W. Sauer, "An iterative approach to calculating dynamic ATC," *Proc. Bulk Power Syst. Dyn. Control IV-Restructuring*, pp. 585–590, 1998.
13. E. De Tuglie, M. Dicorato, M. La Scala, and P. Scarpellini, "A static optimization approach to assess dynamic available transfer capability," *Proc. 21st Int. Conf. Power Ind. Comput. Appl. Connect. Util. PICA 99. To Millenn. Beyond (Cat. No. 99CH36351)*, pp. 269–277, 1999.
14. A. Kumar, S. C. Srivastava, and S. N. Singh, "Available transfer capability assessment in a competitive electricity market using a bifurcation approach," *IEE Proc.-Gener. Transm. Distrib.*, vol. 151, no. 2, pp. 133–140, 2004.
15. Y. Cheng, T. S. Chung, C. Y. Chung, and C. W. Yu, "Dynamic voltage stability constrained ATC calculation by a QSS approach," *Int. J. Electr. Power Energy Syst.*, vol. 28, no. 6, pp. 408–412, 2006.
16. S. Impram, S. V. Nese, and B. Oral, "Challenges of renewable energy penetration on power system flexibility: A survey," *Energy Strateg. Rev.*, vol. 31, pp. 1–12, 2020.
17. M. Sandhu, and T. Thakur, "Issues, challenges, causes, impacts and utilization of renewable energy sources – Grid integration," *Int. J. Eng. Res. Appl.*, vol. 4, no. 3, pp. 636–643, 2014.
18. A. Kalair, N. Abas, M. S. Saleem, A. R. Kalair, and N. Khan, "Role of energy storage systems in energy transition from fossil fuels to renewables," *Energy Storage*, vol. 3, no. 1, pp. 1–27, 2021.
19. Renewable energy country attractiveness index (RECAI). https://www.ey.com/en_gl/recai (accessed Feb. 04, 2022).
20. Energy Statistics 2022. https://www.mospi.gov.in/ (accessed Feb. 14, 2022).

21. Ministry of New and Renewable Energy. https://mnre.gov.in/the-ministry/physical-progress (accessed Feb. 04, 2022).

22. Indian Wind Turbine Manufacturer Association. https://www.indianwindpower.com/news_views.php#tab1 (accessed Jul. 06, 2021).

23. Global Wind Report 2021 (GWEC). https://gwec.net/global-wind-report-2021/ (accessed Jul. 06, 2021).

24. International Renewable Energy Agency (IRENA)." https://www.irena.org/-/media/Files/IRENA/Agency/Publication/2021/Apr/IRENA_RE_Capacity_Statistics_2021.pdf (accessed Feb. 04, 2022).

25. TC 88 – Wind Energy Generation Systems, "Wind energy generation systems – Part 1: Design requirements," *IEC 61400-12019 Stand.*, 2019.

26. Vestas Wind Systems. https://www.vestas.com/ (accessed Feb. 01, 2022).

27. M. G. Khalfallah, and A. M. Koliub, "Suggestions for improving wind turbines power curves," *Desalination*, vol. 209, no. 1–3, pp. 221–229, 2007.

28. M. H. Albadi, and E. F. El-Saadany, "Wind turbines capacity factor modeling – A novel approach," *IEEE Trans. Power Syst.*, vol. 24, no. 3, pp. 1637–1638, 2009.

29. S. H. Jangamshetti, and V. G. Rau, "Site matching of wind turbine generators: A case study," *IEEE Trans. Energy Convers.*, vol. 14, no. 4, pp. 1537–1543, 1999.

30. V. Thapar, G. Agnihotri, and V. K. Sethi, "Critical analysis of methods for mathematical modelling of wind turbines," *Renew. Energy*, vol. 36, no. 11, pp. 3166–3177, 2011.

31. D. Villanueva, and A. Feijóo, "A review on wind turbine deterministic power curve models," *Appl. Sci.*, vol. 10, no. 12, pp. 1–15, 2020.

32. D. Villanueva, and A. Feijóo, "Comparison of logistic functions for modeling wind turbine power curves," *Electr. Power Syst. Res.*, vol. 155, pp. 281–288, 2018.

33. J. A. Carta, S. Díaz, and A. Castañeda, "A global sensitivity analysis method applied to wind farm power output estimation models," *Appl. Energy*, vol. 280, pp. 1–20, 2020.

34. N. Deljouyi, A. Nobakhti, and A. Abdolahi, "Wind farm power output optimization using cooperative control methods," *Wind Energy*, vol. 24, no. 5, pp. 502–514, 2021.

35. S. Díaz, J. A. Carta, and A. Castañeda, "Influence of the variation of meteorological and operational parameters on estimation of the power output of a wind farm with active power control," *Renew. Energy*, vol. 159, pp. 812–826, 2020.

36. F. Meng, A. W. H. Lio, and J. Liew, "The effect of minimum thrust coefficient control strategy on power output and loads of a wind farm," *J. Phys. Conf. Ser.*, vol. 1452, no. 1, pp. 1–12, 2020.

37. E. Yun, and J. Hur, "Probabilistic estimation model of power curve to enhance power output forecasting of wind generating resources," *Energy*, vol. 223, 2021.

38. C. Draxl, R. P. Worsnop, G. Xia, Y. Pichugina, D. Chand, J. K. Lundquist, J. Sharp, G. Wedam, J. M. Wilczak, and L. K. Berg, "Mountain waves can impact wind power generation," *Wind Energy Sci.*, vol. 6, no. 1, pp. 45–60, 2021.

39. G. W. Qian, and T. Ishihara, "Wind farm power maximization through wake steering with a new multiple wake model for prediction of turbulence intensity," *Energy*, vol. 220, pp. 1–17, 2021.

40. F. González-Longatt, P. P. Wall, and V. Terzija, "Wake effect in wind farm performance: Steady-state and dynamic behavior," *Renew. Energy*, vol. 39, no. 1, pp. 329–338, 2012.

41. B. Bilal, M. Ndongo, K. H. Adjallah, A. Sava, C. M. F. Kebe, P. A. Ndiaye, and V. Sambou, "Wind turbine power output prediction model design based on artificial neural networks and climatic spatiotemporal data," *IEEE Int. Conf. Ind. Technol.*, pp. 1085–1092, 2018.

42. X. Ge, Q. Chen, Y. Fu, C. Y. Chung, and Y. Mi, "Optimization of maintenance scheduling for offshore wind turbines considering the wake effect of arbitrary wind direction," *Electr. Power Syst. Res.*, vol. 184, pp. 1–11, 2020.

43. S. T. Suganthi, D. Devaraj, K. Ramar, and S. H. Thilagar, "An improved differential evolution algorithm for congestion management in the presence of wind turbine generators," *Renew. Sustain. Energy Rev.*, vol. 81, pp. 635–642, 2018.

44. D. F. Farzana, and K. Mahadevan, "Performance comparison using firefly and PSO algorithms on congestion management of deregulated power market involving renewable energy sources," *Soft Comput.*, vol. 24, no. 2, pp. 1473–1482, 2020.

45. A. Abdolahi, J. Salehi, F. S. Gazijahani, and A. Safari, "Probabilistic multi-objective arbitrage of dispersed energy storage systems for optimal congestion management of active distribution networks including solar/wind/CHP hybrid energy system," *J. Renew. Sustain. Energy*, vol. 10, no. 4, pp. 1–21, 2018.

46. S. Gope, A. K. Goswami, and P. K. Tiwari, "Transmission congestion management with integration of wind farm: A possible solution methodology for deregulated power market," *Int. J. Syst. Assur. Eng. Manag.*, vol. 11, no. 2, pp. 287–296, 2020.

47. V. Pattanaik, B. K. Malika, S. Mohanty, P. K. Rout, and B. K. Sahu, "Optimal energy storage allocation in smart distribution systems: A review," *Adv. Intell. Comput. Commun.*, vol. 202, pp. 555–565, 2021.

48. S. van der Linden, "Bulk energy storage potential in the USA, current developments and future prospects," *Energy*, vol. 31, no. 15, pp. 3446–3457, 2006.

49. M. A. Rosen, and S. K. Fayegh, "A review of energy storage types, applications and recent developments," *J. Energy Storage*, vol. 27, pp. 1–23, 2020.

50. V. Vahidinasab, and M. Habibi, "Electric energy storage systems integration in energy markets and balancing services," in *Energy storage in energy markets*, pp. 287–316, Academic Press, Cambridge, Massachusetts, USA, 2021.

51. H. Chen, T. Ngoc, W. Yang, C. Tan, and Y. Li, "Progress in electrical energy storage system: A critical review," *Prog. Nat. Sci.*, vol. 19, no. 3, pp. 291–312, 2009.

52. Highview Power Storage Company, UK. https://highviewpower.com/technology/ (accessed Jul. 01, 2021).

53. H. Zhao, Q. Wu, S. Hu, H. Xu, and C. N. Rasmussen, "Review of energy storage system for wind power integration support," *Appl. Energy*, vol. 137, pp. 545–553, 2015.

54. B. Yang, J. Wang, Y. Chen, D. Li, C. Zeng, Y. Chen, Z. Guo, H. Shu, X. Zhang, T. Yu, and L. Sun, "Optimal sizing and placement of energy storage system in power grids: A state-of-the-art one-stop handbook," *J. Energy Storage*, vol. 32, pp. 1–30, 2020.

55. C. Wang, Z. Du, Y. Ni, C. Li, and G. Zhang, "Coordinated predictive control for wind farm with BESS considering power dispatching and equipment ageing," *IET Gener. Transm. Distrib.*, vol. 12, no. 10, pp. 2406–2414, 2018.

56. U. Datta, A. Kalam, and J. Shi, "A review of key functionalities of battery energy storage system in renewable energy integrated power systems," *Energy Storage*, pp. 1–21, 2021.

57. S. F. Santos, M. Gough, D. Z. Fitiwi, A. F. P. Silva, M. Shafie-Khah, and J. P. S. Catalão, "Influence of battery energy storage systems on transmission grid operation with a significant share of variable renewable energy sources," *IEEE Syst. J.*, pp. 1–12, 2021.

58. H. Mehrjerdi, and R. Hemmati, "Modeling and optimal scheduling of battery energy storage systems in electric power distribution networks," *J. Clean. Prod.*, vol. 234, pp. 810–821, 2019.

59. Z. Hameed, S. Hashemi, H. H. Ipsen, and C. Træholt, "A business-oriented approach for battery energy storage placement in power systems," *Appl. Energy*, vol. 298, pp. 1–13, 2021.

60. A. Jindal, and G. Shrimali, "At scale adoption of battery storage technology in Indian power industry: Enablers, frameworks and policies," *Fram. Policies*, pp. 1–25, 2021.

61. S. Gumpu, B. Pamulaparthy, and A. Sharma, "Review of congestion management methods from conventional to smart grid scenario," *Int. J. Emerg. Electr. Power Syst.*, vol. 20, no. 3, pp. 1–24, 2019.

62. A. Raji, and M. Gwabavu, "Dynamic control of integrated wind farm battery energy storage systems for grid connection," *Sustainability*, vol. 13, no. 6, pp. 1–27, 2021.

63. N. T. A. Nguyen, and D. D. Le, "Day-ahead coordinated operation of a wind-storage system considering wind forecast uncertainty," *Eng. Technol. Appl. Sci. Res.*, vol. 11, no. 3, pp. 7201–7206, 2021.

64. G. Pulazza, N. Zhang, C. Kang, and C. A. Nucci, "Transmission planning with battery-based energy storage transportation for power systems with high penetration of renewable energy," *IEEE Trans. Power Syst.*, pp. 1–12, 2021.

65. A. A. R. Mohamed, D. J. Morrow, R. J. Best, I. Bailie, A. Cupples, and J. Pollock, "Battery energy storage systems allocation considering distribution network congestion," *IEEE PES Innov. Smart Grid Technol. Eur.*, pp. 1015–1019, 2020.

66. C. Straub, J. Maeght, C. Pache, P. Panciatici, and Rajagopal, "Congestion management within a multi-service scheduling coordination scheme for large battery storage systems," *IEEE Milan PowerTech, PowerTech*, pp. 1–6, 2019.

67. C. Straub, S. Olaru, J. Maeght, and P. Panciatici, "Zonal congestion management mixing large battery storage systems and generation curtailment," *IEEE Conf. Control Technol. Appl.*, pp. 988–995, 2018.

68. S. Wen, H. Lan, Q. Fu, D. C. Yu, and L. Zhang, "Economic allocation for energy storage system considering wind power distribution," *IEEE Trans. Power Syst.*, vol. 30, no. 2, pp. 644–652, 2015.

69. A. D. Del Rosso, and S. W. Eckroad, "Energy storage for relief of transmission congestion," *IEEE Trans. Smart Grid*, vol. 5, no. 2, pp. 1138–1146, 2014.

70. X. Sun, Z. Tian, Y. Rao, Z. Li, and P. Tricoli, "Probabilistic available transfer capability assessment in power systems with wind power integration," *IET Renew. Power Gener.*, vol. 14, no. 11, pp. 1912–1920, 2020.

71. A. Narain, S. K. Srivastava, and S. N. Singh, "An ATCDF approach for ATC enhancement using charging/discharging of wind-battery system considering wind direction and wake effect of WTs," *Int. Trans. Electr. Energy Syst.*, pp. 1–18, 2021, doi: https://doi.org/10.1002/2050-7038.13086.

72. A. Kumar, S. C. Srivastava, and S. N. Singh, "Available transfer capability (ATC) determination in a competitive electricity market using AC distribution factors," *Electr. Power Components Syst.*, vol. 32, no. 9, pp. 927–939, 2004.

73. A. Narain, S. K. Srivastava, and S. N. Singh, "A novel sensitive based approach to ATC enhancement in wind power integrated transmission system," *SN Appl. Sci.*, vol. 3, no. 5, 2021, doi: 10.1007/s42452-021-04559-8.

74. K. Gaur, H. Kumar, R. P. K. Agarwal, K. V. S. Baba, and S. K. Soonee, "Analysing the electricity demand pattern," *Natl. Power Syst. Conf. NPSC*, pp. 1–6, 2016, doi: 10.1109/NPSC.2016.7858969.

75. T. H. Mehr, M. A. S. Masoum, and N. Jabalameli, "Grid-connected lithium-ion battery energy storage system for load leveling and peak shaving," *Australas. Univ. Power Eng. Conf. AUPEC*, 2013, doi: 10.1109/aupec.2013.6725376.

6 Concentrated Solar Power

A Promising Sustainable Energy Option

Soumitra Mukhopadhyay
Doosan Power Systems (I) Pvt. Ltd.,
Kolkata, West Bengal, India

CONTENTS

6.1   INTRODUCTION

It is a well-known fact that energy and electricity consumption are increasing globally day-by-day due to the growth in various sections like industrialization, urbanization, and population. Till now, mankind is mainly dependent on major conventional fossil fuel energy sources like coal, oil, natural gas, etc., which emit harmful emissions and these are depleting in nature. Regarding the reserve of fossil fuels, there are several data available with us. As per BP Statistical Review of World Energy 2021, at the present consumption rate, it is estimated that the reserve of world's coal,

DOI: 10.1201/9781003369554-6

petroleum, and natural gas is 153, 50, and 52.8 years, respectively [1]. In another model, which is a modified Klass model, depletion times for fossil fuel reserves for coal, oil, and natural gas are approximately 107, 35, and 37 years, respectively, which means that coal reserves will only be available after 2042 and up to 2112 [2]. Burning of fossil fuels results in the emission of greenhouse gases (GHG) like CO_2 due to which mankind is experiencing severe challenges in providing a cleaner, greener, and sustainable energy supply to populations [3–5].

To eliminate or reduce the catastrophic scenario like global warming, air pollution, various diseases, etc., due to GHG emissions for the burning of fossil fuels, shift towards renewable energy options have been felt globally. It has been observed from global awareness for various renewable energy options and simultaneous development of the renewable energy infrastructure that reduction in carbon di-oxide emission will be approximately 30% by 2050, in comparison to the year 2012 [6–8]. In this context, renewable energy options which are considered as no or low carbon emission technologies have become very much pertinent due to following reasons:

   i. **To contribute to a greater CO_2 reduction effort:** As already discussed, fossil fuels have various harmful impacts like global warming, refinery and oil rig explosions, oil spills and water pollution, air pollution and smog, acid rain, etc. Due to these harmful impacts of fossil fuels, use of fossil fuels cannot be relied upon.
  ii. **To satisfy the energy requirement of society:** Energy demand is increasing day-by-day and to meet this energy demand, alternative and sustainable energy sources should be found out.
 iii. **To obtain health benefits from renewable energy options:** Since burning of fossil fuels pollutes air, exposure to this polluted air can cause various kinds of health issues like heart disease, respiratory infections, and lung cancer, which can be avoided by the use of renewable energy options.

This chapter emphasizes on the following main objectives:

A. Understand the basics of concentrated solar thermal power;
B. Understand the disadvantages of photo-voltaic (PV) system, which is also considered as a mature technology nowadays;
C. Discusses various types of concentrated solar thermal power;
D. Discussion on thermal storage considered as an important part of concentrated solar power (CSP) plant along with its various types;
E. And, finally, a Brayton cycle-based power block has been elaborated to find out various parameters from energetic point of view.

6.1.1 ORGANIZATION OF THE CHAPTER

This chapter is organized as follows: Section 6.2 elaborates the concentrated solar thermal power in detail along with its advantages over PV system. Section 6.3 focusses on the working principle of CSP system and its types. Then thermal energy storage system along with its types has been elaborated in Section 6.4 and the sub-sections discuss regarding the working of various types of thermal energy storage systems along with schematic arrangements. Parametric analysis of Brayton cycle-based power block has been studied in Section 6.5 based on certain input parameters and also elaborates mathematical formulations and results. Section 6.6 concludes the chapter with future scope.

6.2 CONCENTRATED SOLAR THERMAL POWER

Out of various renewable energy alternatives available, solar energy is considered as one of the most effective alternative due to its inherent features of being green, low-cost, and availability throughout the globe [9]. Whenever harnessing electricity from solar energy is talked about, what comes into mind first is the PV system which is a mature technology. But the major disadvantages of PV technology are as follows:

i. PV panels require inverters to convert direct current to alternating current to use the current in power network, which is an additional investment cost.
ii. For ground-mounted PV installations, they require large areas for deployment.
iii. The efficiency of PV systems is low.

Another technology which is capable in producing utility scale electricity from the same energy source, that is solar energy is CSP or solar thermal electricity (STE) and it has the capacity to offer dispatchable power by integrating it with thermal energy storage or in hybrid mode. It has higher efficiency compared to PV technology and due to its higher energy saving and efficiency, it is predicted that CSP plants will contribute 7% by the year 2030 and 25% by the year 2050 of global electricity demand [10, 11]. CSP has also considerable potential for job creation, boosting local economy, and more significantly, in CO_2 reduction capability. As per Global Outlook, 2016 [12, 13], 1.2 billion tonnes of CO_2 will be saved in 2050.

The primary requirement of solar thermal power generation is sunlight, generally called Direct Normal Irradiation (DNI). Though the best CSP plants should maintain a DNI level in the range of 2000–2800 kWh/m^2/year [12], commercial CSP plants can be developed with a DNI level at least greater than 1800 kWh/m^2/year. The "Sun Belt" is the region where very high solar radiation level is available. These include the Middle East, North Africa, the Mediterranean, and wide areas in the United States including California, Nevada, Arizona, and New Mexico. The other promising areas for the development of CSP plants are desert areas of India, Pakistan, China, Australia, etc.

It is expected that that CSP plants with a total capacity of 1504 GW will be installed in the United States by 2050 [14,15]. It has also been assessed that with the present deployment policy of CSP, a total of 83 GW will be installed by 2030 in the sunniest regions and will reach to 342 GW by 2050. This expected power generation will be coming from the Northern Africa (30%), Middle East (55%), and the remaining 15% from European countries. It is also predicted that power generated through CSP technologies in desert regions of the Middle East and North Africa (MENA) will be exported to Europe [10].

6.3 TYPES OF CONCENTRATED SOLAR THERMAL SYSTEMS

Solar thermal system works by harnessing the solar energy, that is photons; a sun-facing surface called solar thermal collector concentrates sun's heat energy into a receiver/absorber and this heat energy is converted to a working fluid such as water and air, which in turn generates power [16]. Schematic arrangement of CSP system is shown in Figure 6.1.

Though various concentrated solar plants are at the research, design, and development stages, following four different CSP plants are mostly found [17]:

   i. Solar parabolic dish
  ii. Parabolic trough collector
 iii. Solar power tower or central receiver system and
  iv. Linear Fresnel reflector (LFR)

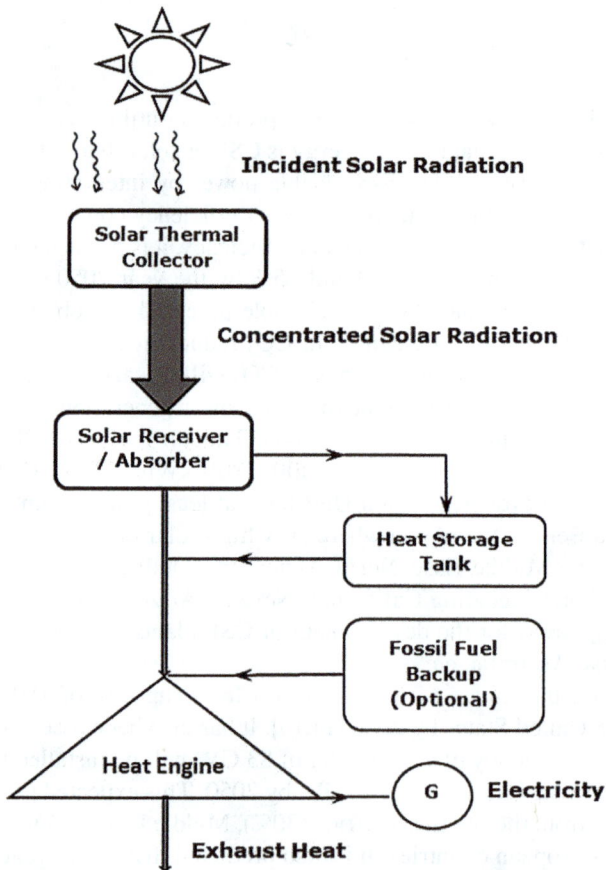

FIGURE 6.1   Schematic diagram of concentrated solar thermal system.

6.3.1 SOLAR PARABOLIC DISH

In the solar parabolic-dish-based CSP system, a parabolic-dish-shaped point-focus concentrator is used, which reflects the solar radiation onto a receiver placed at the focal point of the dish with a concentration ratio of approximately 2000 at the focal point. Two-axis tracking system is used to utilize the maximum solar radiation. At the focal point of the dish, power conversion unit is coupled with an electrical generator to convert the concentrated heat energy into electricity.

The power conversion unit may be a Stirling engine or a Brayton cycle. The schematic arrangement of solar dish system is shown in Figure 6.2. The temperature and pressure of the working fluid generally reach around 700–750°C and 200 bar, respectively [18–20]. The diameter of the dishes may vary from 5 to 10 m with surface area of 40–120 m². The reflecting surface of the dish is constructed of aluminium or silver, coated on glass or plastic. Higher performance can be attained while silver surface having a thickness of 1 μm is used coated on a glass surface. Further to increase the reflectivity, a certain percentage of iron is added in the glass. Solar reflectance can reach as high as 94% with power generation from a single

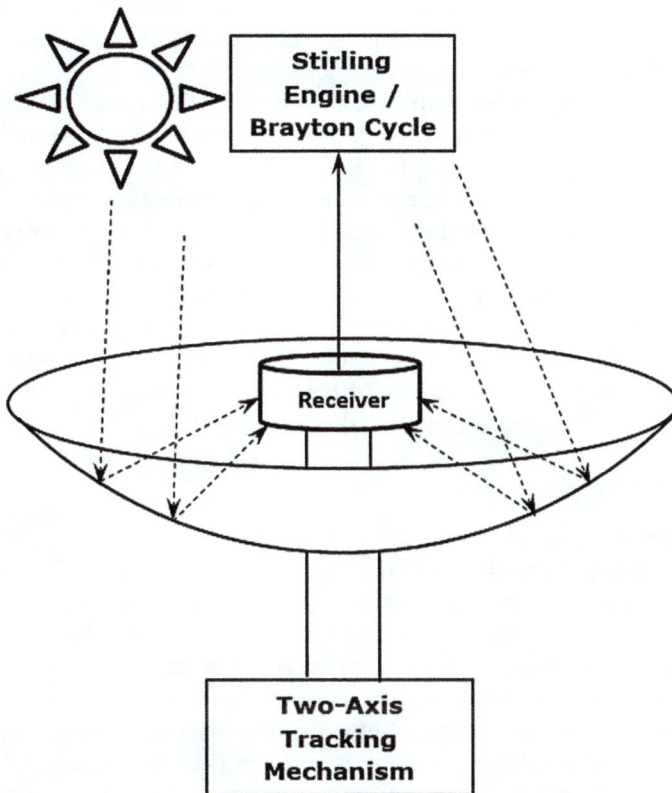

FIGURE 6.2 Schematic diagram of solar dish concentrator system.

parabolic-dish CSP system varying from 0.01 to 0.5 MW [21–23]. Various types of solar dish systems have been explained later.

- A. **Solar Dish Stirling System:** The major components of solar dish Stirling systems are as follow:
  - i. Solar Concentrator: The dish, called solar concentrator, directs the beam of concentrated sunlight into a thermal receiver placed at the focal point of the dish. The dish is mounted on a structure which tracks the sunlight all over the day to reflect the maximum percentage of sunlight possible onto the thermal receiver.
  - ii. Power Conversion Unit: The power conversion unit consists of the thermal receiver and the engine-generator combination. The receiver absorbs the concentrated solar radiation, converts the solar energy to heat through a suitable fluid, and the heat energy is transferred to the engine. The fluid used in the receiver, which is also the working fluid of the engine, is generally hydrogen or helium.

    The engine, generally a Stirling engine, receives the hot fluid from the thermal receiver. Mechanical power is generated due to the linear movement of piston of the engine. The linear movement of the piston is converted to rotational movement through a crankshaft, which drives a generator. The generator produces electrical energy. Thus, solar energy is ultimately converted into electrical energy.
- B. **Solar Dish Brayton System:** Like solar dish Stirling system, the parabolic dish concentrates the sun's rays onto the receiver so that the solar heat can be absorbed by the working fluid passing through the tubes in the inner walls of the receiver. A compressor and recuperator are used to compress and preheat the working fluid (generally air), respectively, before entering into receiver. Hot exhaust air from the turbine is used to preheat compressed air. The hot air from the receiver expands in the turbine, which produces mechanical power through the rotational movement of the shaft. A generator is coupled with the turbine to convert the mechanical power to electrical power. Generally, the generator, turbine, and the compressor are mounted on the same shaft and all rotate at the same speed.

6.3.2 Solar Trough Collector

Out of various concentrating solar power technologies available, parabolic troughs are the most mature technology and they are commercially viable. The first system was installed near Cairo in Egypt in the year 1912 to generate steam and this plant was competitive with respect to coal fired steam generation system [24].

In the parabolic trough system, sunlight is concentrated to 25–100 times on absorber tubes, placed at the focal point of parabolic trough which acts a solar concentrator here. Due to the combined effect of the shape and mirror polished surface of the trough, operating temperatures of the working/heat transfer fluid (HTF) can be achieved from 350°C to 550°C. Generally, the HTF is pumped through the absorber tube and transfers the thermal energy to a Rankine cycle-based steam turbine power cycle. The solar-to-electric efficiency of the system achieved is 15% [25].

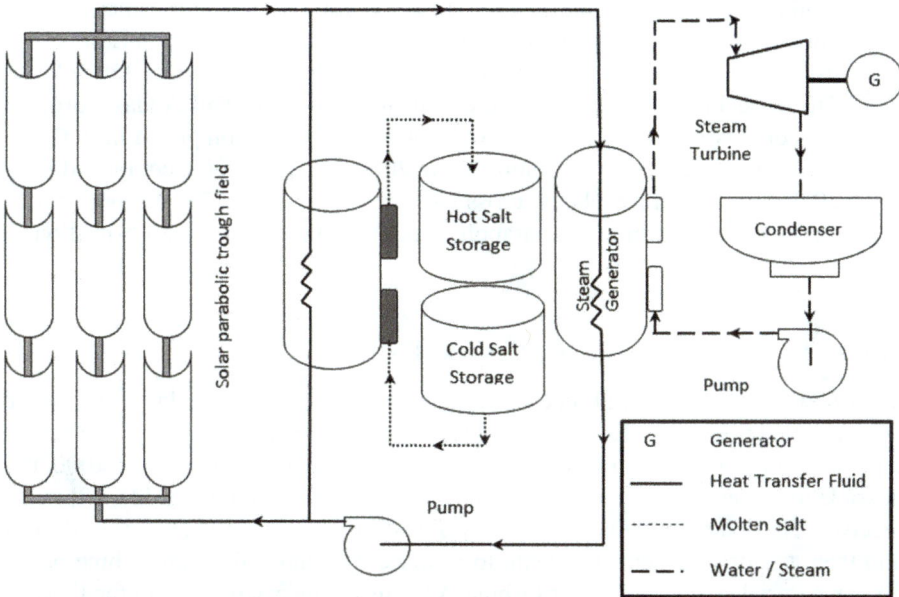

FIGURE 6.3   Schematic diagram of parabolic trough system with molten salt storage.

Though various types of HTFs are available, most plants use synthetic thermal oil as a medium of heat transfer. This hot thermal oil is used to produce superheated steam at high pressure. Depending on the various factors like concentration ratio, solar intensity, flow rate of thermal oil, etc., the temperature of thermal oil can reach up to 400°C [26], which is a limitation of parabolic trough collector system. This thermal oil is used to produce superheated steam which in turn rotates a steam turbine. Generator is coupled with the steam turbine to generate electricity. Direct steam generation parabolic trough power plant can also be developed where steam is generated in the absorber tube. To supply electricity round-the-clock, molten salt storage can be incorporated with the system. Schematic arrangement of parabolic trough power plant with molten salt storage system is shown in Figure 6.3.

Parabolic troughs, being a very mature technology with more than 20 years of operational experience, are considered as low-risk from investment point of view by the entrepreneurs. Examples of some parabolic trough power plants are given as follows:

A. **California Solar Energy Generating System (SEGS):** Nine parabolic trough solar thermal plants were developed in the Mojave Desert of United States by Israeli-American company Luz between 1984 and 1991. Three plants were constructed at separate locations with combined capacity of 354 MW. These plants are collectively called as SEGS. These plants use solar-generated steam with gas back-up of 25% of total heat input. These three plants consist of more than 2 million square metres of parabolic trough mirrors [12].

B. **Andasol Solar Parabolic Trough Power Plants:** The Andasol solar parabolic trough power project, located in Andalusia, Spain, which is Europe's
first commercial plant using parabolic troughs, has a capacity of 150 MW.
The project uses molten salt for thermal energy storage. This Andasol project consists of three plants: (i) Andasol 1, which was completed in 2008;
(ii) Andasol 2, which was completed in 2009; (iii) Andasol 3, completed in
2011. From the three plants, gross electricity output is 525 GWh per year
(approx.) and a combined parabolic trough surface area is of 1.5 million
square metres [12].

6.3.3 SOLAR POWER TOWER OR CENTRAL RECEIVER SYSTEM

Solar power tower or central receiver systems use heliostats – a field of distributed
mirrors – that track the solar energy and focus the sunlight on a receiver placed at the
top of a tower. The concentration ratio being very high in the range of 300–1000, it is
possible to achieve the temperatures from 800°C to well over 1000°C [27, 28] in the
receiver. The solar energy is absorbed by a working fluid or an HTF in the receiver
and then it is used to generate steam to power a conventional steam turbine or hot
air can be produced to run an air turbine. Various media can be used in the thermal
cycle like air, steam, and molten salts. The schematic diagram of power tower with
heliostats and solar receiver is shown in Figure 6.4. A solar central receiver system
consists of the following subsystems:

i. **Collector System or Heliostat Field:** It consists of several thousands of
planar mirrors called heliostats; individual mirror tracks the sun with two-
axis tracking mechanism. The mirrors are generally placed all around the
tower.
ii. **Solar Receiver:** It is mounted at the top of the tower where the concentrated solar energy is absorbed. In the receiver, the working fluid or the
HTF is heated by the help of concentrated solar energy. The fluids generally used in the receiver may be either molten salt of nitrates ($NaNO_3$ and
KNO_3) or water or air. Other fluids, such as sodium or helium, can also be
used. The solar receiver may be of various types like tubular and volumetric
receiver [29].
iii. **Storage System:** Storage system can be placed with the power tower system to provide energy to the working fluid when sufficient sunlight is not
available during night time or cloudy days.
iv. **Power Block:** It consists of turbine – generator system to generate
electricity.

Some of the largest solar power tower plants are given in Table 6.1 [30].

6.3.4 LINEAR FRESNEL REFLECTOR

In LFR [31], flat linear facets solar collector rows or loops form almost a parabolic
shape. These long rows of flat or slightly curved reflectors move independently on a

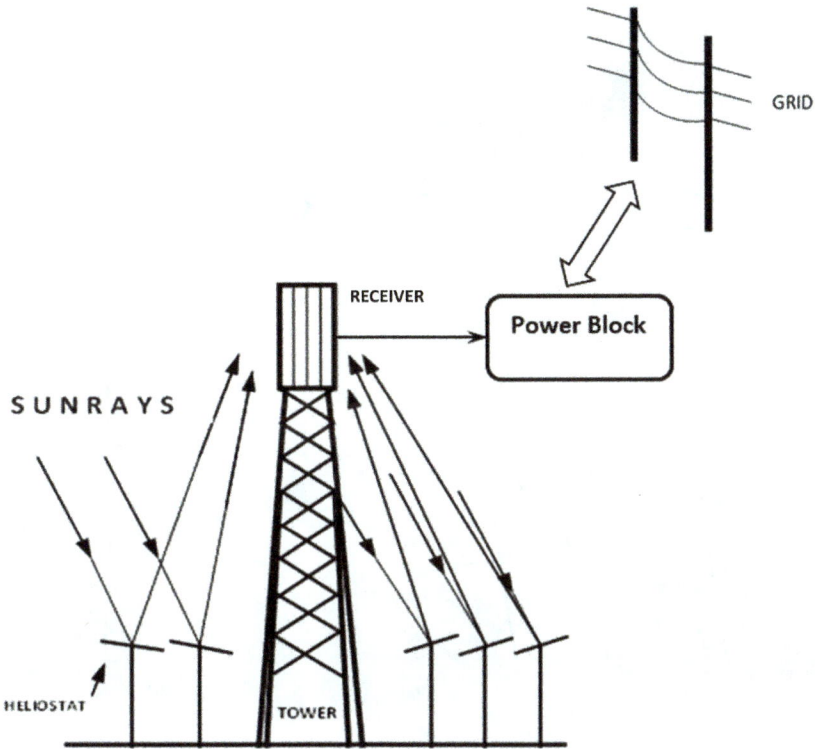

FIGURE 6.4   Schematic diagram of power tower system.

TABLE 6.1
Some Solar Power Tower Plants

Sl. No.	Name	Developer/ Owner	Year of Completion	Country	Height (m)	Installed Maximum Capacity (MW)
1.	Ivanpah Solar Power Facility (three towers)	Bright Source Energy	2014	United States	139.9	392
2.	Ouarzazate Solar Power Station	Moroccan Agency for Sustainable Energy	2009	Morocco	250	150
3.	Ashalim Power Station	Megalim Solar Power	2018	Israel	260	121
4.	Cerro Dominador Solar Thermal Plant	Acciona (51%) and Abengoa (49%)	2021	Chile	243	110
5.	Crescent Dunes Solar Energy Project	Solar Reserve	2016	United States	200	110
6.	Shouhang Dunhuang 100 MW Phase II	Beijing Shouhang IHW	2018	China	220	100

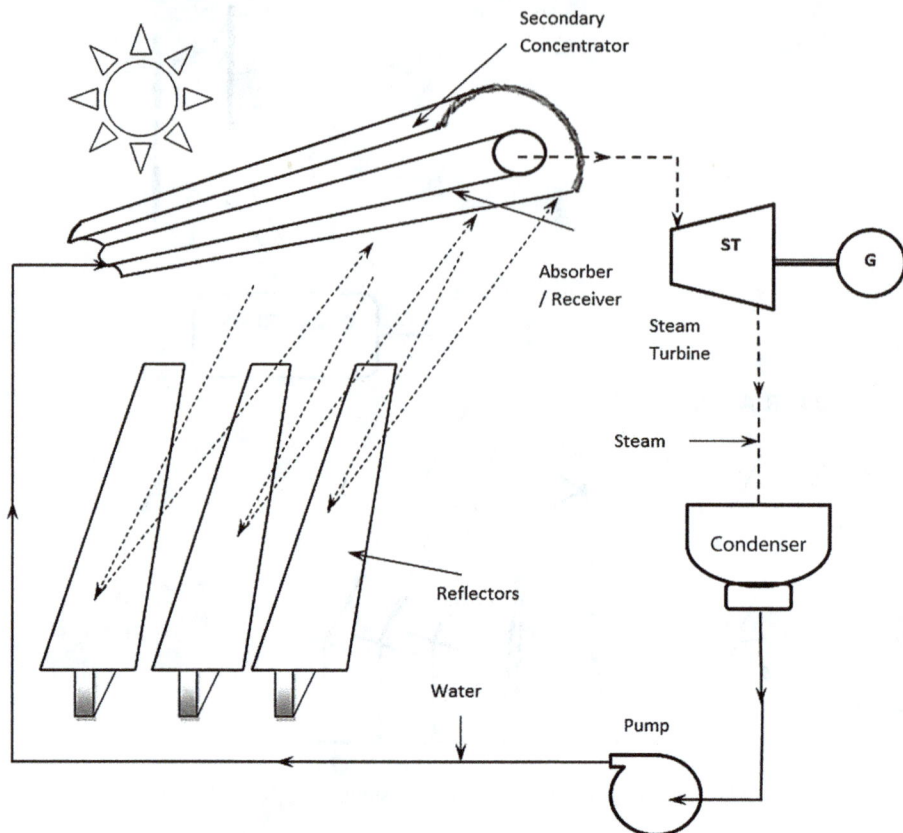

FIGURE 6.5   Schematic diagram of linear Fresnel system.

single axis to reflect the sun's rays onto the stationary receiver/absorber. Sometimes, secondary concentrator is placed above the receiver to increase the temperature of the working fluid. It is possible to generate superheated steam at 500°C.

To produce electricity in the evening or during cloudy weather, linear Fresnel systems may be incorporated with thermal storage. In that case, additional collector fields are required to heat a storage system, which will produce steam while sunlight is not available. LFR plants can also be hybridized with fossil fuel-based power plants. Schematic diagram of linear Fresnel system is shown in Figure 6.5.

Examples of some solar thermal power plants based on LFR technology are given as follows:

A. **Jaisalmer Solar Thermal Project:** This 125 MW STE project is situated at Jaisalmer District, Rajasthan, India. The project was awarded to Reliance Power and was successfully connected to the grid in November, 2014. It is the world's largest STE project based on Compact Linear Fresnel Reflector (CLFR) technology [12].

B. **Puerto Errado 2:** Puerto Errado 2 (PE2), a 30 MW solar thermal power station using linear Fresnel technology, was built by Novatec Solar in Murcia, Spain, and it has been operating since August, 2012. Direct super-heated steam is produced from the plant at a temperature of up to 270°C and a pressure of 55 bar. The generated power can supply electricity to 15,000 Spanish homes [12].

6.4 THERMAL STORAGE

Out of various renewable energy options available for power generation, concentrated solar thermal power generation is becoming a very attractive renewable energy production system, as it has a better potential for dispatchability. This dispatchability demands an efficient and cost-effective thermal storage system. Thermal energy storage for concentrating solar thermal power (CSP) plants is required in avoiding the intermittency of the solar resource and it also reduces the levelized cost of energy (LCOE) by utilizing the turbine-generator system, that is power block while sufficient sunlight is not available. Thus, thermal storage is a key one in CSP system. However, more research and development are required in this field and a few plants only in the world have high-temperature thermal energy storage systems. The following points need to be considered during designing a thermal energy storage system:

   i. Cost of the system
   ii. Operating range of temperatures
  iii. Maximum load to be handled, that is maximum energy to be stored
   iv. The ease with which the system can be integrated into the power plant

Several thermal energy storage technologies are available such as [32, 33]:

   i. Two-tank direct system
   ii. Two-tank indirect system
  iii. Single-tank thermocline system

6.4.1 TWO-TANK DIRECT SYSTEM

In this system, for storing solar thermal energy and for heat transfer, same fluid is used. As the name indicates, two tanks are used here – one is for high-temperature fluid storage and the other is for low temperature. During power generation, steam is produced while fluid from the high-temperature storage tank flows through a heat exchanger. After transferring heat to water, the fluid exits the heat exchanger at a lower temperature and accumulates in the low-temperature storage tank. During the charging period, low-temperature fluid from the storage tank flows to the solar receiver/absorber and gains solar energy to achieve high temperature. Then the hot fluid flows to the high-temperature storage tank for storage. The schematic diagram of the system is shown in Figure 6.6.

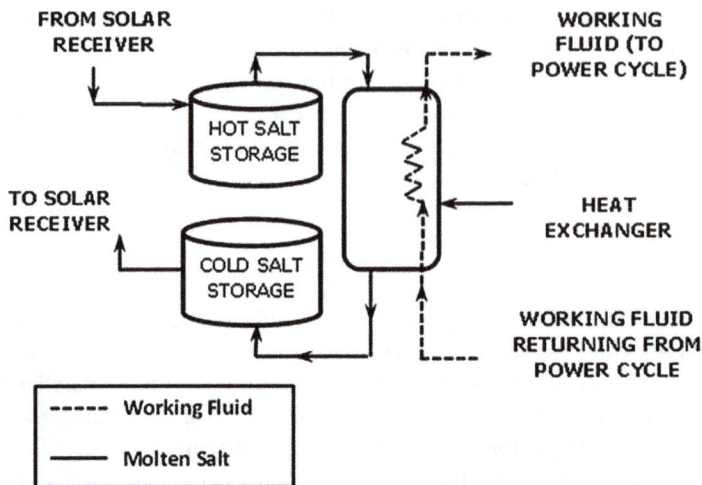

FIGURE 6.6   Schematic diagram of two tank direct storage system.

6.4.2   TWO-TANK INDIRECT SYSTEM

While same fluid is used for heat transfer and storage media in two-tank direct system, different fluids are used as the heat transfer and storage media in case of two-tank indirect systems. This system is used in CSP plants while either the HTF is too expensive or it cannot be suited for the use as the storage media. In this system, another heat exchanger is incorporated and storage media from the low-temperature tank is heated by the high-temperature HTF coming from solar receiver/absorber. After discharging heat to storage media, the HTF from the heat exchanger exits at a lower temperature and returns back to the solar receiver/absorber to gain energy. After getting heated, the high-temperature storage fluid is stored in the high-temperature storage tank. Steam is generated while hot storage fluid flows from the hot-storage tank through a heat exchanger, as in the case of two-tank direct systems. The schematic diagram of a two-tank indirect storage system is shown in Figure 6.7.

6.4.3   SINGLE TANK THERMOCLINE STORAGE

It is another method of thermal storage system where molten nitrate salts in a single-tank thermocline storage system are used as direct HTF. The thermocline storage system generally uses a single tank with hot and cold fluid, and it is based on the principle of thermal buoyancy to maintain thermal stratification due to which the hot fluid remains on the top, and the cold fluid remains at the bottom. A low-cost filler material, which can be used to pack the single-storage tank, may also act as the primary thermal storage medium, which is heated by the hot fluid during the charging process. The most commonly used solid filler materials like silica sand, pebbles, rocks, etc. are located in a single tank. At any time, the top

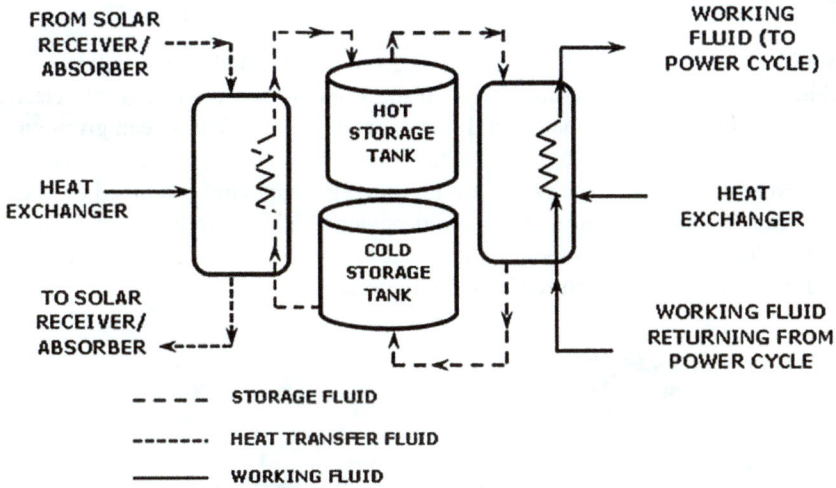

FIGURE 6.7   Schematic diagram of two-tank indirect storage system.

portion of the tank is generally at high temperature, and the bottom portion is at low temperature. The hot- and cold-temperature zones are separated by a temperature gradient or thermocline. The HTF [34] at high temperature flows into the top of the thermocline and exits at the bottom. This is called charging. During discharging or during power generation, the flow direction of HTF is reverse. The schematic of single-tank thermocline storage is shown in Figure 6.8. While the solid arrows indicate the direction of hot HTF during charging, the dotted arrows indicate the direction of cold HTF flow during discharging [32].

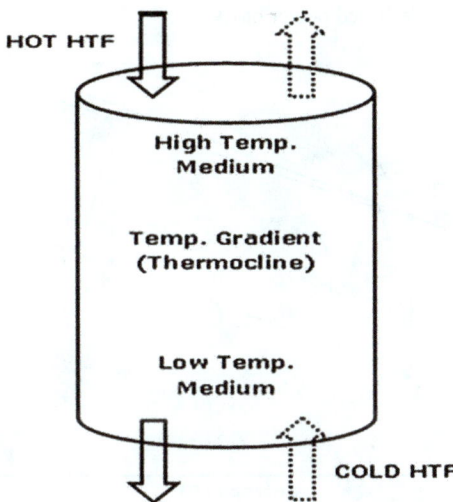

FIGURE 6.8   Schematic diagram of thermocline storage.

6.5   MODEL OF CONCENTRATED SOLAR POWER PLANT

Power block is the "heart" of the CSP plant. Here, the thermal energy delivered to the working fluid either from storage or from the solar field is converted into electrical energy via turbine and alternator. In this section, a closer look has been given on how the thermal energy is converted into electrical energy. For CSP plants, main thermodynamic cycles used to produce power are Rankine, Brayton, and combined cycles [35].

Figure 6.9 shows Brayton cycle-based power block which has also been taken for study for thermal analysis in the present section. Figure 6.10 shows the corresponding temperature-entropy (T vs. S) diagram.

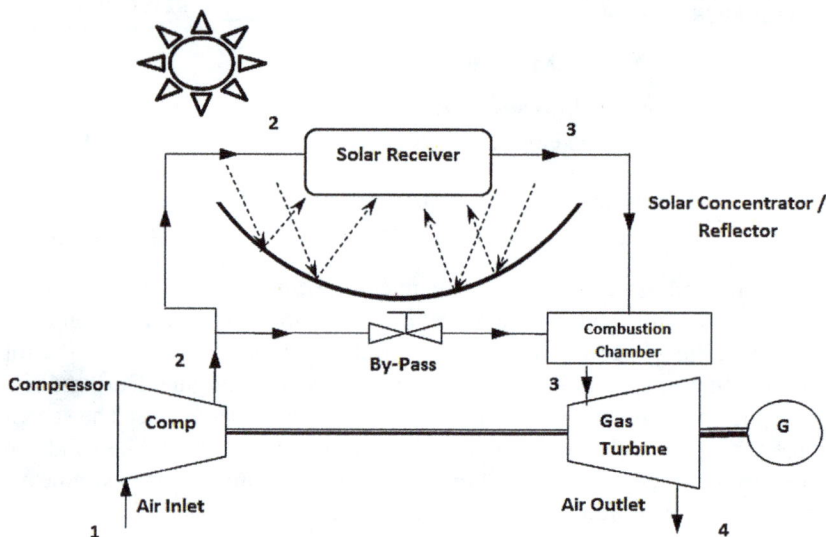

FIGURE 6.9   Brayton cycle-based power block.

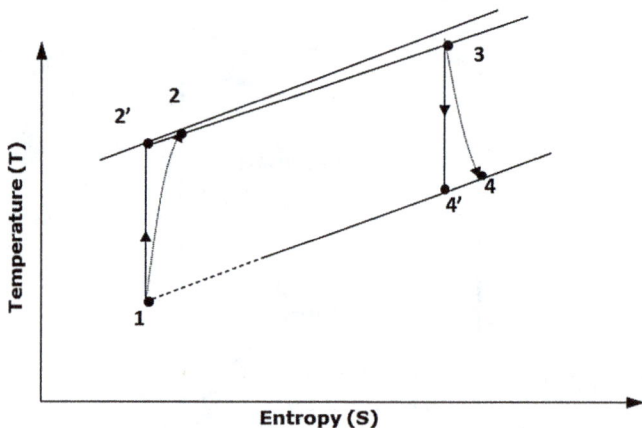

FIGURE 6.10   Temperature-entropy diagram of the Brayton cycle-based power block.

TABLE 6.2
Main Assumptions of Brayton Cycle-Based Power Block

Characteristics	Value
Pressure ratio of compressor	4–22 (varied at the interval of 2)
Mass flow rate of air (m_a)	5 kg/s
Gas turbine inlet temperature (TIT) of air	1000°C
Compressor inlet temperature	25°C
Compressor inlet pressure	1.0132 bar
Isentropic efficiency of compression (η_C)	0.88
Isentropic efficiency of expansion	0.90
Mechanical efficiency (η_{mech})	0.95
Generator efficiency (η_G)	0.97
Field optic efficiency $(\eta_{Concentrator})$	0.90
Receiver efficiency $(\eta_{Receiver})$	0.82
Pressure drop in the solar receiver	0.12 bar

Various processes involved with the Brayton cycle-based power block are as follows:

i. **Process 1–2:** Compression of air in compressor
ii. **Process 2–3:** Heat input to the cycle (Q_{in})
iii. **Process 3–4:** Gas turbine work
iv. **Process 4–1:** Heat rejection (Q_{out})

Assumptions considered in the energy analysis are given in Table 6.2.

6.6 MATHEMATICAL FORMULATION

The temperature of air after compression is given by Equation (6.1):

$$T_2 = T_1 + \frac{1}{\eta_c}(T_{2'} - T_1) \tag{6.1}$$

Rate of solar energy required on concentrator for heating air to the required turbine inlet temperature (TIT) can be obtained from Equation (6.2).

$$Q_{in} = \frac{m_a \int_{T_2}^{T_3} C_{pa}\, dT}{\eta_{Concentrator} * \eta_{Receiver}} \tag{6.2}$$

where C_{pa} is the specific heat of air.

Net power obtained from the gas turbine cycle is obtained from the following relationship, that is Equation (6.3):

$$(W_{net})_{GT} = (W_{GT} - W_{comp})\eta_{mech}\eta_G \tag{6.3}$$

Thermal efficiency of the Brayton cycle is obtained from Equation (6.4).

$$\eta_{th} = \frac{(W_{net})_{GT}}{Q_{in}} \tag{6.4}$$

6.7 RESULTS AND DISCUSSION

In the present analysis of the model, various parameters have been obtained against varying pressure ratios of the compressor, and these are shown in Figures 6.11–6.13. Values corresponding to graphs are presented in Tables 6.3–6.5, respectively.

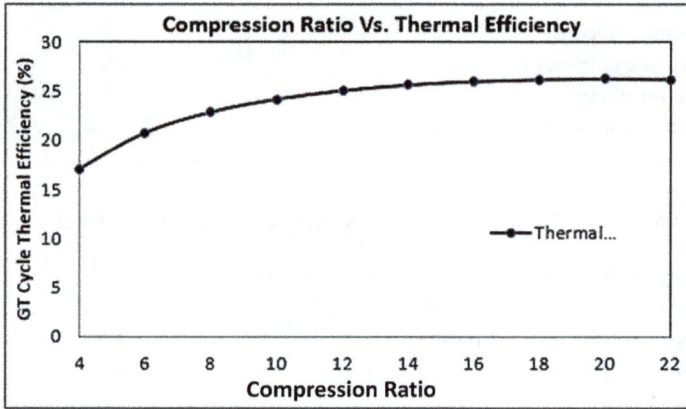

FIGURE 6.11   Compression ratio versus gas turbine cycle thermal efficiency.

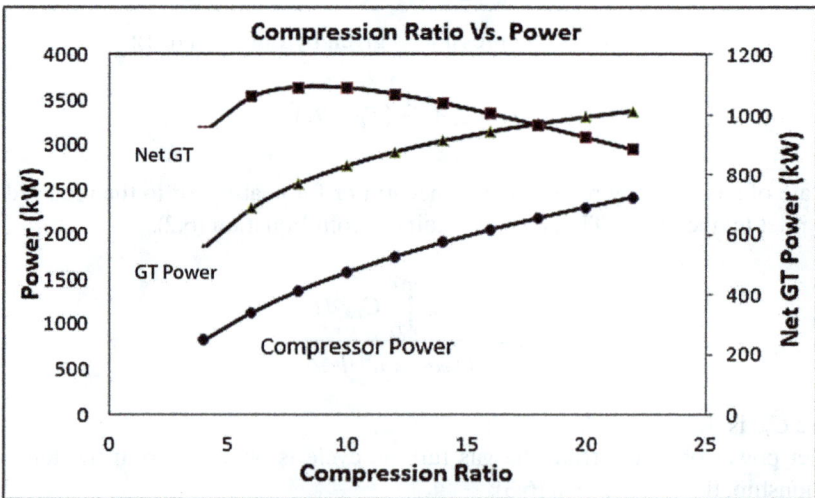

FIGURE 6.12   Compression ratio versus compressor power, gas turbine power, and net gas turbine power.

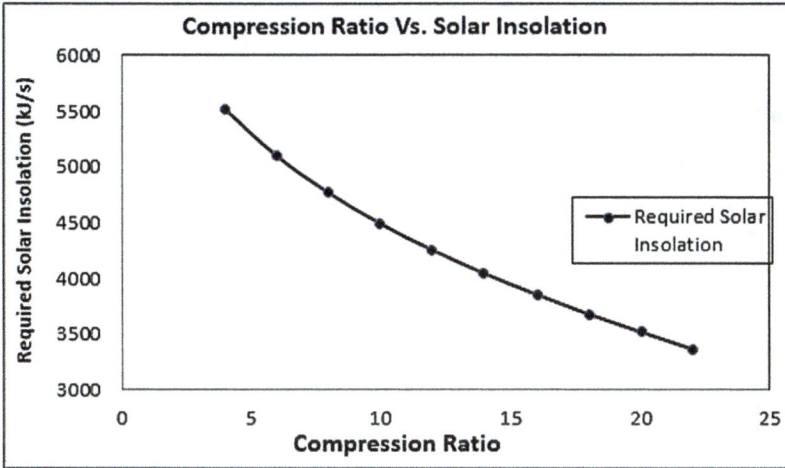

FIGURE 6.13 Compression ratio versus solar insolation.

Major findings from the analysis are as follows:

1. With the increase of pressure ratio of the compressor, thermal efficiency of the gas turbine initially increases, reaches a maximum value, and then becomes almost flat.
2. With the increase of pressure ratio of the compressor, both the compressor power and the gas turbine power increase but net power from Brayton cycle initially increases, reaches a maximum value at pressure ratio 8, and then decreases.
3. With the increase of pressure ratio of the compressor, required solar insolation to heat the air at solar receiver decreases since temperature of air after compression increases and TIT has been fixed at 1000°C.

TABLE 6.3

Compression Ratio versus Gas Turbine Cycle Thermal Efficiency

Compression Ratio	Thermal Efficiency (%)
4	17.07
6	20.78
8	22.9
10	24.23
12	25.11
14	25.68
16	26.03
18	26.23
20	26.29
22	26.24

TABLE 6.4

Compression Ratio versus Net Gas Turbine (GT) Power, GT Power, and Compressor Power

Compression Ratio	Net GT Power	GT Power	Compressor Power
4	942.2	1849	827
6	1059	2287	1138
8	1092	2565	1381
10	1089	2765	1584
12	1068	2919	1759
14	1038	3042	1915
16	1003	3144	2056
18	964.8	3231	2184
20	924.8	3307	2303
22	883.9	3373	2414

TABLE 6.5

Compression Ratio versus Required Solar Insolation

Compression Ratio	Required Solar Insolation (kJ/s)
4	5518
6	5097
8	4768
10	4493
12	4255
14	4044
16	3853
18	3679
20	3518
22	3368

6.8 CONCLUSION AND FUTURE SCOPE

It is a fact that dependence on fossil fuels for the generation of energy or power should be reduced since energy or electricity production from conventional sources creates several impacts such as emission of major GHG like CO_2, depletion of fossil fuel sources, and a threat to overall environmental sustainability.

But renewable- energy sources are clean, abundant, and sustainable in nature, which will definitely be at the forefront in terms of delivering an inexhaustible source of energy. CSP is capable of utility scale electricity generation and various nations are investing largely in CSP technology. Among four different technologies available in the CSP system, the solar power tower and parabolic trough collector are the two popular technologies while parabolic trough collector is already a mature technology. Thermal storage can be incorporated with the CSP technology to obtain electricity round the clock. Various conventional cycles can be used to obtain power such as Brayton, Rankine, and Brayton–Rankine combined cycle.

R&D is going on the various aspects of Concentrated Solar Thermal Power – to improve its efficiency, to improve the storage system, to invent better HTF, and a lot more. The Brayton cycle shown earlier can be integrated with other cycles like conventional Rankine cycle, organic Rankine cycle, etc. to utilize the solar power in a more efficient way.

REFERENCES

1. BP Statistical Review of World Energy 2021, https://www.bp.com/content/dam/bp/business-sites/en/global/corporate/pdfs/energy-economics/statistical-review/bp-stats-review-2021-full-report.pdf [accessed August 2, 2021].
2. S. Shafiee, E. Topal, 'When will fossil fuel reserves be diminished?' Energy Policy, pp. 181–189, Volume 37, Issue 1, 2009.
3. M. T. Islam, S. A. Shahir, T. M. I. Uddin, A. Z. A. Saifullah, 'Current energy scenario and future prospect of renewable energy in Bangladesh,' Renewable and Sustainable Energy Reviews, pp. 1074–88, Volume 39, 2014.
4. S. Mekhilef, R. Saidur, A. Safari, 'A review on solar energy use in industries,' Renewable and Sustainable Energy Reviews, pp. 1777–90, Volume 15, 2011.
5. T. M. Pavlović, I. S. Radonjić, D. D. Milosavljević, L. S. Pantić, 'A review of concentrating solar power plants in the world and their potential use in Serbia,' Renewable and Sustainable Energy Reviews, pp. 3891–902, Volume 16, 2012.
6. International Energy Agency, Energy Technology Perspectives 2015: Mobilising Innovation to Accelerate Climate Action, Paris, https://iea.blob.core.windows.net/assets/3f901e93-c083-4649-a9e6-c591e28a7b70/ETP2015.pdf, 2015 [accessed September 2, 2021].
7. Z. Zhang, Y. Yuan, N. Zhang, Q. Sun, X. Cao, L. Sun, 'Thermal properties enforcement of carbonate ternary via lithium fluoride: a heat transfer fluid for concentrating solar power systems,' Renewable Energy, pp. 523–31, Volume 111, 2017.
8. S. Ahmed, M. T. Islam, M. A. Karim, N. M. Karim, 'Exploitation of renewable energy for sustainable development and overcoming power crisis in Bangladesh,' Renewable Energy, pp. 223–35, Volume 72, 2014.
9. J. Sun, Q. Liu, H. Hong, 'Numerical study of parabolic-trough direct steam generation loop in recirculation mode: characteristics, performance and general operation strategy,' Energy Conversion and Management, pp. 287–302, Volume 96, 2015.
10. S. Izquierdo, C. Montanes, C. Dopazo, N. Fueyo, 'Analysis of CSP plants for the definition of energy policies: the influence on electricity cost of solar multiples, capacity factors and energy storage,' Energy Policy, pp. 6215–21, Volume 38, 2010.
11. A. Ummadisingu, M. Soni, 'Concentrating solar power–technology, potential and policy in India,' Renewable and Sustainable Energy Review, pp. 5169–75, Volume 15, 2011.
12. Solar Thermal Electricity Global Outlook 2016, European Solar Thermal Electricity Association (ESTELA), Greenpeace International and SolarPACES, https://www.estelasolar.org/wp-content/uploads/2016/02/GP-ESTELA-SolarPACES_Solar-Thermal-Electricity-Global-Outlook-2016_Full-report.pdf, 2016 [accessed September 2, 2021].

13. R. E. Sims, H. H. Rogner, K. Gregory, 'Carbon emission and mitigation cost comparisons between fossil fuel, nuclear and renewable energy resources for electricity generation,' Energy Policy, pp. 1315–26, Volume 31, 2003.

14. C. Richter, S. Teske, R. Short, Concentrating Solar Power Global Outlook 09, Greenpeace International/European Solar Thermal Electricity Association (ESTELA)/ IEA SolarPACES, http://www.solarpaces.org/wp-content/uploads/concentrating-solar-power-2009.pdf, 2009 [accessed September 2, 2021].

15. EIA. U.S. Energy Information Administration, International Energy Outlook, https:// www.eia.gov/outlooks/ieo/pdf/0484(2017).pdf, 2017 [accessed September 2, 2021].

16. H. Müller-Steinhagen, F. Trieb, F. Trieb. Concentrating Solar Power: A Review of the Technology, https://www.dlr.de/tt/Portaldata/41/Resources/dokumente/institut/system/ publications/Concentrating_Solar_Power_Part_1.pdf [accessed September 2, 2021].

17. SolarPACES, CSP Projects Around the World, http://www.solarpaces.org/csp-technology/ csp-projects-around-the-world, 2016 [accessed September 2, 2021].

18. K. Kaygusuz, 'Prospect of concentrating solar power in Turkey: the sustainable future,' Renewable and Sustainable Energy Review, pp. 808–14, Volume 15, 2011.

19. F. G. Braun, E. Hooper, R. Wand, P. Zloczysti, 'Holding a candle to innovation in concentrating solar power technologies: a study drawing on patent data,' Energy Policy, pp. 2441–56, Volume 39, 2011.

20. R. B. Affandi, C. K. Gan, A. Ghani, M. Ruddin, 'Performance comparison for parabolic dish concentrating solar power in high level DNI locations with George Town, Malaysia', Applied Mechanics and Materials, pp. 570–76, 2015.

21. F. Cavallaro, 'Multi-criteria decision aid to assess concentrated solar thermal technologies,' Renewable Energy, pp. 1678–85, Volume 34, 2009.

22. A. Poullikkas, G. Kourtis, I. Hadjipaschalis, 'Parametric analysis for the installation of solar dish technologies in Mediterranean regions,' Renewable and Sustainable Energy Review, pp. 2772–83, Volume 14, 2010.

23. J. H. Peterseim, S. White, A. Tadros, U. Hellwig, 'Concentrated solar power hybrid plants, which technologies are best suited for hybridisation?,' Renewable Energy, pp. 520–32, Volume 57, 2013.

24. W. Fuqiang, C. Ziming, T. Jianyu, Y. Yuan, S. Yong, L. Linhua, 'Progress in concentrated solar power technology with parabolic trough collector system: a comprehensive review,' Renewable and Sustainable Energy Review, pp. 1314–28, Volume 79, 2017.

25. A. F. García, E. Zarza, L. Valenzuela, M. Perez, 'Parabolic-trough solar collectors and their applications,' Renewable and Sustainable Energy Review, pp. 1695–721, Volume 14, 2010.

26. A. Mohamad, J. Orfi, H. Alansary, 'Heat losses from parabolic trough solar collectors,' International Journal of Energy Research, pp. 20–8, Volume 38, 2014.

27. S. A. Kalogirou, 'Solar thermal collectors and applications,' Progress in Energy and Combustion Science, pp. 231–95, Volume 30, 2004.

28. O. Behar, A. Khellaf, K. Mohammedi, 'A review of studies on central receiver solar thermal power plants,' Renewable and Sustainable Energy Review, pp. 12–39, Volume 23, 2013.

29. C. K. Ho, B. D. Iverson, 'Review of high-temperature central receiver designs for concentrating solar power,' Renewable and Sustainable Energy Review, pp. 835–46, Volume 29, 2014.

30. Solar Power Tower, https://en.wikipedia.org/wiki/Solar_power_tower [accessed October 12, 2021].

31. V. Kumar, R. L. Shrivastava, S. P. Untawale, 'Fresnel lens: a promising alternative of reflectors in concentrated solar power,' Renewable and Sustainable Energy Review, pp. 376–90, Volume 44, 2015.

32. Thermal Storage System Concentrating Solar-Thermal Power Basics, https://www.energy.gov/eere/solar/thermal-storage-system-concentrating-solar-thermal-power-basics [accessed October 12, 2021].

33. U. Pelay, L. Luo, Y. Fan, D. Stitou, M. Rood, 'Thermal energy storage systems for concentrated solar power plants,' Renewable and Sustainable Energy Review, pp. 82–100, Volume 79, 2017.

34. K. Vignarooban, X. Xu, A. Arvay, K. Hsu, A. M. Kannan, 'Heat transfer fluids for concentrating solar power systems – a review,' Applied Energy, pp. 383–96, Volume 146, 2015.

35. M. T. Dunham, B. D. Iverson, 'High-efficiency thermodynamic power cycles for concentrated solar power systems,' Renewable and Sustainable Energy Review, pp. 758–70, Volume 30, 2014.

7 Efficient and Effective Techniques for Intensification of Renewable Energy (Wind) Using Deep Learning Models

Kavita Arora[1], Sailesh Iyer[2], and Mariya Ouaissa[3]
[1]Department of Computer Applications, Manav Rachna International Institute of Research & Studies, Faridabad, Haryana, India
[2]Department of Computer Science and Engineering, Rai School of Engineering, Rai University, Gujarat, India
[3]Computer Science and Networks, Moulay Ismail University Meknes, Morocco

CONTENTS

DOI: 10.1201/9781003369554-7

7.1 INTRODUCTION

The continuous industrial development of the last 150 years has been linked to a progressive increase in energy consumption. In return, the main sources of energy that have been used to make this development possible have been fossil fuels and their derivatives due to which large amounts of greenhouse gases have been emitted and thus are held responsible for the global warming the world is experiencing.

On the other hand, the continuous increase in energy demand in developed and developing countries is leading to a strong energy dependence for many countries, mainly in Asia. The strong dependence on energy, social pressure, and awareness on the part of governments in the fight against climate change has led to the adoption of regulatory frameworks that favor the development and use of renewable energy resources in a cleaner and sustainable manner. These renewable resources are obtained from natural resources and waste and include, among others, hydraulic, wind, solar, biomass, geothermal energy, and the use of solid urban and industrial waste. Furthermore, in addition to the environmental benefits, the push for renewable energy brings other benefits. On the one hand, due to the great dispersion in the location of the generating facilities that use renewable resources, their promotion has led to the creation of local jobs, regional development, and substantial economic performance for the municipalities in which the generating plants are installed. This type of dispersed generation, which in most cases is of relatively low power, carries other advantages, such as the reduction of transmission losses in the network, thanks to the fact that generation is closer to consumption [1].

The strong boost that has been given to renewable energy sources has led to a large increase in installed power in recent years and among them, wind energy has been the one that has had the greatest boom, mainly due to the great maturity of the technology, resulting in a lower and lower cost per megawatt installed. The great advantage of wind energy, which has driven its development to a greater extent compared to other renewable energy sources, is its relatively high geographical availability due to the fact that to a greater or lesser extent, there are wind currents in almost any region on the planet. In return, wind power generation has certain drawbacks. On one hand, the great variability of the wind complicates its integration into the electrical system. In addition, some generators can disconnect from the system due to disturbances such as voltage sags, sudden drops in voltage that take place when a short circuit occurs in the system. These disturbances can cause a sudden loss of wind generation in the system, which can jeopardize the security of supply. However, the recent incorporation of technological improvements in the behavior of

State	Total Capacity
Tamil Nadu	9231.77
Gujarat	7203.77
Maharashtra	4794.13
Karnataka	4753.4
Rajasthan	4299.73
Andhra Pradesh	4077.37
Madhya Pradesh	2519.89
Telangana	128.1
Kerala	62.5
Others	4.3
Total	37074.96

Source: https://en.wikipedia.org/wiki/Wind_power_in_India

wind turbines in the face of disturbances in the grid begins to allow a high degree of penetration of wind energy without compromising excessively with the security of the electricity supply. Thanks to improvements in the behavior of wind turbines, to the continuous decrease in the cost per installed megawatt of wind power, and to advances in the methods of supporting the programming and management of this energy, wind power has become the source of renewable energy that has been developed the most in recent years.

Renewable energies are becoming more relevant worldwide every day due to the great pollution generated by conventional energies and the search for new sustainable models. One of the problems associated with the exclusive use of renewable energies is the complication of foreseeing whether their production will be able to cover the required energy expenditure, since otherwise, conventional energies would have to be resorted to in any way. Therefore, the prediction of the production (in this case wind) is of vital importance to contribute to the model being sustainable and thus to be able to try to get to do without conventional energies. So far, there are many production prediction models, but is it possible that predictions that are in principle more inaccurate a priori can give us more information and be more optimal when it comes to knowing how much wind energy will be produced in a day or at a certain time? From this question the present study is born, which will try to approach from basic models that, although they could be sophisticated to try to minimize the prediction error as much as possible and give predictions as accurate as possible, it is not the objective of the same, since it is only trying to propose an approach to the reasonable doubt previously exposed for a possible later deepening in the subject once the results of this study have been obtained.

The objectives of this chapter are as follows:

1. To carry out a study of the time distance of a weather forecast while predicting energy outputs from wind power of certain regions;
2. To propose the model built with the short-term data and to make it better and more accurate predictions since lower errors are associated with the prediction;
3. And, to evaluate the performance of two prediction models, that is, feed-forward and long short-term memory (LSTM).

7.1.1 ORGANIZATION OF THE CHAPTER

This chapter is organized as follows: Section 7.2 enlightens the need for wind energy prediction. Section 7.3 elaborates existing technique of numerical weather forecast. Sections 7.4–7.7 highlight model definition, output statistics, atmosphere as a chaotic system, and model prediction. Section 7.8 gives general background and concepts definition, and prediction models are elaborated in Section 7.9. Section 7.10 concludes the chapter with future scope.

7.2 NEED FOR WIND ENERGY PREDICTION

A statistical prediction problem consists of analyzing the past values of a variable and other related variables to look for significant patterns, intending to be able to know or extrapolate the values that said variable will take in the future. In recent decades, a great advance has taken place in prediction systems due to the enormous increase in computing power of current computers, which allow to store, analyze, and relate large amounts of variables and their values in a very short time. That is how prediction algorithms are used as tremendously developed tools today [2]. The importance of predictions lies in the help they provide to plan and anticipate future values that will affect a system, helping to manage the acquisition of the necessary resources well in advance, or serving as a tool to maximize profitability by taking decisions that maximize the benefits of an activity. The great boom that renewable energy sources have had in recent years has forced us to consider a new forecasting problem of knowing in advance the energy that these plants are going to generate.

Given that the greatest boom in the field of renewable energy has been wind energy, it is in this sector that greater efforts have been made to create reliable and efficient energy prediction tools that help to integrate this form of energy into the grid. Wind power is a non-programmable form of generation since energy is only produced when the wind blows, which can become highly variable even in the short term, with the possibility of intermittence and large changes in short intervals of time [3]. For this reason, it is difficult to know in advance and with sufficient precision the amount of wind energy that we can count on at all times. This variability makes its operation especially complex, so its future production has to be estimated or predicted, this forecast of future power being inevitably affected by a prediction error or uncertainty. If the wind decreases, the power generated in the wind farms also decreases and that lack of power must be replaced by other sources of generation with a sufficient reserve in magnitude and response speed so that the electricity demand is not affected. On other occasions, it may happen that all available wind energy production cannot be integrated into the system, since wind energy is not generated according to consumer needs and it is necessary to reduce the supply of this energy source. For all this, the prediction of wind generation has become a key issue to make the development and implementation of wind energy feasible with its integration into the electrical system. From the point of view of wind generation or any other source of renewable energy, its forecast is useful both for the system operator and for market agents or park owners. Thus, the electrical system operator needs to know in advance the amount of wind energy that will be injected into the network to manage the power that conventional power plants must generate, to cover the total demand of the system [4]. Meanwhile, market agents will be interested in knowing with the greatest possible certainty the power that their wind farms will generate to follow the most profitable strategies in the electricity market. The value of the wind generation forecast in economic terms has two perspectives. On one hand, we have the reduction in operating costs in the system caused by the reduction in the necessary reserve. On the other, there are possible economic penalties that are applied to the agents, due to deviations in their generation commitments acquired in the electricity market.

In recent years, various methods have been applied in the field of wind forecasting. This work will focus on short-term wind predictions and usually, for this horizon, two basic approaches have been used: physical models and statistical models. Physical models take physical considerations into account to adapt predictions of wind in an area to the specific conditions of the site where measurements are made. To make this adaptation, mesoscale or microscale models are used, starting from the initial and boundary conditions obtained from a larger-scale atmospheric model, calculating the incident wind speed in the wind turbines of the park to later calculate the power prediction using the power curve [5].

Furthermore, using the statistical models, we can find the family of time series which use past values of the variables as input data of the model, and in addition to past values, use the meteorological prediction values of models as inputs, atmospheric values, relating them to historical power values or other measured historical values. This is the approach highlighted by the ARIMA or Box Jenkins models, which is useful for prediction of certain industrial processes, and in the context of wind, energy prediction provides reasonably good results for horizons up to 6 hours. Different types of regressive models have also been used, such as AR and ARMA models and models based on Support Vector Regression, which have given very good results in recent years. The objective behind this is to present two other prediction models, based on a series of architectures belonging to machine learning and deep learning. Specifically, different artificial neural network architectures are used as prediction models, such as feedforward neural networks (FFNN) and LSTM networks.

7.3 EXISTING TECHNIQUE OF NUMERICAL WEATHER FORECAST (WIND ENERGY)

Weather prediction uses mathematical models of fluid mechanics (air in the case of the atmosphere, water in the case of the oceans) in order to predict their future state. This mathematical dependency is quite problematic for a number of reasons.

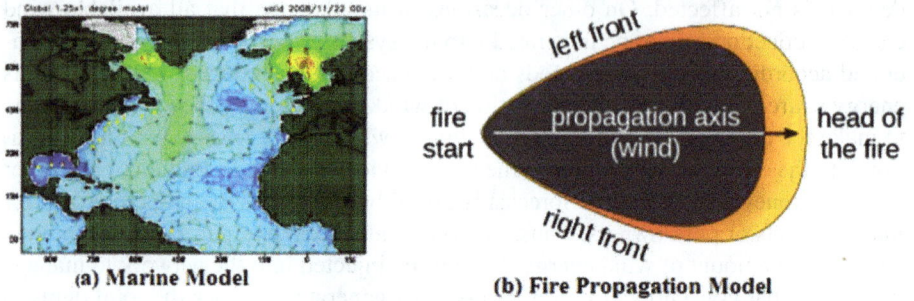

(a) Marine Model   (b) Fire Propagation Model

FIGURE 7.1   Modeling examples of ocean surface state prediction and modeling prediction of fire spread using NWP. (a) Marine Model, (b) Fire Propagation Model. (Images taken from Wikipedia.)

Among other problems, the equations are formulated analytically, which is quite difficult for a computer to deal with in terms of computing time.

This is where the numerical weather prediction (NWP) model or numerical weather prediction comes in as showin in Figure 7.1 (a) and Figure 7.1 (b). It manages to eliminate the treatment of the equations in analytical form of fluid mechanics to treat it in the form of numerical integration, taking as reference previous states of the fluid to predict later states. This is why an initialization of said numerical integration is necessary, for which the current state of the fluid in question is usually used.

The first approach to the model dates around 1920s by the hand of Lewis Fry Richardson but it was not until the 1950s that the specifications of computers (The Electronic Numerical Integrator and Computer [ENIAC] more specifically) were good enough to simulate such a model and produce useful results [11]. Computers need to be powerful enough to be able to get really interesting results since, for example, in the 1920s it took at least six weeks of work to produce a six-hour forecast from the initial state which was uninteresting in terms of productive use [7].

7.4   MODEL DEFINITION

The model is based on dividing the fluid in question into a mesh that is easily treatable for an automated system, an example of coordinates being longitude, latitude, and height or pressure. In each of these cells is where we will find the different predictions of the model for the region represented by each cell.

To establish this three-dimensional mesh, it is necessary to choose the coordinate system to use. The most common system for horizontal coordinates is usually latitude and longitude coordinates. To establish the vertical coordinate, there are different options such as pressure or height according to the particular interests of the model. In our particular case, Figure 7.2 shows the schematic 3 Dimensional division of the atmosphere in the NWP model the horizontal coordinates will be latitude and longitude and the vertical differences will be established as different components under the same cell.

The atmosphere is a fluid and, as we have seen, there are models to given some initial conditions and predict future states of the fluid in question. The problem is that when establishing the initial conditions, they carry an error associated with

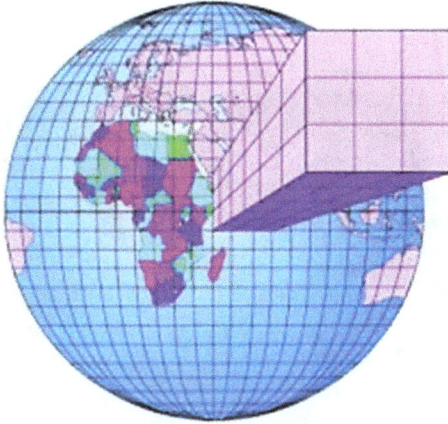

FIGURE 7.2 Schematic image of the three-dimensional division of the atmosphere in the NWP model. (Image taken from Wikipedia.)

them (either from a direct measurement of the component introduced or from a previous not completely coincident prediction). While treating these values numerically, this error is seen increased as we use these initial values to operate.

7.5 MODEL OUTPUT STATISTICS

Previously, we have commented that the prediction of the state of any fluid has an associated error, which increases as we distance ourselves from the initial situation presented to prediction. This is why it is necessary to find error handling methods for the numerical prediction methods.

Model Output Statistics (MOS) is a type of statistical post-processing of data based on multiple linear regressions in order to establish statistical relationships between the variables that make up the system to be studied in order to try to correct possible predictive errors of any of the components and thus try to fit the prediction more closely to the future real data and, as a consequence, improve the following predictions.

7.6 THE ATMOSPHERE AS A CHAOTIC SYSTEM

One of the prediction complications of meteorology is that it is what is called a chaotic system. A chaotic system or complex system is one that is extremely sensitive to small variations in the initial conditions, which make them quite difficult to predict in the long term. One of the best-known chaotic systems is that of the "Lorenz attractor" presented in the article "Non-periodic deterministic flow" [8]. This system interests us to a great extent since it is a system based on simplified equations derived from the dynamic equations of the Earth's atmosphere.

With the system represented in Figure 7.3, Edward Norton Lorenz tried to explain the chaotic behavior of some unstable systems, including weather forecasting. Thus, he introduced the widely known concept of the "Butterfly Effect", according to which he asked whether "the simple flapping of a butterfly in Brazil could make

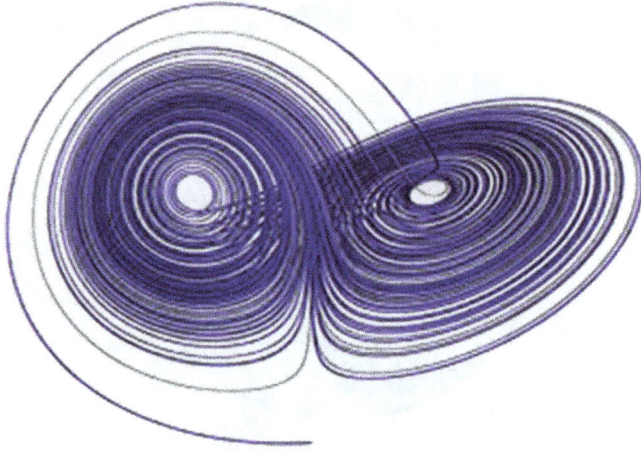

FIGURE 7.3   Lorenz attractor represented three-dimensionally.

a tornado appear in Texas", title of a conference of the American Association for the Advancement of Science of which he was the author [6].

Observing that the atmosphere attends to such a delicate and unstable system and given that the prediction of its state is quite useful, techniques are necessary to try to make the prediction more accurate. One of these techniques is ensemble forecasting [9]. This technique consists of making several predictions with the same model but varying the initial conditions minimally between one execution and another. This establishes a set of possible future states under which the atmosphere can end (in the case that this is the study fluid). In this way, a greater range of possibilities can be given and even try to reduce the errors associated with statistical techniques such as MOS seen in Section 7.5.

In Figure 7.4 we can see schematically how a series of predictions are made with slightly varied initial conditions and it is capable of obtaining several future states

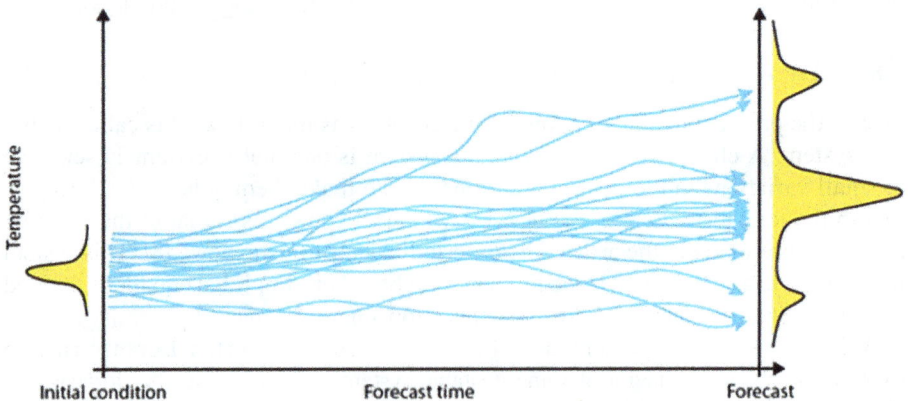

FIGURE 7.4   Schematic image of the ensemble forecasting technique.

with different probabilities depending on the number of predictions made and how many have ended up in similar states.

7.7 MODEL PREDICTION

Although there are a large number of models to predict meteorology, each one has its mathematical characteristics and are more useful for solving one type of problem than other. Figure 7.5 shows the comparison of the precision of the prediction according to the temporal distance of some of the most important models.

We use NWP modeling due to the specifications of the problem to be dealt with, since we will always be dealing with predictions of the same day in question or at most three days ahead and, at first glance, it is seen that it is the most optimal for our problem in particular [10].

In the concept of Figure 7.5, we can find the motivation for this work, since we have to observe at what temporal distance is the most optimal prediction data for solving prediction problems. Likewise, a series of experiments are carried out with these models to study their prediction capacity and analyze the learning process by varying the different parameters that these networks have. An exhaustive evaluation of errors of the models is also carried out following a series of protocols and guidelines already established in the field of wind forecasting.

7.8 BACKGROUND AND CONCEPTS DEFINITION

The objective of this chapter is to guide the reader in the field of deep learning and define a series of concepts that will be related to the two neural network architectures used to make the different predictions of the meteorological data.

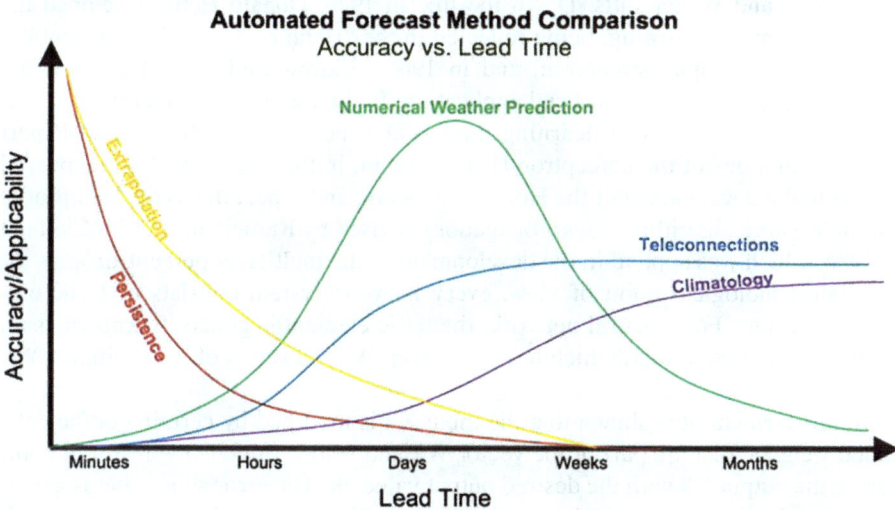

FIGURE 7.5 Comparison of accuracies of different models according to the prediction distance.

7.8.1 MACHINE LEARNING

Machine learning is the division of artificial intelligence that has emerged to evolve methodologies that permit computers to learn. More specifically, it is about developing algorithms capable of generalizing behaviors and recognizing patterns from data patronized as examples. Thus, it is a process of induction of knowledge that permits to obtain by the generalization of a general statement that describes particular cases. When all the particular cases have been observed, the induction is considered complete, so the generalization it gives rise to is considered to be valid. However, many a time, it is tough to obtain a complete induction, so the statement to which it gives rise is subject to a certain degree of uncertainty and consequently cannot be considered a formally valid inference scheme, nor can it be considered to be justified empirically.

In many cases, the field of action of machine learning overlaps with Data Mining, since the two disciplines are focused on data analysis. However, machine learning emphasizes more on the study of the computational complexity of problems to make them feasible from a practical point of view.

7.8.2 NEURAL NETWORKS

One of the most successful methods of learning machine learning is artificial neural networks. They are a paradigm of machine learning and their processing is inspired by the nervous system of humans. It is an interconnection of neurons that collude with each other to produce a resultant stimulus.

7.8.2.1 History

The first models of neural networks connect back to 1943 by neurologists Warren McCulloch and Walter Pitts. Down the line in 1949, Donald Hebb developed his ideas about neural learning, being reflected in the "Hebb rule". In 1958, Rosenblatt developed the simple perceptron, and in 1960, Widrow and Hoff developed the ADALINE, the first real industrial application. In the following years, research was reduced, due to the lack of learning models and the study of Minsky and Papert on the limitations of the perceptron [12]. However, in the 1980s, ANNs reappeared thanks to the development of the Hopfield network, and especially to the backpropagation learning algorithm (backpropagation) devised by Rumelhart and McClelland in 1986, which was applied in the development of the multilayer perceptron's.

From a biological point of view, every nervous system consists of basic elements, neurons. For a neural network, the basic element is generally known as an adaptive linear combiner, which has an answer Sk for a series of vector inputs Wk (Figure 7.6).

In this element, it is shown that the input Xk is modified by certain coefficients, called weights, that are part of the vector Wk and whose value is the result of comparing the output Yk with the desired output value dk. The error signal that is generated is used in turn to update the weights W1k, W2k, etc., in such a way that through an iterative process, the output approaches the desired value and an error εk from zero. The general structure of the element in its entirety is shown in Figure 7.6.

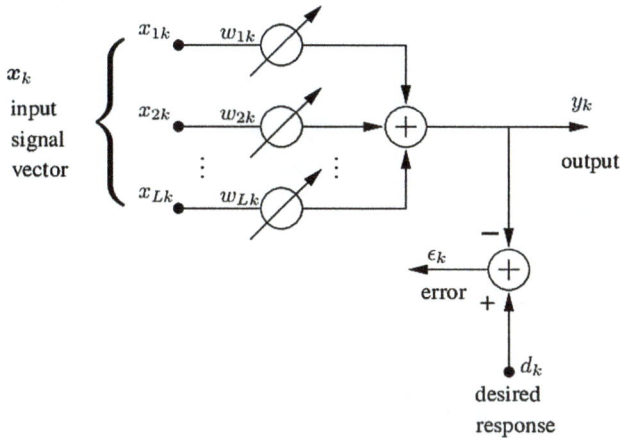

FIGURE 7.6   Adaptive linear combiner.

This type of linear element can be modified by certain functions called activation functions, which are in charge of relating the input information of the neuron with the next activation state that the neuron has. There are two models of activation function:

- Bounded models: The neuron activation value can be any within a continuous range of values.
- Unbounded models: There is no limit to the activation values.

While designing a network, it must be established that how the activation values of each neuron and the activation function with which each neuron processes the inputs, must be decided. Once the drive function has been included in the output signal path, the linear combiner becomes an adaptive linear element (Adaline). This is the true fundamental unit of all neural networks, which performs the sole function of producing a bounded output for any input presented to it, for classification purposes.

7.8.2.2 Feedforward Networks

Depending on the architecture and interconnection of all the neurons in a network, it can be classified into different categories. The first one is the feedforward network. As its name indicates, in this type of network we start with a vector of inputs that is equivalent in magnitude to the number of neurons in the first layer of the network [13]. The information, modified by the multiplicative factors of the weights in each neuron, is transmitted forward passing through the hidden layers to be processed by the output layer. Feedforward networks are the simplest in terms of implementation and simulation. Each input vector presented as training for this type of network is an entity isolated from the rest and, at the end of the said test period, the network will be ready to begin to identify and classify patterns, recognize images, or any other application that you want to give.

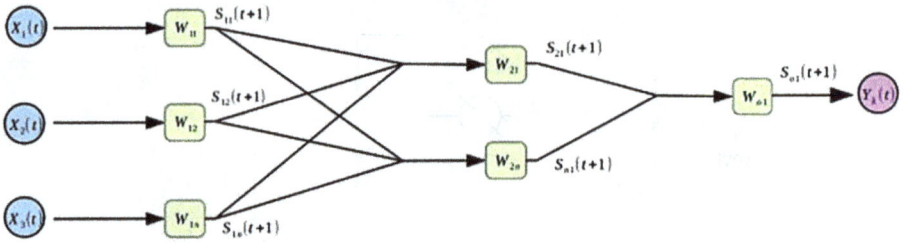

FIGURE 7.7 Feedforward network.

Compared to other networks, the FFNN is an option whose cost-speed and cost-accuracy balance is such that it gives a greater cost advantage than the other parameters. Figure 7.7 shows an FFNN where all the neurons of one layer are interconnected with the neurons of the next layer, starting with the main layer and the elements of the vector. Xk, providing your information (the outputs Smn [t + 1]) propagating it forward within the network. The weight Wk is updated as the times pass while the training continues, and at the end of the same the weights individual eleven …W1n, W2n, … Wn, etc., assume their final values to start the FFNN work with new input data once one or more outputs have been reached global Son (t + 1). The sigmoid trigger function is located at the output of the network to convert Son (t + 1) in the final value AND k.

7.8.2.3   Recurrent Neural Networks

Recurrent neural networks (RNNs) have feedback paths between all the elements that make them up. A single neuron is connected to the posterior neurons in the next layer, the neurons passed from the previous layer, and to themselves through vectors of varying weights that undergo alterations at each epoch to achieve operating parameters or goals [14].

RNNs are far more complex as compared to a feedforward network, for example, the latter network is capable of transmitting information to the following layers, resulting in a propagation effect backward in time. On the contrary, RNNs carry out the exchange of information between neurons in a much more complex way, and due to their characteristics, depending on the type of training algorithm chosen, they can propagate the information forward in time.

The basic architecture of an RNN is shown in Figure 7.8. An important feature is the inclusion of delays (Z-1) at the output of neurons in the layers intermediate; partial departures Smn (t + 1) become values Smn (t), a previous instant of time, and thus feedback is given to all the components of the network, saving information from previous instants of time. It can be seen how all the nodes are interconnected with each other and also with the nodes before them through direct connections and also delays before each layer or temporary memories. The diagram has been simplified to avoid excessive complexity, but each of the layers is represented by several neurons [15].

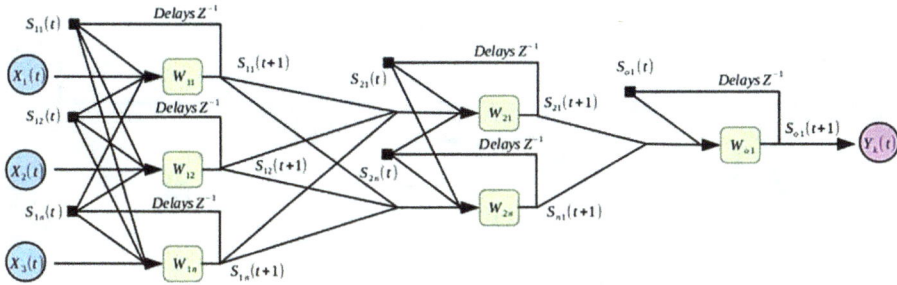

FIGURE 7.8   Recurrent neural networks.

RNNs are most effective in solving problems with remarkable temporal non-linearities. These are by and large effective in application areas vis-á-vis recognition of sequential patterns, changing over time since the prediction and mapping capabilities of RNNs allow this. The RNN has variable behavior over time and this offers possibilities of solving problems different from those with traditional architecture.

In biological systems, which were the conceptual bases of neural networks, the number of interconnections between all neurons is very large [16]. RNNs can be closer to this behavior than other types of networks, but in general, the intrinsic complexity of these requires much longer processing times. The variable characteristics of its internal states over time also make it very important to consider when the weights should be updated: at the end of each epoch (epoch-wise training) or continuously. In the second case, finding the right time to update is a significant challenge.

7.8.3   Long Short-Term Memory Networks

In traditional RRNs, during the backpropagation phase, the gradient signal can be multiplied many times with the weight matrix associated with the neurons of the hidden layer. This can have a great effect on the network learning process since the magnitude of the matrix weights is affected.

If the weights in this matrix are paltry (weights are less than 1.0), it may account for a situation known as evanescent gradient, where the gradient signal becomes so small that learning becomes too slow or stops working altogether. This can also make it cumbersome to learn long-term dependencies on the data. If, on the other hand, the matrix weights are large (greater than 1.0), it may account for a situation wherein the gradient signal is so large that it can cause the learning to never converge. This increase in the gradient signal is also known as a gradient burst [17]. These issues are what motivated the development of the LSTM model, which brings in a new structure termed a memory cell to correct these problems.

The memory cell or state of the cell corresponds to the horizontal line that runs through the upper part of the diagram. It could be said that it would be equivalent to a conveyor belt capable of storing information for long periods.

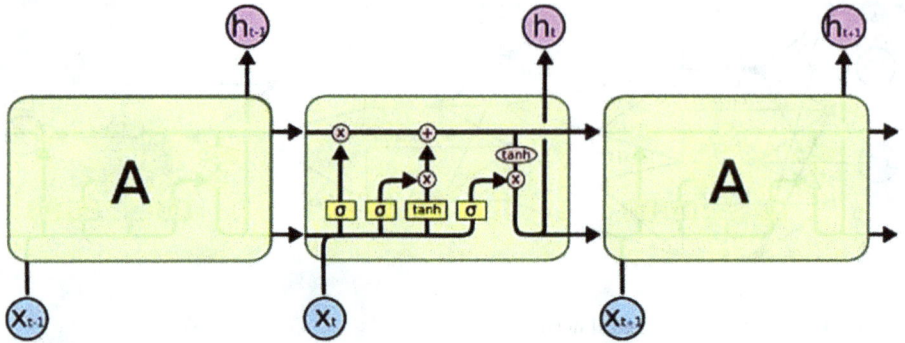

FIGURE 7.9   Representation of an LSTM cell. (Ref. [2].)

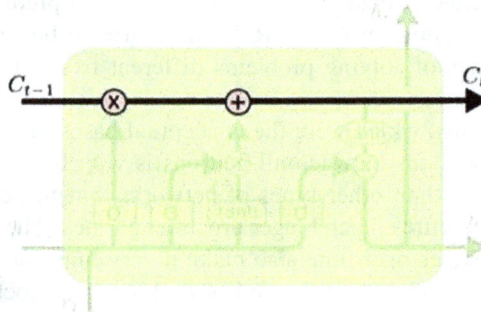

FIGURE 7.10   A memory cell. (Ref. [2].)

Figure 7.9 shows representation of LSTM Cell and Figure 7.10 shows a memory cell. LSTM has the potential to remove or add information to memory. These operations are carefully regulated by internal structures called gates. These doors allow information to be added or removed from memory at any given time. They are composed of a neural network (usually with sigmoid as the activation function) of a layer together with an arithmetic operation (multiplication or addition).

Figure 7.11 shows the representation of Gates and the outputs. The outputs of the network with sigmoid will be values between zero and one, which allows deciding

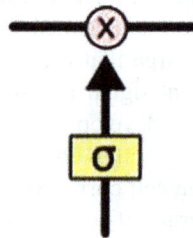

FIGURE 7.11   Gates. (Ref. [2].)

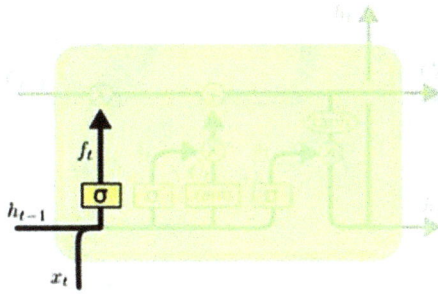

$$f_t = \sigma\left(W_f \cdot [h_{t-1}, x_t] + b_f\right)$$

FIGURE 7.12   Forget gate.

the amount of information that the component should let through. Zero would mean "let nothing go", while one would mean "let everything go". The LSTM reaches out to these gates to protect and control the state of memory. The following will explain step by step the internal operation and the information path in an LSTM:

- The first thing the LSTM does is to decide which information is to be discarded from memory. The conclusion is reached by "Forget gate". This is quite visible in Figure 7.12 where ht−1 and xt are concatenated and the result will be the entrance to the small network that makes up the Forget gate. As we saw before, the result of the network will be used to decide if the memory state will be left as it is or will be altered by eliminating some element.
- The next step that the LSTM performs is to figure out new information is to be kept in a memory state. It is done in stages. Firstly, a gate called "Input Gate" determines whichever values are to be updated. Then a small network with tanh creates a vector of new value candidates, $\hat{C}_t$, which can be added to the state. In the next step, both results are merged for status updation.
- At this time, we can update the memory status by changing C_{t-1} for the new state C_t. As in the earlier steps, it has already been decided what to do, we only have to apply the operations of each door. Figure 7.13 denotes the Input Gate. We multiply the state previous by F_t, forgetting a certain amount of

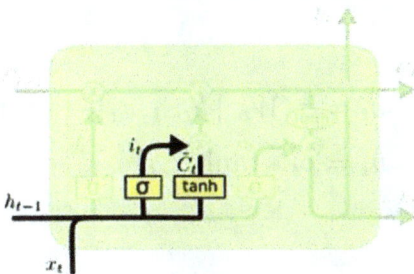

$$i_t = \sigma\left(W_i \cdot [h_{t-1}, x_t] + b_i\right)$$
$$\tilde{C}_t = \tanh(W_C \cdot [h_{t-1}, x_t] + b_C)$$

FIGURE 7.13   Input gate. (Ref. [2].)

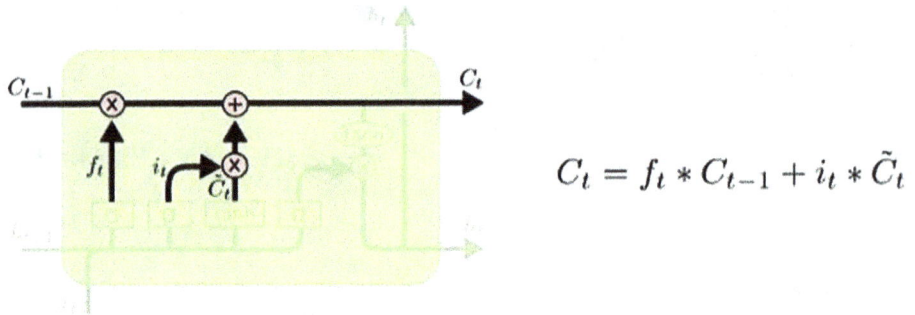

$$C_t = f_t * C_{t-1} + i_t * \tilde{C}_t$$

FIGURE 7.14 Memory upgrade. (Ref. [2].)

information based on F_t. Then we add it to $\hat{C}_t$ memory, updating it with new values that could serve us in the future. Figure 7.14 shows memory upgrade on changing the state.

- Lastly, we have to determine the LSTM cell output. This will be obtained through the product of two elements. The first one will be the output of the network with sigmoid, which will be used to decide which memory elements will be combined. The second element will be the filtering of data from memory by a tanh (to push the values to be between 1 and 1). These two elements will multiply, resulting in the new output from the cell.

7.8.4 Description of the Tools Used

All neural network architectures described in the previous section can be really difficult to implement; therefore, there are various other frameworks specially developed so that the user does not have to delve too deeply into the design and implementation of the different machine learning structures [18]. Among the most important are Theano, Torch, or Caffe, but recently another framework has been released that in theory offers even more facilities for the user when using the different architectures, not only to implement them but also to study their behavior and perform experimental studies. Figure 7.15 shows the output.

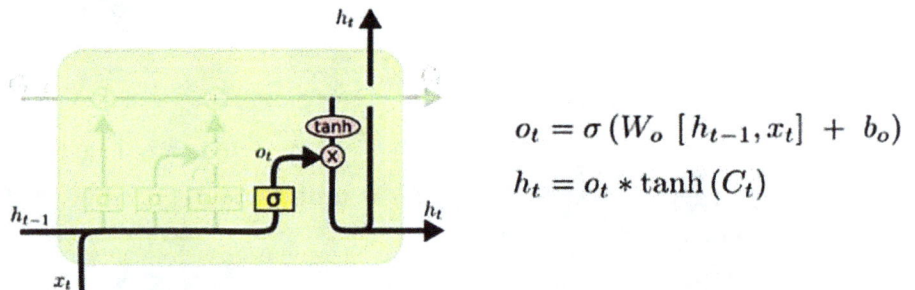

$$o_t = \sigma \left(W_o \left[h_{t-1}, x_t \right] + b_o \right)$$
$$h_t = o_t * \tanh \left(C_t \right)$$

FIGURE 7.15 Output. (Ref. [2].)

The framework is called TensorFlow, primitively conceived by researchers and engineers from Google's Intelligence research group to develop deep learning machines based on neural networks. It is open-source and is intended for numerical calculation using data flow graphs. These diagrams describe mathematical computation with a directed graph where we have nodes and edges. Nodes usually represent math operations, but also data feeding points results in removal, or reading/writing persistent variables.

The edges express the I/O association between the nodes. They transmit compound data sets or tensors that can be assigned to computing devices to run asynchronously or in parallel. Another important feature is that, thanks to its architecture, it permits to implementation of computing to multiple Central Processing Units or Graphics Processing Units on a desktop computer, server, or mobile device with the help of a single Application Programming Interface (API). It has APIs available in several languages, both for the construction and execution of a TensorFlow chart. The Python API is currently the most comprehensive and the easiest to use, but the C ++ API can offer some performance benefits in graphical execution and supports the deployment of small devices such as Android [19].

7.9   PREDICTION MODELS

As specified earlier, in this work two prediction models based on neural networks are proposed. The first using an FFNN and the second using an LSTM recurrent neural network.

FFNNs could be considered as Adaptive Information Processing structures, where processing is carried out through the interconnection of their neurons. They are capable of capturing complex non-linear relationships between different variables, and for this reason, they can be used to solve certain types of very complex problems where the existing relationships between the explanatory variables and the output, such as prediction, are not known. On the other hand, RNNs present one or more cycles in the graph defined by the interconnections of their processing units. The existence of these cycles allows them to work innately with temporal sequences. They are non-linear dynamic systems capable of discovering temporal regularities in the processed sequences and can therefore be applied to a multitude of processing tasks of this type of sequence [20].

Its main advantage lies in the possibility of storing a representation of the recent history of the sequence, which allows, unlike what happens with non-RNNs, that the output before a given input vector can vary as a function of current internal network configuration.

The structure of the two networks is as follows:

1. Feedforward network
   - Input layer: 32
   - Inner layers: 1 inner layer of 30 neurons output
   - Layer: 1 of a neuron
   - Learning algorithms: Gradient down activation
   - Function: Sigmoid
   - Stop algorithm: Number of iterations = 500

2. LSTM recurring network
   - Input layer: 32
   - Inner layers: 1 inner layer of 50 neurons
   - Output layer: 1 of a neuron
   - Learning algorithms: Gradient down trigger function: None
   - Actual output values Stop algorithm: Number of iterations = 100

The reason for choosing these configurations (number of inputs, neurons, iterations, and so on) will be addressed in the following sections.

7.9.1 TRAINING DATA

The data that will be used to carry out the different tests, training sessions, and, of course, the prediction will be the different powers of wind speed (m/s) captured by an anemometer located in the Muppandal Wind Farm, specifically powers taken at a height 40 meters during July and August.

Wind speed measurements are taken every minute, therefore there are about 80,000 samples, but at work, different trainings will be carried out taking the data every 10 minutes or every hour to study different aspects of the prediction. Tests will also be done by taking the average of the data in different ranges (average every ten samples for example). The consummation of a model is related to its predictability in novel data and unfettered training. The evaluation of this performance is very crucial as it helps us in measuring the caliber of the implicit prediction model. Therefore, evaluation of error measures in data is of utmost importance, specifically unused data for the construction of the prediction model or to adjust remarkable parameters of the method. To execute this, data will be divided as shown in Figure 7.16.

7.9.2 HYPERPARAMETERS

In any neural network training, there are a series of hyperparameters that the programmer must adjust depending on how the training was developed, the evolution of the cost function, the quality of the network output, or the execution time [21]. Of the same, specifically, the hyperparameters that will have to be adjusted in the different neural networks are the following:

- The number of epochs: Learning in this study is done by periods. The network it evolves during a period (epoch) in which the different batches of the data set are passed to it. Once the end of an epoch has been reached,

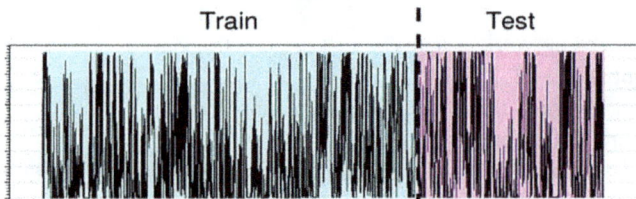

FIGURE 7.16   Data division.

the network will have already been trained with all the data and will have reached a certain cost error that will be tried to reduce in the following periods.

- Batch size: The input of the network is nothing more than a vector of numbers (from wind speeds). TensorFlow is developed in such a way that, when training, the programmer can pass more than one input to the network, that is, more than one vector of data. This batch of data is known as the batch name, and the size of this batch is the crucial criterion of the network as it can not only influence the execution time of the network but also vary the quality of the network's predictions.
- Neurons in hidden layer: The number of neurons in the network is not data that are known in advance, it depends on the input data. Therefore, it is necessary to carry out a little experimentation with the number of neurons to be used to optimize the functioning of the network.
- Learning factor: The learning factor is directly related to the speed of network learning. This parameter indicates the percentage by which the net weights are allowed to vary in each training season. If the value is high, the modification of the weights from one era to another can be very large, while if the value is low, the weights can only vary in a small proportion:
  - In the learning period, high learning rate values favor a rapid approach to the optimal values of the weights but do not allow the fine adjustment of these weights causing a continuous movement around the optimum.
  - The low values of the learning factor suppose a slow adjustment of the value of the weights, which can cause the fall to local minimums from which it is not possible to get out due to the low value of the learning rate.
  It is necessary to locate an appropriate learning factor compromise value for each specific problem, high enough to approach the optimum in a reasonable time and avoid the risk of falling to local minimums, and small enough to be able to have guarantees of locating a nearby value to the optimum.
- Window size: The window size refers to the length of the vector of network input, that is, the number of data that the network will take and that it will evaluate to predict future data. Figure 7.17 shows the window representation related to predict the future of data.

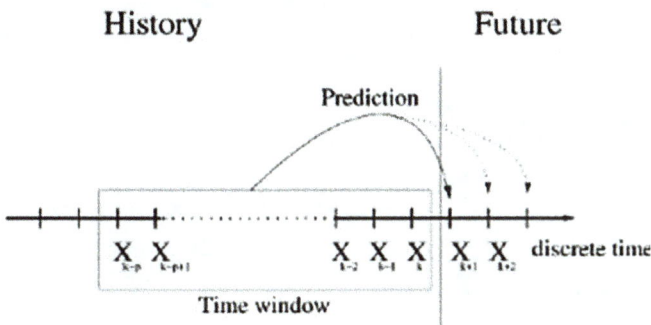

FIGURE 7.17   Window. [Ref. 18].)

7.9.3 TRAINING

Training a neural network is a process that modifies the value of the weights and biases associated with each neuron so that the RNA can generate an output. In the case of supervised learning, there is a set of data but the function or mathematical relationship that represents them is not known. By propagating each of these patterns forward, a response is obtained in the RNA output which is compared with the desired output, allowing to obtain the network performance error [22].

The cost function is a fundamental part of training a network, as it is responsible for estimating the quality of the output by calculating an error (usually mean square error) using the resulting output from the network and the desired output. This error should decrease as the training progresses, which would mean that the network is learning and therefore the output from it is increasingly looking like the desired output [23].

This error can be collected and then represented in a graph that can give us some information on how the network has been training. For example, in the graph of Figure 7.18 we can see that the evolution of the cost in recent times continues to decline, that is, it does not remain more or less constant, which means that the cost of the network can continue to decrease. Therefore, it is advisable to increase the number of epochs to obtain better results.

Another example could be the one we see in Figure 7.19, where the cost does not decrease constantly but has risen in the early stages. This may be because perhaps we have trained data with a learning factor that is too high, and that causes descent made through the gradient. The approach to the optimal minimum function is not precise, causing that the error increases and that it takes longer to adjust to reach that minimum.

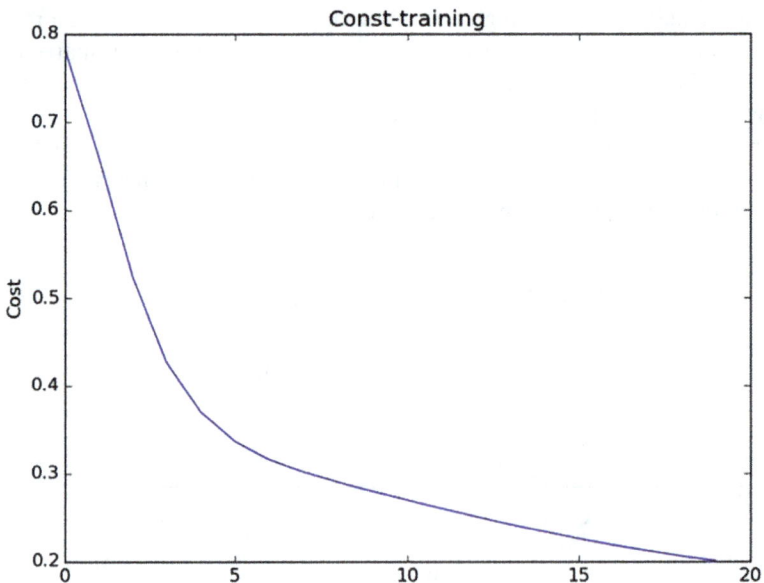

FIGURE 7.18   Example of cost graph.

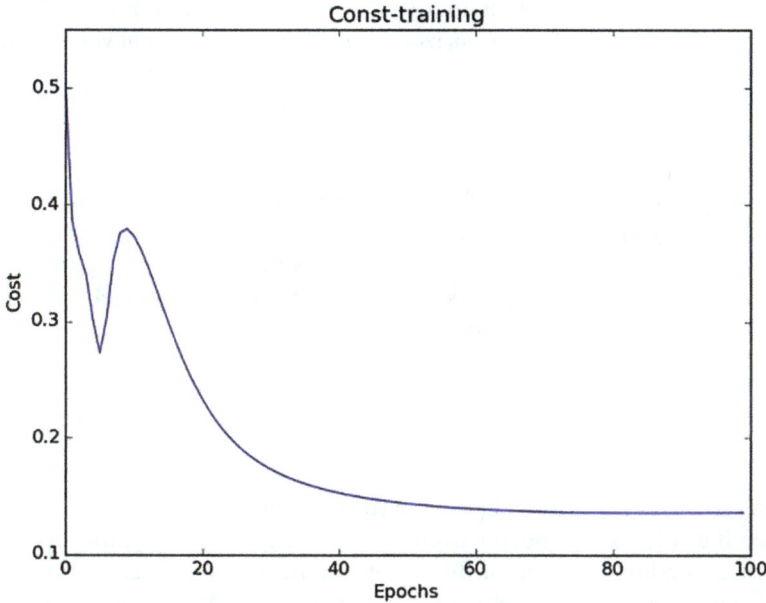

FIGURE 7.19   Example of cost graph.

To correct this, we would simply have to reduce the learning factor until we observe that the cost graph descends progressively.

7.9.4   EVALUATION OF THE MODELS

There are two evaluation modes already proposed in the literature for the comparison between different models with different case studies, and to know the goodness of the different proposed calculation procedures for aggregate predictions, and to be able to compare its accuracy. To quantify the prediction, gain of the models proposed to the models of reference, the following formula will be used:

$$Imp_{ref}, EC(k) = \frac{EC_{ref} - EC(k)}{EC_{ref}(k)}$$

wherein (k) is the evaluation criteria to be considered, that is, the error measures calculated in the validation stage (BIAS, MAE, MSE, etc.), and the term EC_{ref} corresponds to the reference models.

- Reference models

In the field of prediction, some simple models have been defined that serves as a reference when evaluating the quality of the new prediction models developed. These reference models are based on simple time series models. One of the most widely used reference models is the Persistence model. According to this model, future power predictions, for any horizon, coincide with the current value of the variable, in this case, the wind power:

$$\hat{P}_P(t) = P(t)$$

wherein $\hat{P}_P(t)$ is the prediction for time horizon t + h, calculated at the current instant t, and $P(t)$ is the current measure of power recorded in the park or the study region. The persistence model is a concrete case of the moving average model of the last n observations, with n = 1, which is the second reference model which may be adhered to for wind energy evaluation.

$$\hat{P}_{MA,n}(t) = \frac{1}{n} \sum_{i-0}^{n-1} P(t-i)$$

wherein $P(t-i)$ is a measure of power at time $t-i$ of age i.

When the order of the moving average model tends to infinity, then we have the global or unconditional mean of the entire historical wind power registered in the park or region, also called the climatic mean, which would be a very good long-term prediction model.

$$\hat{P}_0(t) = \underline{P(t)}$$

As a result of combining the goodness of the predictions of the persistence model for a very short term, and the relative goodness of the predictions of a global mean power model for long-term predictions, we obtain a new reference model that results from a weighting of both methods, whose weights depend on the prediction horizon.

$$\hat{P}_{NR}(t) = ._{ak} P(t) + (1 - ._{ak}) \underline{P(t)}$$

wherein $._{ak}$ is weighting factor to be estimated, with values between 0 and 1 depending on the prediction horizon, and that will take values close to unity for very short forecast horizons, giving much weight to the model of persistence, and values close to zero for very high prediction horizons, in those whose expected power, will be closer to the average power recorded historically. Furthermore, even though the latest proposed reference model is more elaborated and provided better results than the persistence model, still the reference model is used more due to its simplicity and the fact that it does not require the estimation of any parameter.

7.10 CONCLUSION AND FUTURE SCOPE

To tackle this study, first of all, it was necessary to deepen the knowledge of the architectures of machine learning such as artificial neural networks. Learning and understanding the feedforward network is relatively easy.

The case of the recurring network LSTM was very different. As the model has been developed recently, the high complexity of its operation and scarce information about its architecture has made learning much slower and more difficult than expected. There is no doubt that both the architecture of the LSTM network and the TensorFlow framework will continue to advance and continue to develop, as it is evident that there is great potential in these architectures. So, in the future, whoever enters the world of LSTM networks and intends to use TensorFlow as a framework to implement the different aspects of it, will surely have it much easier since there will be more examples and more documentation to facilitate their learning

In short, it has been empirically demonstrated that Feedforward and LSTM networks, with the proposed configuration, exhibit very similar behavior to the persistence model. Furthermore, it seems that both networks generate their outputs based on the last observed value, a behavior that was not expected with these networks and it could be a matter of investigation in the future to determine the reason for this procedure.

The recommendations for future research work suggest that the result obtained would open up new questions to answer. It would be interesting to continue with the experiment by extending the days of both training and testing, as well as the testing of new topologies or prediction strategies. A convolutional network could be studied when dealing with spatially located data or also a recurring network to try to reduce the error in those cases that, due to unpredictable human interference, the performance of wind energy production has been modified. For the latter case, a more rigorous statistical comparison would also be required. Another line of research that would emerge would be to continue the investigation with a smaller and more specific area.

REFERENCES

1. M.R. Bachute & J.M. Subhedar, "Autonomous Driving Architectures: Insights of Machine Learning and Deep Learning Algorithms", *Machine Learning with Applications*, 6, 100164, 2021.
2. P. Baldi & R. Vershynin, "The Capacity of Feedforward Neural Networks", *Neural Networks*, 116, 288–311, 2019.
3. F. Beaufays & E.A. Wan, "Relating Real-Time Backpropagation and Backpropagation-Through-Time: An Application of Flow Graph Interreciprocity", *Neural Computation*, 6(2), 296–306, 1994.
4. F.M. Bianchi, E. Maiorino, M.C. Kampffmeyer, A. Rizzi, & R. Jenssen, *Recurrent Neural Networks for Short-Term Load Forecasting: An Overview and Comparative Analysis*, Springer, 2017.
5. F. Chollet, *Deep Learning with Python* (2nd ed.), Manning, 2021.
6. E.N. Lorenz, *Predictability; Does the Flap of a Butterfly's Wings in Brazil Set Off a Tornado in Texas*, Cambridge, 1972.
7. E. Pelikán et al., "Wind Power Forecasting by an Empirical Model Using NWP Outputs", *9th International Conference on Environment and Electrical Engineering*, pp. 45–48, doi: 10.1109/EEEIC.2010.5490019, 2010.
8. N. Edward, "Deterministic Nonperiodic Flow", *Journal of the Atmospheric Sciences*, 20, 130–141, 1963.
9. M. Ekman, *Learning Deep Learning: Theory and Practice of Neural Networks, Computer Vision, Natural Language Processing, and Transformers Using TensorFlow* (1st ed.), Addison-Wesley Professional, 2021.

10. G. Röth, "Tutorial 1: NVIDIA's Platform for Deep Neural Networks", *IEEE International Conference on Data Science and Advanced Analytics (DSAA)*, pp. XXXVII–XXXIX, 2015, doi: 10.1109/DSAA.2015.7344778.

11. H. Goldstine & A. Goldstine, "The Electronic Numerical Integrator and Computer (ENIAC)". *IEEE Annals of the History of Computing*, 18(1), 10–16, 1996, doi: 10.1109/85.476557.

12. X. Han & L. Clemmensen, "On Weighted Support Vector Regression", *Quality and Reliability Engineering International*, 30(6), 891–903, 2014, doi: 10.1002/qre.1654.

13. I.Z. Memon, S. Talpur, S. Narejo, A.Z. Junejo, & E.F. Hassan, "Short-Term Prediction Model for Multi-currency Exchange Using Artificial Neural Network", *3rd International Conference on Information and Computer Technologies (ICICT)*, pp. 102–106, 2020, doi: 10.1109/ICICT50521.2020.00024.

14. J. Korstanje, *Advanced Forecasting with Python: With State-of-the-Art-Models Including LSTMs, Facebook's Prophet, and Amazon's DeepAR* (1st ed.), Apress, 2021.

15. P. Laguna, R. Jané, E. Masgrau, & P. Caminal, "The Adaptive Linear Combiner with a Periodic-Impulse Reference Input as a Linear Comb Filter", *Signal Processing*, 48(3), 193–203, 1996.

16. A. Lawi & E. Kurnia, *Accurately Forecasting Stock Prices Using LSTM and GRU Neural Networks: A Deep Learning Approach for Forecasting Stock Price Time-Series Data in Groups*, LAP LAMBERT Academic Publishing, 2021.

17. P. Nerurkar, *Convolutional Neural Networks and Recurrent Neural Networks: Convolutional Neural Networks and Recurrent Neural Networks*, Independently Published, 2020.

18. P. Saatwong & S. Suwankawin, "Short-Term Electricity Load Forecasting for Building Energy Management System", *13th International Conference on Electrical Engineering/Electronics, Computer, Telecommunications and Information Technology (ECTI-CON)*, pp. 1–6, 2016, doi: 10.1109/ECTICon.2016.7561477.

19. Q. Xu, L. Chen, P. Zeng, & X. Xu, "Correlation Modeling among Multi-Wind Farms Based on Copula-ARMA Wind Speed Model", *China International Conference on Electricity Distribution (CICED)*, pp. 543–546, 2014, doi: 10.1109/CICED.2014.6991768.

20. S. Mukhopadhyay, P.K. Panigrahi, A. Mitra, P. Bhattacharya, M. Sarkar, & P. Das, "Optimized DHT-RBF Model as Replacement of ARMA-RBF Model for Wind Power Forecasting", *IEEE International Conference ON Emerging Trends in Computing, Communication and Nanotechnology (ICECCN)*, 2013, pp. 415–419, 2013, doi: 10.1109/ICE-CCN.2013.6528534.

21. R. Saunders, "The Use of Satellite Data in Numerical Weather Prediction", *Weather*, 76(3), 95–97, 2021.

22. M. Scheuerer & D. Mölle, "Probabilistic Wind Speed Forecasting on a Grid Based on Ensemble Model Output Statistics", *The Annals of Applied Statistics*, 9(3), 2015.

23. N. Son, S. Yang, & J. Na, "Hybrid Forecasting Model for Short-Term Wind Power Prediction Using Modified Long Short-Term Memory", *Energies*, 12(20), 3901, 2019.

8 Machine Learning for Renewable Energy Applications

Dhanasekaran Arumugam[1], Christopher Stephen[2,3], Richa Parmar[4], Vishnupriyan Jegadeesan[5], Ajay John Paul[6], and Ibrahim Denka Kariyama[7]
[1]Center for Energy Research, Department of Mechanical Engineering, Chennai Institute of Technology, Chennai, Tamil Nadu, India
[2]Department of Mechanical Engineering, Vel Tech Rangarajan Dr. Sagunthala R&D Institute of Science and Technology, Chennai, Tamil Nadu, India
[3]Teaching Associateship for Research Excellence (TARE) Fellow under SERB, National Institute of Solar Energy (NISE), Gurugram, Haryana, India
[4] Solar Water Pump Laboratory, National Institute of Solar Energy, Gurugram, Haryana, India
[5]Center for Energy Research, Department of Electrical and Electronics Engineering, Chennai Institute of Technology, Chennai, Tamil Nadu, India
[6]School of Mechanical Engineering, Kyungpook National University, Daegu, Gyeongbuk Province, South Korea
[7]Department of Agricultural Engineering, Dr. Hilla Limann Technical University, Wa, Upper West Ghana, Ghana

CONTENTS

DOI: 10.1201/9781003369554-8

8.1   INTRODUCTION

Energy consumption is rapidly increasing due to various industrial activities and progress in all countries. It is expected that the global energy demand will increase significantly in the next few decades. This is mainly due to the estimated growth in population and the economic and industrial developments in the countries, especially in China and India [1]. Energy demand projections show an increasing trend globally with annual consumption requirement of 778 Etta Joule by the year 2035. Achieving the growing energy demand while mitigating the effects of pollution is one of the biggest issues in the modern era [2]. One of the most promising energy sources to meet the energy demand is fossil fuel. These fuels are basically hydrocarbons and their derivatives. It took millions of years for the fossil fuel to form and now they are being depleted in a much faster rate than new fuels are being made. Consequently, they emit greenhouse gases, which accelerate climatic change and pollution.

Hence, renewable energy sources have received much attention globally. Renewable energy sources are solar energy, wind energy, hydropower, biomass energy, marine energy, and geothermal energy (each discussed briefly later in the chapter). These energy resources are abundant, inexhaustible, clean, green and will emit very low carbon. So, they are beneficial for the protection of environment. Figure 8.1 shows global installed capacity of various renewable energy sources [3].

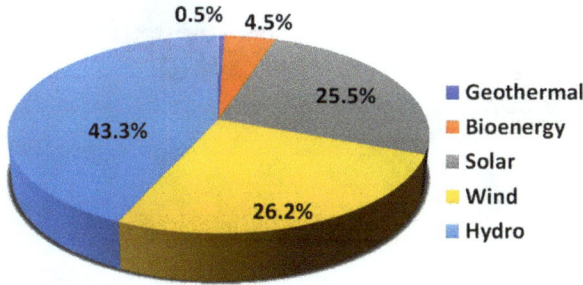

FIGURE 8.1 Global installed capacity of renewable energy sources.

Moreover, the use of renewable energy can reduce the utilization of fossil fuels and reduce soil pollution. For these reasons, renewable energy is globally emphasized and developed rapidly. Figure 8.2 shows the classifications of various renewable energy resources.

Solar energy: It is enormous and can be used to produce electricity and heat. In 2018, around 100 GW of solar energy was added, causing the total solar energy usage to 505 GW. The global irradiance [4] is shown in Figure 8.3. Regions of high diffuse radiation are Northern Europe, South-East China and the tropical belt surrounding the equator.

Wind energy: It is the second largest renewable energy source. It has produced more than 5% of global electricity in the year 2018 with 591 GW of global energy capacity. The kinetic energy of wind is converted into electricity by using wind energy generators.

Hydropower: The potential energy of water is converted into electrical energy through hydraulic turbines. Majority of electric power generation is produced by hydropower plants which produced an estimated 4210 TWh of the 26,700 TWh total global electricity in 2018. Small hydropower projects and micro-hydropower have a lower environmental impact than large conventional hydropower plants.

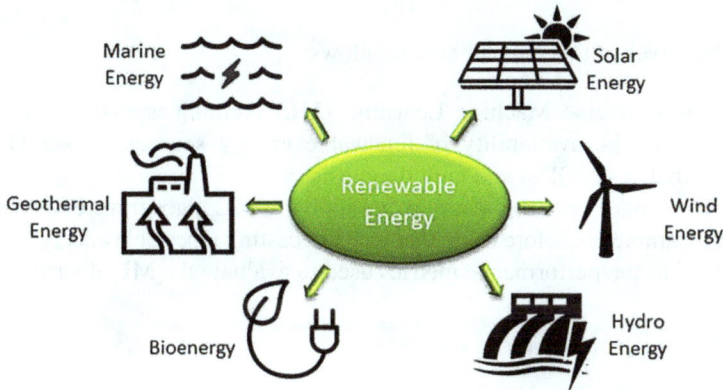

FIGURE 8.2 Classification of renewable energy resources.

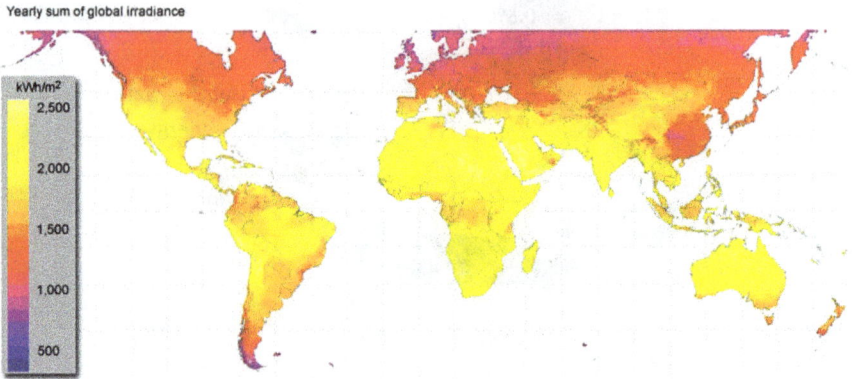

Yearly sum of global irradiance

FIGURE 8.3 Annual global irradiance.

Biomass energy: Energy obtained from plant and animal wastes can be used to produce electricity and heat. Global biomass electric power capacity is 130 GW in the year 2018. One ton of waste can produce 550–750 kWh of electrical energy. Landfill gas has methane which can be used in to produce energy.

Marine energy: Offshore wind, tides, ocean currents, waves, thermal differences, salinity gradients and biomass are all sources of marine energy. The estimated technical potential of off-shore wind energy is 16,000 (TWh) ·a–1 by 2050. It is estimated that the total ocean energy resource has been two million (TWh) ·a–1, the tidal energy potential is 26,000 (TWh)·a–1 and the theoretical wave energy is about 32,000 (TWh) ·a–1 having a technical potential of 5600 (TWh)·a–1. The global ocean thermal energy conversion (OTEC) potential is huge with a theoretical potential of about 44,000 (TWh) a–1. The salinity gradients in the ocean have technical potential of 1650 (TWh)·a–1 [5].

Geothermal Energy: In 2018, geothermal energy provided 175 TWh globally, half of which was in the form of electricity and the other half in the form of heat. In 2018, the United States generated 16 billion kWh of geothermal energy [6].

The objectives of this chapter are as follows:

- To study diverse Machine Learning (ML) techniques that can be used to predict the availability of renewable energy sources, Power Quality Disturbances (PQDs) and fault diagnosis;
- And, to understand the methods for applying ML algorithms and selecting input parameters before using them for forecasting renewable energy systems, as well as the performance metrics used to evaluate the ML algorithms.

8.1.1 ORGANIZATION OF THE CHAPTER

This chapter is organized as follows: Section 8.2 elaborates the methodology for forecasting Numerical Weather Prediction (NWP), statistical models and soft computing models. Section 8.3 highlights in depth towards ML – definition, groupings, types as well as Deep Learning (DL) and neural networks. Section 8.4 gives a detailed overview of implementation of ML in renewable energy sources. Section 8.5 concludes the chapter with future scope.

8.2 METHODOLOGY FOR FORECASTING

Renewable Energy (RE) is intermittent in nature that affects the reliability and stability of large-scale integration, reserve capacity and the cost of power generation. The intermittency of renewable energy is an area of concern for the power system planners. The technical challenges caused by the penetration of renewable energy sources can be reduced if renewable energy resource and its energy output can be accurately predicted.

Moreover, renewable energy systems have power electronic devices, which reduce the rotational inertia and stability of the system. Hence to mitigate the uncertainties and to effectively plan and operate energy systems forecasting the renewable energy is essential [7]. But accurate prediction of renewable energy sources is a challenging task due to its randomness and intermittency. Several algorithms have been developed in recent years to accurately predict renewable energy sources for the next few minutes to few days. Physical models, statistical models, artificial intelligence (AI) techniques and hybrid techniques are the four types of algorithms [8].

8.2.1 NUMERICAL WEATHER PREDICTION

Numerical Weather Prediction (NWP) data are widely used to model weather data. It processes current weather observations for predicting the weather conditions in the future. Output is based on current weather observations, which are incorporated in the model and to predict temperature, precipitation and several other meteorological variables. The main input parameters for the NWP are meteorological data, geographical information, temperature, pressure and orography. Though physical methods are effective at forecasting atmospheric dynamics, they necessitate more computational resources in order to calibrate more data [9]. Hence, physical methods may not be suitable for short-term forecasting horizons.

8.2.2 STATISTICAL MODELS

Statistical models like Autoregressive Moving Average (ARMA) models, Autoregressive (AR) models, Bayesian approach, Kalman filter, Markov chain model, grey theory and linear prediction methods can be used for wind power forecasting due to the availability of a large amount of historical data. These models establish a mathematical relationship between time series data of renewable energy sources. When compared to physical models, statistical models perform satisfactorily for short-term wind power forecasting. Despite the fact that statistical models assume a linear relationship between data, they are incapable of accurately forecasting non-linear wind power.

8.2.3 SOFT COMPUTING MODELS

Due to the rapid advancement in the soft computing techniques, ML-based forecasting models can provide reliable performance than their counterparts because

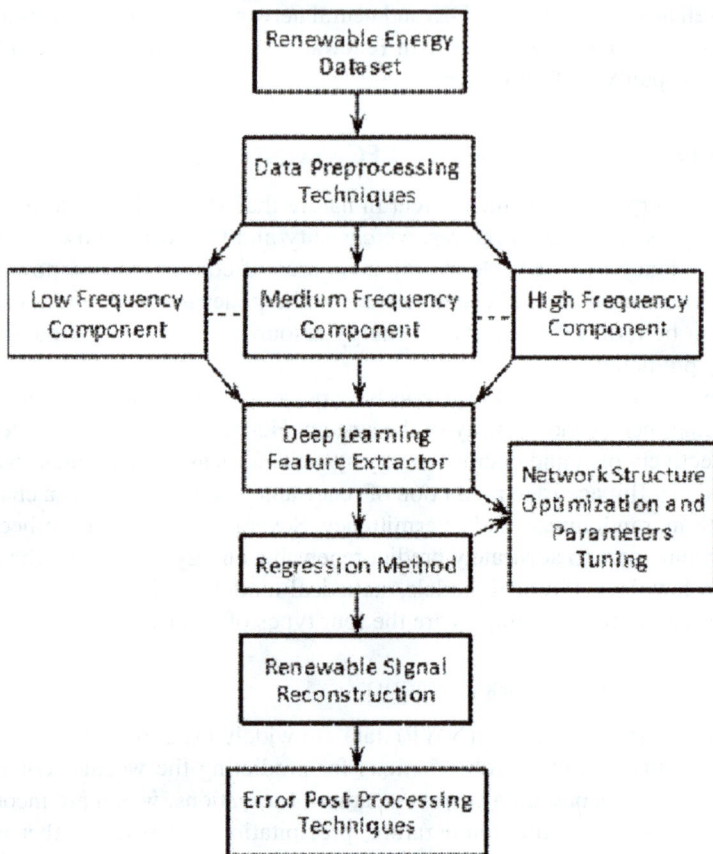

FIGURE 8.4   The general architecture of renewable energy forecasting.

of their ability in data mining and feature extraction. Figure 8.2 depicts possible areas where the ML techniques can be employed for improvements and better performance management of renewable energy systems. The figure's right side depicts consumers and prosumers (i.e., those who consume energy from the grid as well as produce small-scale renewable energy and feed the excessive energy to the grid). The left side shows large-scale renewable energy producers. Traditional power plants can still be used in the grid to keep demand and supply balanced and to ensure power quality (PQ). The general architecture of renewable energy forecasting [10] is shown in Figure 8.4.

8.3   OVERVIEW OF MACHINE LEARNING

Machine Learning (ML) techniques can be used to forecast the sectors of renewable energy generation and demand. Depending on the requirements, it can be used as a stand-alone or grid-connected renewable energy resource. Figure 8.5 illustrates the various sectors where ML can be applied.

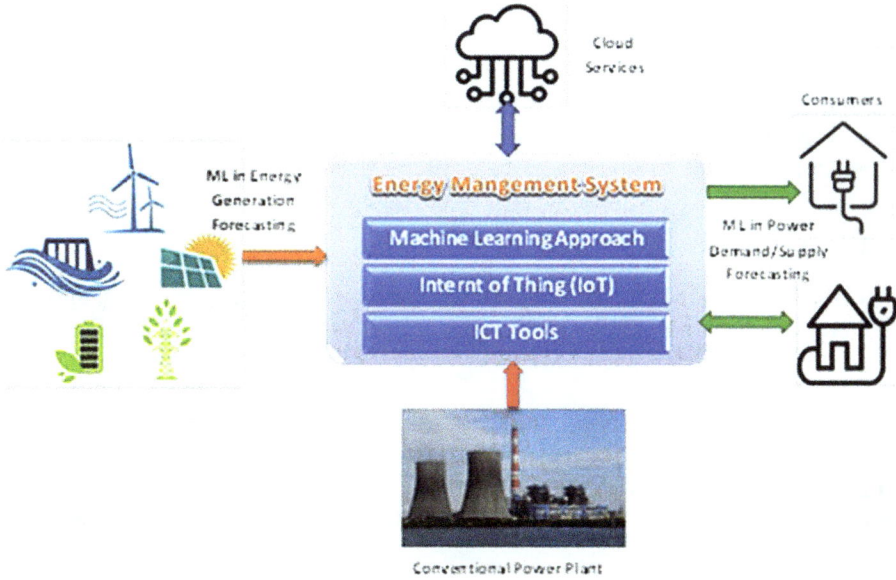

FIGURE 8.5   Usages of modern technologies in hybrid-renewable-energy systems.

The ML is a branch of science that studies the theory, performance and properties of learning systems and algorithms. It incorporates concepts from a variety of scientific fields, including AI, optimization theory, information theory, statistics and so on [11]. Figure 8.6 shows the Venn diagram of ML.

Generally, ML algorithms learn from data rather than through programming. The algorithms iteratively learn from data to improve, describe and to predict outcomes. Due to this fact, ML can produce more accurate models. It can be used for a variety

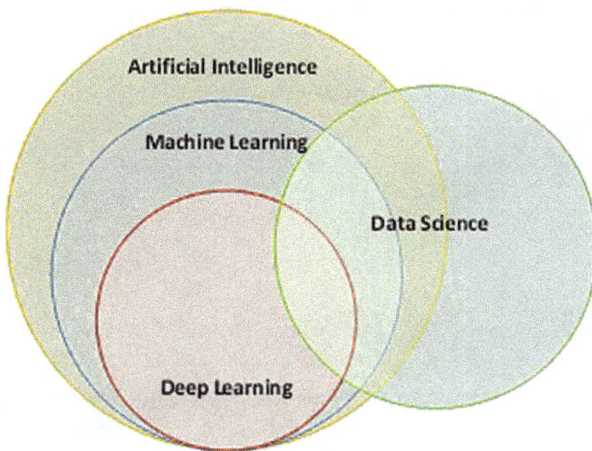

FIGURE 8.6   Venn diagram of machine learning.

of problems, including data mining, autonomous control systems, pattern recognition, classification and forecasting problems. The three main parts of an ML algorithm are:

- Decision Process: To make a prediction or classification.
- Error Function: To evaluate outcome of the model.
- Optimization: To adjust the weights till the required accuracy has been met.

8.3.1 GROUPINGS OF MACHINE LEARNING

Machine Learning (ML) is divided into five categories: supervised learning, unsupervised learning, semi-supervised learning, reinforcement learning and DL. Figure 8.7 shows the classification tree of ML.

8.3.2 SUPERVISED LEARNING

In supervised learning, machines are trained using labelled training data (i.e. some inputs are tagged with the right output) and from the data, it predicts the output. The

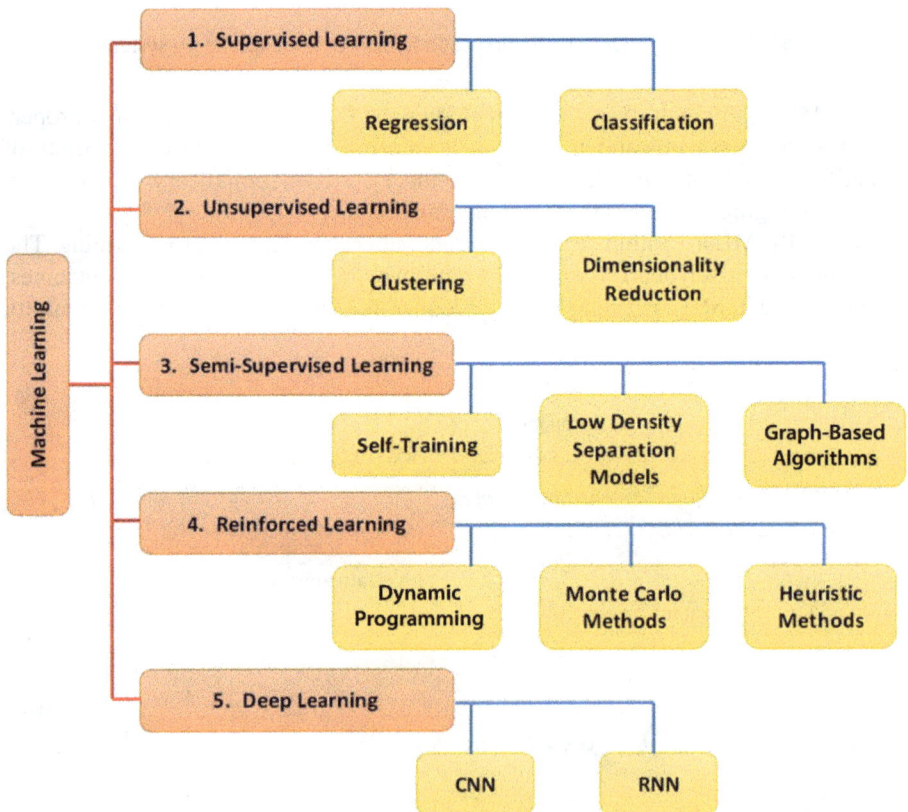

FIGURE 8.7 Classifications of machine learning.

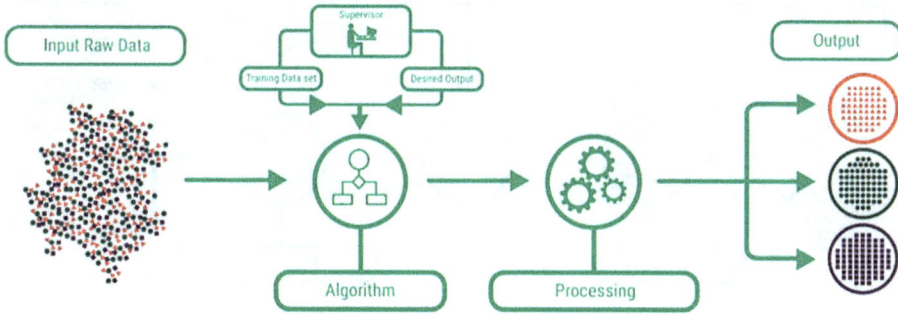

FIGURE 8.8 Process flow of supervised learning.

training data act as the supervisor which trains the ML to predict the correct output. Figure 8.8 shows the process flow of supervised learning [12]. Supervised learning is divided into regression algorithms and classification algorithms. Regression algorithms are used to establish a relationship between the input and the output variables. Linear regression, regression trees, non-linear regression, Bayesian linear regression and polynomial regression are the various types of regression algorithms.

When the output variable is categorical, such as Yes–No, Male–Female, True–False, etc., classification algorithms are used. Random forest, decision trees, logistic regression and support vector machines are the different types of classification algorithms [13]. The inputs and outputs of supervised learning can be described mathematically as follows. If $x_j \in R$ represents a set of training data points along with their labels y_j for each point where $j = 1,2, \ldots m$, then when test data are provided, the algorithm generates labels for test data.

Input

$$\text{data} \left(x_j \in R_{n,j} \in Z := \{1, 2, \cdots, m\} \right)$$

$$\text{labels} \left(y_j \in \{\pm 1\}, j \in Z' \subset Z \right)$$

Output

$$\text{labels} \left(y_j \in \{\pm 1\}, j \in Z \right).$$

8.3.3 UNSUPERVISED LEARNING

Unsupervised learning is an ML technique in which models are created without the use of a training dataset. Models, on the other hand, discover hidden patterns and insights in the given data. Figure 8.9 shows the typical processes of unsupervised learning [12].

It is used to discover the underlying structure of a dataset, group it based on similarities and represent it in a compressed format. It is divided into two categories: clustering and association. Clustering is a method of categorizing objects

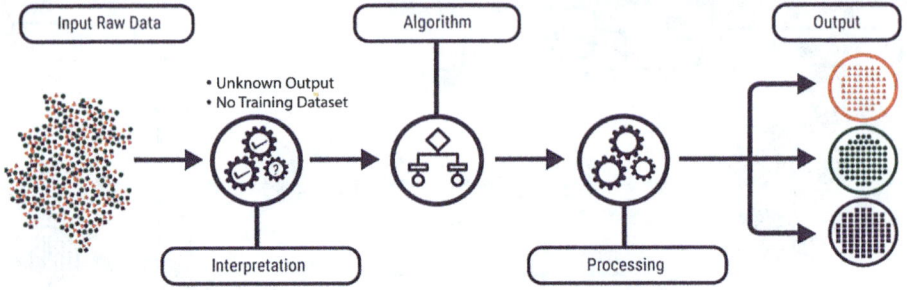

FIGURE 8.9 Process flow of unsupervised learning.

into groups, in which data with the most similarities remain in one group and data with few or no similarities remain in another. Association technique finds it useful relations in the parameters of a large dataset. K-means clustering, K-Nearest Neighbours (KNN), hierarchal clustering, anomaly detection, neural network, principal component analysis, independent component analysis, a priori algorithm and singular value decomposition are some popular unsupervised learning algorithms. In the simplest form, unsupervised learning can be described mathematically using the following form. This form of learning generates labels y_j for input data x_j.

Input

$$\text{data} \left(\mathbf{x_j} \in R_{n,j} \in Z := \{1, 2, \cdots, m\} \right)$$

Output

$$\text{labels} \left(\mathbf{y_j} \in \{\pm 1\}, j \in Z \right)$$

8.3.4 SEMI-SUPERVISED LEARNING

Semi-supervised learning is a type of ML that falls somewhere in between supervised and unsupervised learning. It includes a small number of labelled examples as well as a large number of unlabelled examples. Self-training, generative models, graph-based algorithms, semi-supervised support vector machines and multi-view algorithms are examples of semi-supervised learning algorithms.

8.3.5 REINFORCEMENT LEARNING

Reinforcement learning is a type of ML in which the algorithm learns by interacting with its surroundings. The learning agent gains knowledge from its own actions. It chooses current actions based on previous experiences and new options. A signal received by the agent in the form of a numerical reward value can be used to determine the success of a specific action. The agent learns to choose

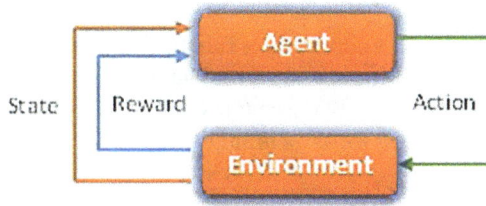

FIGURE 8.10   Reinforcement learning.

actions that maximize the numerical reward's value. The elements of a reinforcement learning are environment, state, reward, policy and value which are shown in Figure 8.10.

Markov Decision Process (MDP), a mathematical framework for defining an environment, and Q learning, a value-based method of supplying information to convey the action to be selected by the agent, are two important learning models in reinforcement learning.

8.3.6 Deep Learning and Neural Networks

Deep Learning (DL) is a subfield of ML that is based on algorithms that learn from multiple levels to provide a model that represents complex data relations. DL employs artificial neural networks to process large amounts of data. A neural network is similar to the human brain in that it is made up of artificial neurons known as nodes. These nodes are stacked in three layers on top of each other. The input layer, the hidden layer(s) and the output layer are all present. Figure 8.11 shows the configuration of DL [14].

The DL algorithms can work with any type of data, including large amounts of computational power and complex problems. Convolutional Neural Networks (CNNs), Long Short-Term Memory Networks (LSTMs), Recurrent Neural Networks (RNNs), Generative Adversarial Networks (GANs), Radial Basis Function Networks (RBFNs), Multi-layer Perceptron's (MLPs), Self-Organizing Maps (SOMs), Deep Belief Networks (DBNs), Restricted Boltzmann Machines (RBMs) and Autoencoders are various algorithms on DL.

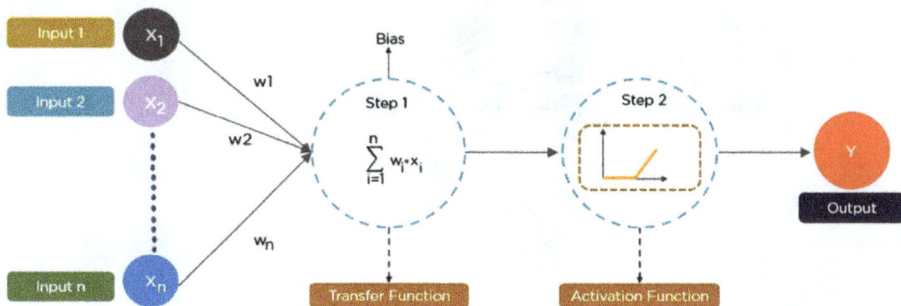

FIGURE 8.11   Deep learning architecture.

8.3.6.1 Basic Concepts of Neural Networks

A neural network is designed to optimize a function of other nested functions and is usually composed of inputs, weights and a regularization term. The neural networks are solved using stochastic gradient descent (SGD) and backpropagation algorithms. Mathematically, in its simplest form, neural networks can be represented as:

$$\text{argmin } A_j\left(f_M\left(A_M, \cdots, f_2\left(A_2, f_1(A_1, x)\right)\cdots\right) + \lambda g\left(A_j\right)\right)$$

The matrix A_k describes the weight given to the connection from the kth layer to $(k+1)^{th}$ layer. As it is an underdetermined system, a regularization term $g(A_j)$ is used.

In its simplest form, a neural network can be constructed with only one layer consisting of only one output with a linear function $AX = Y$ (A denotes the weights, X denotes the inputs and Y the outputs). A single layer network is shown in Figure 8.12.

More complex forms, according to the neural network application, may be more appropriate. Non-linear functions $y = f(A,x)$ are usually used in contemporary neural networks in place of linear functions where $f(.)$ is the activation function.

Here non-linear functions can take the form:

$$x^{(1)} = f_1\left(A_1, x\right)$$

$$x^{(2)} = f_2\left(A_2, x^{(1)}\right)$$

$$y = f_3\left(A_3, x^{(2)}\right)$$

A multi-layer neural network is shown in Figure 8.13 and common activation functions are shown in Table 8.1.

Two algorithms are commonly used for enabling the nature of neural networks: SGD and backpropagation. Backprop, as it is commonly called, uses the nested feature of neural networks to create an optimization problem to find the values of the weights. Backprop uses the chain rule for differentiation. While backprop allows for better computation, SGD enables a quicker evaluation of best weights that can be achieved.

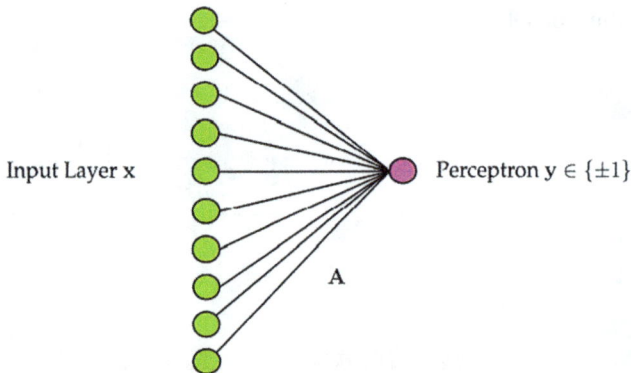

FIGURE 8.12　Illustration of a single layer network.

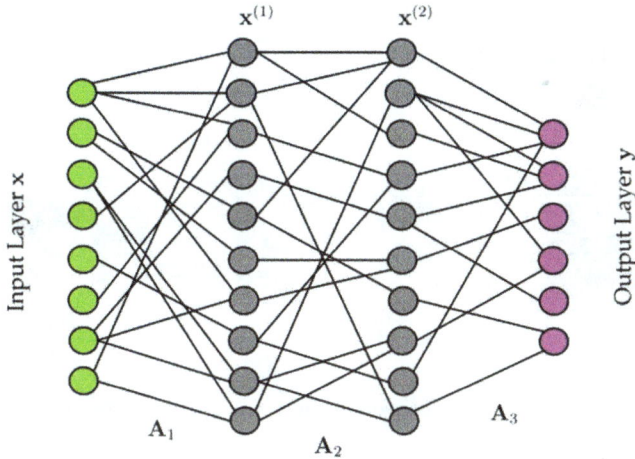

FIGURE 8.13   A multi-layer neural network.

TABLE 8.1

Common Activation Functions

$f(x) = x$	Linear
$f(x) = \{ \ 0 \quad x<0 \ 1 \ x>0$	Binary step
$f(x) = \{ \ \dfrac{1}{1+ \exp(-x)}$	Logistic (soft step)
$f(x) = \tanh(x)$	TanH
$f(x) = \{ \ 0 \quad x \leq 0 \ x \quad x> 0$	Rectified linear unit (ReLU)

Deep convolutional neural nets (DCNNs) are the basis of DL methods. They find application in time series data which is typical of renewal energy sources. Even though it can be mentioned that the construction of these networks requires a structure or a method, in actuality, it is more of an art and it requires instinct which is developed by experience. Design questions commonly asked regarding neural networks include the number of layers to be used, the dimension of the layers, the design of the output layer and the type of connections and mapping between the layers. Some of the important elements of the DCNNs include convolution layers, fully connected layers, pooling layers and dropout.

In convolution layers, only a small part of the full high-dimensional input space is removed and used for feature engineering. Figure 8.14 shows deep convolutional neural nets architecture in which the windows (the darkly shaded boxes) that move across the entire layer (lightly shaded boxes). Each convolution window converts the data into a new node through a given activation function, as shown in Figure 8.14 [15]. Feature spaces are then built from the smaller patches of the data.

In a DCNN architecture, pooling layers are frequently inserted between successive convolutional layers. Pooling layers are used in networks to reduce the number of parameters and computations. Pooling layers are used to: (1) manage overfitting

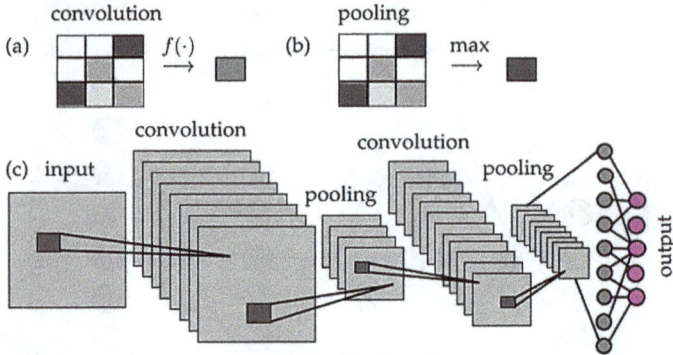

FIGURE 8.14 A typical deep convolutional neural nets architecture (a) convolution, (b) pooling, and (c) input.

and (2) improve memory management. Pooling layers act on every depth slice of the input and resize them. Using the maximum value for all the nodes in its convolutional window is called max pooling.

Fully connected layers are inserted into the DCNN so that different regions can be connected. The pooling and convolutional layers are local connections only, while the fully connected layer restores global connectivity. This is another commonly used layer in the DCNN architecture, providing a potentially important feature space to improve performance.

Overfitting is an important problem in DCNNs. Overfitting is the key reason of DCNN's failure in demonstrating good generalizability properties. Larger DCNNs are typically slow and tricky to deal with overfitting by combining the predictions of many different neural networks for online implementation. Hence, dropout helps in addressing this issue. During training, the trick is to randomly remove nodes from the network (along with their connections). This prevents units from co-adapting excessively. Dropout samples form an exponential number of different "thinned" networks during training. It is simple to approximate the effect of averaging the predictions of all these thinned networks at test time by simply using a single unthinned network with smaller weights. This significantly reduces overfitting and has been shown to outperform other regularization methods.

8.4 IMPLEMENTATION OF MACHINE LEARNING IN RENEWABLE ENERGY SOURCES

The main objectives of ML and DL methods in renewable energy sources are [16]:

- To forecast the output generation of renewable energy
- To determine the geographic location, configuration and sizing of renewable energy systems
- To oversee the overall operation of the renewable energy-integrated smart grid
- Forecasting the energy demand
- Developing renewable energy materials

TABLE 8.2

Pre-Processing Techniques Used in Machine Learning

Technique	Description
Pre-processing	Removing data with missing values which come from the abnormal data collection.
Splitting	Divide the original data into three datasets: training, validation and testing.
Decomposition	The high-dimensional dataset is divided into several low-dimensional sub-datasets.
Discretization	Converting continuous data into discrete data.
Selection	Identifying appropriate independent variables and removing irrelevant attributes.
Imputation	Replacing the missing value with substituted values if null data presents.
Normalization	Modification of datasets expressed in different scales to allow machine-learning algorithms to handle the various datasets.
Standardization	Converting data from various sizes to the same size.
Transformation	A method to convert the state of the data.

8.4.1 DATA PRE-PROCESSING

ML is implemented in following steps: (1) data collection, (2) pre-processing, (3) feature selection, (4) extraction, (5) model choice and (6) model validation [17]. The modelling process in renewable energy systems can be divided into various stages: (1) data pre-processing, (2) determining appropriate hyper-parameters model, (3) training model, (4) testing and (5) forecasting issues [18]. Table 8.2 shows the pre-processing techniques and their description.

8.4.2 DATA REQUIRED FOR MACHINE LEARNING

The input variables have a strong influence in the performances of ML models. Most of the models contain more than two parameters. Some of the commonly used input parameters for the forecasting of renewable energy are given in Table 8.3.

8.4.3 FEATURE SELECTION IN MACHINE LEARNING

The Feature Selection Process (FSP) is a crucial task in the ML-based forecasting since irrelevant features can unnecessarily increase cost, time and error in the results. The FSP for a learning problem from data can be defined as follows: given a set of labelled data samples $(X_1, y_1), ..., (X_1, y_1)$ where $X_i \in R^n$ and $y_i \in R$ (or $y_i \in \{\pm 1\}$ in the case of classification problems) choose a subset of m features ($m < n$) that achieves the lowest error in the prediction of the variable y_i. Several algorithms are available to solve an FSP. In general, they are divided into two different groups [27].

Wrapper algorithm [28] searches for a good subset of features by including the classifier/regressor in the evaluating function, which runs on the training dataset with different subsets of features. In an independent but representative test set,

TABLE 8.3

Input Parameters Required for Various Types of Renewable Energy Systems

S. No.	Energy Source	Input Parameters
1	Solar PV	Sunshine duration, maximum and minimum temperatures, long-term radiation, water vapour pressure, relative humidity, extra-terrestrial GSR on a horizontal surface, wind speed, evaporation, precipitation, clearness index, global solar radiation, upper-level cloudiness, mid-level cloudiness, low-level cloudiness, atmospheric transmissivity data, sky cover, dew point, visibility and a descriptive weather summary [19].
2	Solar Thermal	Weather conditions from recorded data, actual weather forecasts from meteorological services, solar flux, inlet and outlet temperature and flow rate of heat transfer fluid. Solar radiation, wind speed and direction [20].
3	Wind	Meteorological and geographical data such as air temperature, pressure, relative humidity, longitude and latitude to determine wind direction and speed, and Global Horizontal Irradiance are used (GHI) [21]. The National Renewable Energy Laboratory (NREL) in the United States has a large wind dataset that is open to the public [22].
4	Hydro	Hydroelectric Energy (MWH), River Inflow at month, precipitation at month [23], daily water level and rainfall.
5	Geothermal	Drilling parameters, hydraulics, drilling fluid properties, location, weather, water availability, rig conditions, depth, formation properties, hole problems, bottom hole temperature, and crew efficiency, pore pressure gradient, equivalent circulation mud density at the bottom hole, flow rate, mud viscosity, weight-on-bit-to-diameter ratio, bit nozzle diameter and topographic and surface layer/soil data [24].
6	Biomass	Biomass inlet temperature, air inlet temperature, syngas exiting gasifier temperature, environmental pressure, gasifier pressure, moisture content, molar composition of environmental air, oxygen, oxygen-enriched air, air-steam mixture [25] methane co-feeding, time and heating value, fuel flow rate, equivalence ratio, and temperature distribution, ultimate and proximate values.
7	Tidal	Tidal current speed and direction data, wave height and mean peak period [26].

the final feature set will be the one with the lowest estimated error. An external measure calculated from the data is defined in the filter approach for selecting a subset of features. The best feature subset on the data algorithm is evaluated using the classifier.

8.4.4 NECESSITY FOR PREDICTION

8.4.4.1 Solar Energy

The integration of renewable energy into the energy supply structure is one of the most critical tasks in the future global energy supply. Voltage fluctuations, local power quality and stability issues are caused by the unpredictability of solar energy. As a result, predicting the energy output of solar systems is necessary for the effective operation and management of the power grid. It is also necessary to estimate reserves in order to properly schedule the power system and trade the produced power. The increase in solar power generation necessitates a more accurate prediction [29].

In the prediction of solar energy, solar irradiance is considered as a time series having different time scales. ARMA method is commonly used in the time series forecasting analysis of solar energy [30]. Data-driven forecasting models that use machine-learning and deep-learning models, such as Support-Vector Machines (SVM) and Artificial Neural Networks (ANN) [31], are widely used. CNN, Deep Neural Network (DNN) [32], Long Short-Term Memory (LSTM) and other hybrid models are also used. Figure 8.15 shows block diagram of solar forecasting using ANN [33].

Multiscale Decomposition Techniques (MDT), Empirical Mode Decomposition (EMD), Integrated Empirical Mode Decomposition (EEMD) and Wavelet Decomposition (WD) can be used to predict solar radiation [34]. The parameter selection affects the forecasting accuracy and therefore, to improve the accuracy, optimization methods – like grid search, Genetic Algorithm (GA), Firefly Algorithm (FFA) and Particle Swarm Optimization (PSO) – can be used [35].

For example, Zendehboudi et al. [36] used Least Square Support Vector Machine (LSSVM) with input parameters historical data of atmospheric transmissivity in a novel two-dimensional form, sky cover, relative humidity and wind speed collected between 1991 and 2005 at 1454 locations in the United States to predict short-term solar energy. The results are shown in Figure 8.16. The results showed that the SVM model is more robust than the Radial Basis Function Neural Network (RBFNN) and AR models for estimating short-term solar power. The commonly used model to predict solar energy is given in Appendix A.

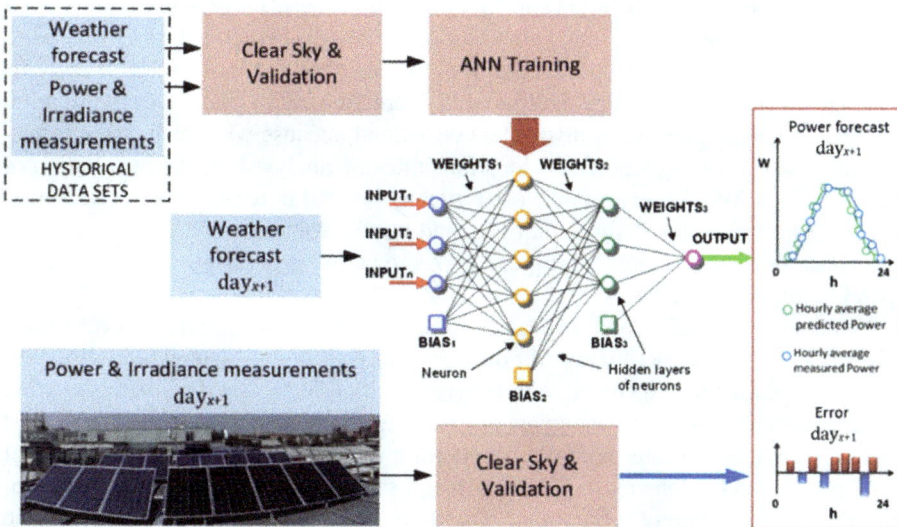

FIGURE 8.15   Block diagram of solar forecasting using artificial neural network.

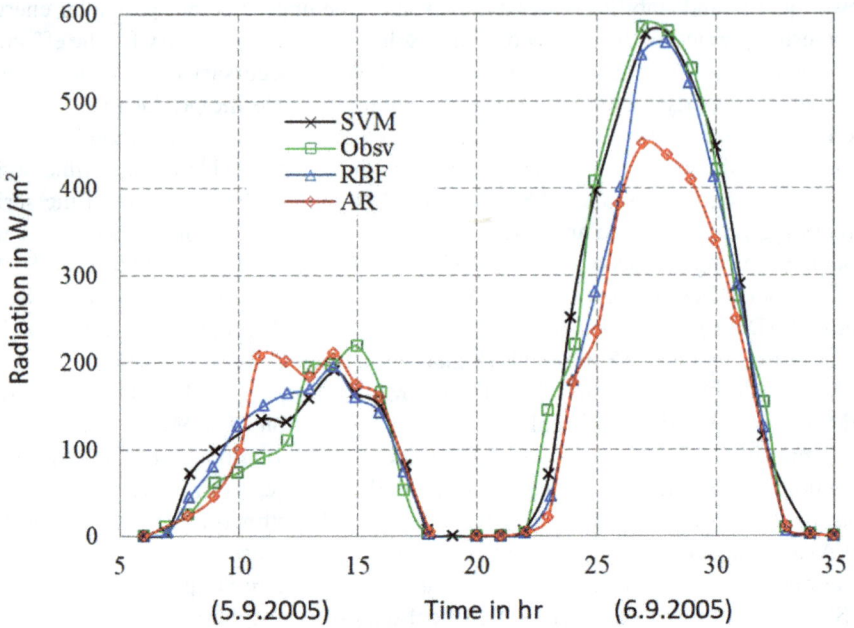

FIGURE 8.16 One hour ahead solar radiation predicted values using machine learning models.

8.4.4.2 Wind Energy

When wind energy penetration into grids increases, it is necessary to forecast the amount of energy to be produced on short-to-medium timescales (one hour to seven days). Since the level of installed capacity of wind power plant grows, so forecasting the wind energy production has gained its importance. Owing to the fact that wind energy has the potential to produce power for each hour, it is suitable for systems which need continuous energy. Despite this, there are several challenges to use wind energy, such as higher initial investment costs, and because wind turbines are massive in size, wind energy potential locations must be analysed for the installation of wind turbines. Wind energy plants are typically located in remote areas, necessitating the installation of transmission lines, and wind turbines can harm local wildlife and cause noise pollution [37]. Figure 8.17 shows the general procedure for wind speed prediction [38].

Wind speed and direction forecasts are the starting point for wind power uncertainty, which affects the quality of power system operations, distribution, dispatching and peak load management. Wind energy prediction depicts the amount of wind power that can be expected at a given point in time in the future. It is one of the most crucial stages in the integration and operation of wind power. The forecasting of wind power can be done on long, medium and short term. Accordingly, energy produced from wind cannot be supplied directly to the electricity demand as it is done in conventional plants. The penetration level of wind energy leads to new challenges

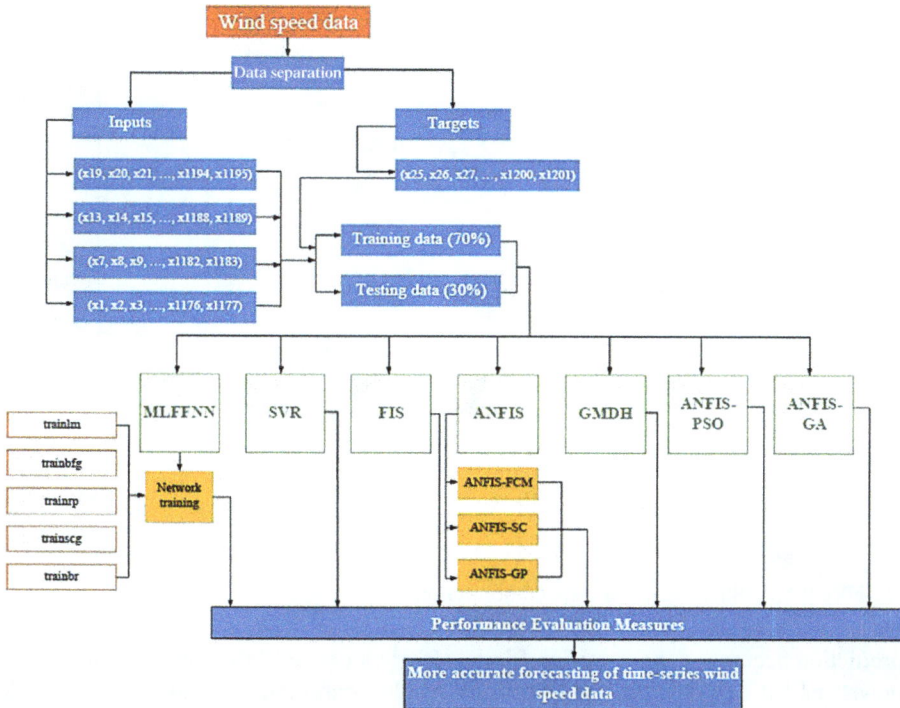

FIGURE 8.17 Wind speed forecasting procedure.

in the power system. Hence, for the successful integration of substantial amount of power produced by wind into the grid, accurate forecasting of the wind power is essential [39]. Two types of wind power forecasting techniques are [40]:

- Time series or univariate: Past data are employed to predict future generation. Data are decomposed into multidimensional form.
- Multivariate: Multiple exogenous inputs are used to predict wind power generation.

Statistical methods were used in the early stages to predict wind energy [41]. SVM, RF classification algorithms, Gradient Boosting Decision Trees (GBDT), Adaptive Neuro-Fuzzy Inference System (ANFIS), ANN [42] and LSTM networks have all been used in recent studies. The ML techniques can capture data trends while forecasting the wind energy. Furthermore, hybrid algorithms that incorporate data processing approaches and optimization algorithms into ML models have been developed to improve forecasting models effectively and efficiently.

Other methods, such as extreme-learning machines (ELM), wavelet decomposition (WD), wavelet packet decomposition (WPD) and ensemble empirical mode decomposition (EEMD), are used to remove noise from data and effectively improve

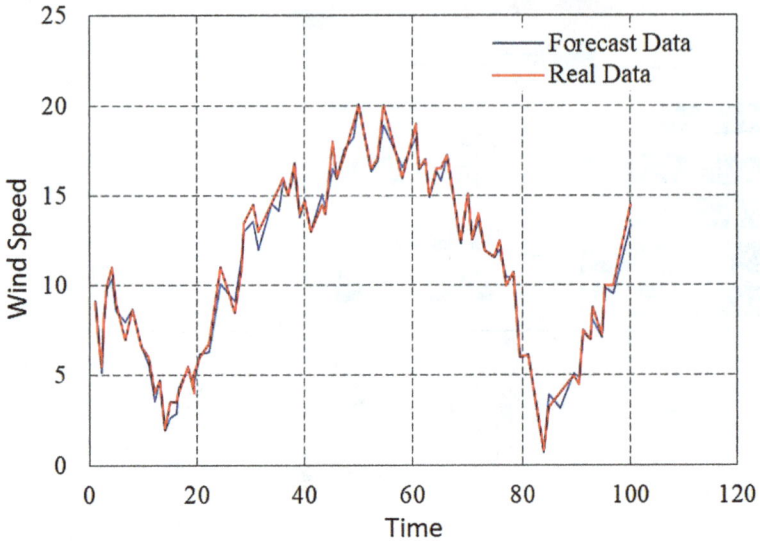

FIGURE 8.18 Short-term wind speed prediction.

prediction accuracy. The Bayesian Model (BM) can be used to forecast hybrid wind power, and it can provide more accurate results than other forecasting models. A hybrid model incorporating WPD, CNN and LSTM can be used to predict wind speed [43].

For example, a prediction of short-term wind speed conducted using the best WPD method and SVM using 1000 points collected over a one-month period [44], of which 900 samples were used for training and the remaining data were used to test the model's accuracy. The values predicted by this model, as shown in Figure 8.18, have shown a good agreement with experimental data. Some of the models commonly used in wind energy forecasting are given in Appendix B.

8.4.4.3 Hydropower

Hydropower is among the cleanest sources of renewable energy. The rate of hydropower generation depends on the dam height and the inflow to the dam. The amount of energy that will be generated in the coming months is necessary to manage the electricity distribution and operation of the dam. The hydropower generation forecasting is a key component in reservoir operation [23]. The following ML models can be used for the prediction of hydropower, ANFIS and Grey Wolf Optimization (GWO) [23]. To predict the water levels, Bayesian linear regression, neural network regression, boosted decision-tree regression and decision forest regression are used. For the inflow predictions with various lead-times intervals, LMS-BSDP can be used [45].

The influence of the seasonal inflow forecast for hydroelectric generation study was conducted by Zhang et al. [46] using long, medium and short-term Bayesian Stochastic Dynamic Programming inflow forecasts (LMS-BSDP) and Two Stage Bayesian Stochastic Dynamic Programming (TS-BSDP). The results are shown in Figure 8.19.

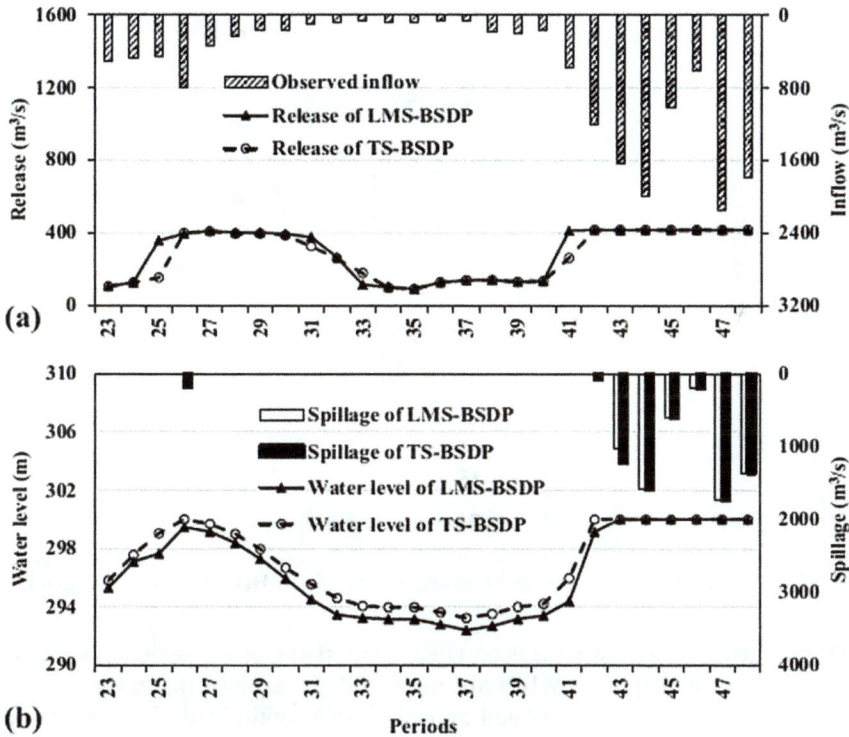

FIGURE 8.19   (a and b) Seasonal inflow forecast for hydroelectric generation.

Figure 8.19a shows the inflows and releases for hydroelectric generation, and Figure 8.19b shows water levels and spillage. It may be noted that there is a difference in the release at period 25 between LMS-BSDP and TS-BSDP. The LMS-BSDP model increases release to reduce storage due to the high seasonal inflow forecasted at period 26. Some other models used in hydropower prediction are given in Appendix C.

Energy from biomass has been used by mankind for centuries. Today, the biomass industry has huge market potential. The development of new technologies like carbonization, liquefaction, pyrolysis and gasification has augmented to use the wood. The higher heating value (HHV) is an important parameter to determine biomass economy and to design the biomass conversion technologies. Determining the heating value by bomb calorimeter is a time-consuming and costlier process. Therefore, to overcome this issue, several mathematical models were developed to predict HHV. These models have shown the reliance of HHV with the other theoretical and experimental analysis [47]. Biomass and hydrogen can increase energy sustainability and reduce greenhouse gas emissions [48].

8.4.4.4   Biomass Energy

To model gasification products, ML models like linear regression (LR), KNN, SVMR and DTR can be used [49]. To forecast HHV of biomass, gradient-boosted regression trees (GBRT) can be used. To predict CO, CO_2, CH_4, H_2 during gasification and

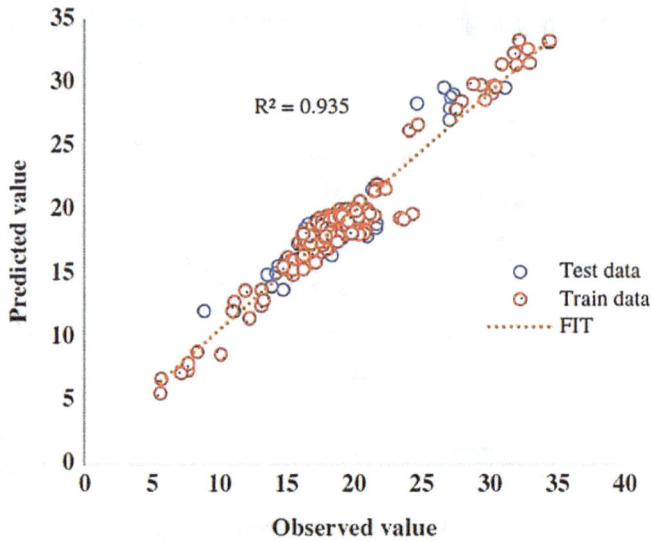

FIGURE 8.20   Comparison between observed and predicted HHV for GBRT-5 model.

HHV outputs, polynomial regression (PR), support-vector regression (SVR), DTR and multi-layer perceptron (MLP) are used [50]. To analyse the baking process, combination of SVM and simulated-annealing (SA) optimization technology can be used. To forecast the HHV in a roasting process, PSO-SVM and PSO is developed [51].

The prediction of HHV of biomass materials based on proximate analysis using GBRT method was proposed by Samadi et al. [47] is shown in Figure 8.20. It can be seen that the GBRT model can predict HHV with high precision. Some other models used in biomass energy prediction is given in Appendix C.

8.4.4.5 Marine Energy

Tidal waves are produced by changes in the gravitational forces of the sun and the moon. Prediction of tidal currents are necessary for navigation, protection from flooding, coastal management and energy extraction [52]. The accurate prediction of wave features plays a vital role in various ocean engineering activities. Marine energy is one of the most promising sources of renewable energy in several offshore islands. The accurate estimation of the wave energy flux is essential to characterize the wave energy production. Therefore, it is necessary to identify the regions to implement wave energy farms in a faster and more inexpensive manner [26]. To get the highest efficiency from tidal power, accurate modelling and prediction is important [53]. Prediction of tidal currents output power is based on the prediction of tidal currents speed and direction. The accuracy of models used for tidal currents forecasting is very important as the integration of the tidal currents power into the grid depends on its forecasting accuracy [54].

In wave-power-generation, the prediction algorithms are employed to predict the height of waves in a short period of time. Ensemble EMD-ELM model can be

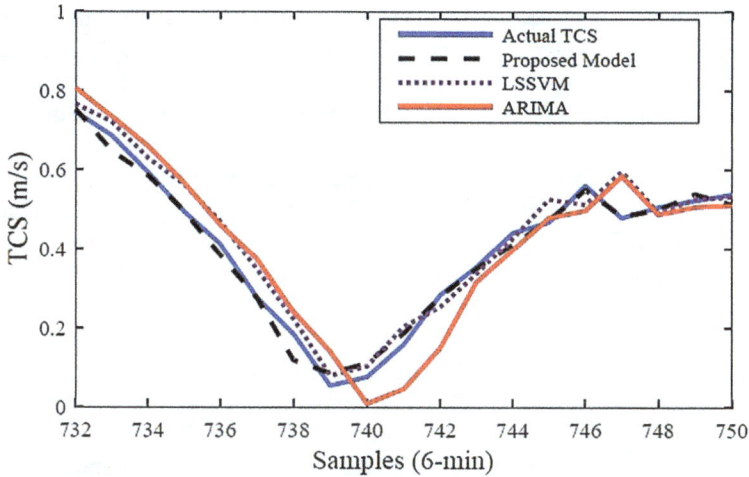

FIGURE 8.21   Comparison of different prediction models for TCS.

used to predict the wave heights [55]. Similarly, Bayesian Optimization with hybrid Grouping Genetic Algorithms and an Extreme-Learning Machine (BO-GGA-ELM) model can be used to forecast the wave height and the wave energy flux [56]. In predictions of tidal-power generation to forecast the tide velocity and directions with high accuracy, a univariate prediction method based on Wavelet Transform (WT) and SVR can be used. To obtain harmonic power flows, a power flow prediction method based on clustering technology has been developed. These models have contained Wavelet and Artificial Neural Networks (WNN and ANN) and Fourier Series based on Least Square Method (FSLSM) [57].

To predict wave energy faster, Fuzzy inference systems (FIS) and ANN can be used. These forecasting models can predict wave power efficiently in deep oceans [58]. Gaussian process (GP) can be used to forecast short-term waves, which is essential for the wave energy converters [59]. A study was carried out using decomposition-based LSSVM to forecast the short-term tidal current and direction by Safari et al. [60]. Figure 8.21 shows a result obtained from different prediction models.

The prediction results obtained from ARIMA-based and LSSVM-based models show deviation from the actual TCS. Whereas the decomposition-based LSSVM accurately tracked the changes in the actual speed. The other models used in marine energy prediction are given in Appendix C.

8.4.4.6   Geothermal Energy

The amount of data in the geothermal energy industry is incredibly large, and with the growing importance of geothermal as an energy source with lesser fluctuations, it is necessary to develop models to predict the geothermal energy and heat-flow forecasting. ML can be used to accurately spot geothermal identifiers from land maps and detect sites fit for geothermal energy production. ML can solve a variety of complex problems such as site drilling locations in the industry in a fraction of second.

Determining thermal breakthrough is important to manage a sustainable geothermal reservoir which involves characterizing the geothermal reservoir, fractures, flow paths and constructing a geological model. Since fractures control the mass and heat flow, hence, understanding the layout of fractures help to establish optimal injection strategies in the geothermal reservoirs. Present mathematical models can negatively be impacted by the unknowns like unexpected gases or indistinct reservoir boundaries. This conventional way of prediction is time-consuming and computationally expensive. ML models can be used to accurately understand the current and evolving nature of each geothermal reservoir to support the geothermal energy industry's efficiency and mitigation of risk [61].

For geothermal energy predictions, LSTM encoder–decoder model is employed [62]. Multiple regression and ANN can be used to predict drilling parameters and rates of penetration (ROP) [63]. Random forest algorithms and the data collected from geographic information systems can be used to predict very shallow geothermal potential [24]. In the study [64] of geothermal heat exchanger energy prediction based on time series and monitoring sensors optimization using non-linear autoregressive (NAR), non-linear autoregressive with external input (NARX) and Ridge model and regression values obtained for several of the models, calculated for a time horizon of one hour is shown in Figure 8.22. It shows values obtained from each model along with the real value to predict. The other models used in geothermal energy prediction are given in Appendix C.

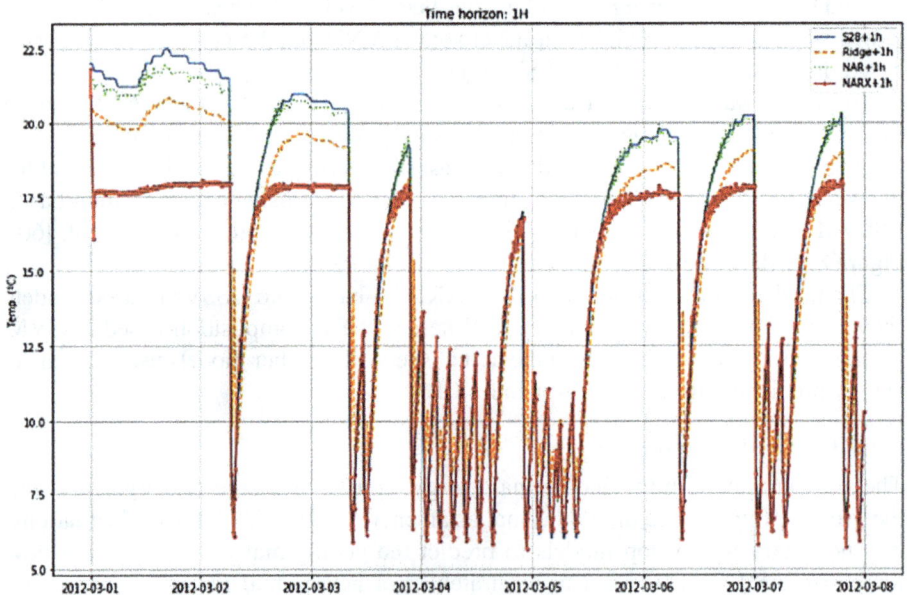

FIGURE 8.22 Geothermal heat exchanger energy prediction using machine learning models.

8.4.5 POWER QUALITY DISTURBANCES

The huge investment in power distribution systems and the increased integration of renewable energy resources are influencing the power quality, grid reliability, stability and safety issues. Intermittent variations in voltages, voltage sags, voltage harmonics, short-period shortage in voltages and temporary incidents can introduce disturbances in voltages, affect the performance of electrical power networks and sustainable energy supply. The source of PQDs are electromagnetic transients caused by lightning strokes, switching actions, self-clearing faults and switching of end-user equipment.

The identification and interpretation of such random and dynamic events that affect power quality are essential tasks to address and mitigate PQDs. Generally, disturbances in PQ are associated with the integration of generators which are operated by renewable energy and any non-linear features caused from the connected load. In particular, DC microgrids have large number of renewable energy sources, energy efficient loads and energy storage systems. Hence the hybrid power system which are incorporating renewable energy sources into the utility networks will experience challenges in the power quality. Therefore, PQD detection is a crucial task to achieve a resilient and sustainable power [65].

International standards are available to provide guidelines to the manufacturers and PQ monitoring community. the Institute of Electrical and Electronics Engineers (IEEE), International Electrotechnical Commission (IEC) and European Committee for Electrotechnical Standardization (CENELEC) are some of the globally accepted standards. IEC 61000 and EN 50160 are the generally applicable standards for PQ disturbances [66].

The PQDs' identification using ML can be done in three steps: signal analysis, feature selection and classification. The steps involved in PQ monitoring with renewable energy sources are shown in Figure 8.22. PQ disturbances can be identified from the deviations in frequency, variations voltage, transient, the occurrence of flickers and harmonics. In PQD's analysis, ML techniques are used to extract the feature eigenvectors that enable disturbance identification. Figure 8.23 shows various stages involved in analysing power quality.

They are as follows: short-time Fourier transform (STFT), Stockwell transform (ST), WT, Hilbert-Huang transform, Kalman filter, strong trace filter (STF), sparse signal decomposition (SSD), Gabor–Wigner transform and empirical mode decomposition (EMD).

The techniques used in the selection of features are the principal component analysis, K-means-based a priori algorithm, classification and regression tree algorithm, multi-label extreme learning machine, random forest model, sequential forward selection and bionic algorithms. The bionic algorithms are genetic algorithms (GA) and swarm-based approaches like ant colonies and particle swarm optimizers (PSO) and a combination of PSO and support vector machine (PSO-SVM).

The techniques used for classifying PQDs are ANN, K-NN algorithm, SVM, DT methods, multiclass SVM (M-SVM), directed acyclic graph SVMs (DAG-SVMs) and radial basis function kernel SVM (RBF-SVM). PQD's classifiers based on DT include fuzzy decision tree and regression tree algorithm (CART) [67]. In addition,

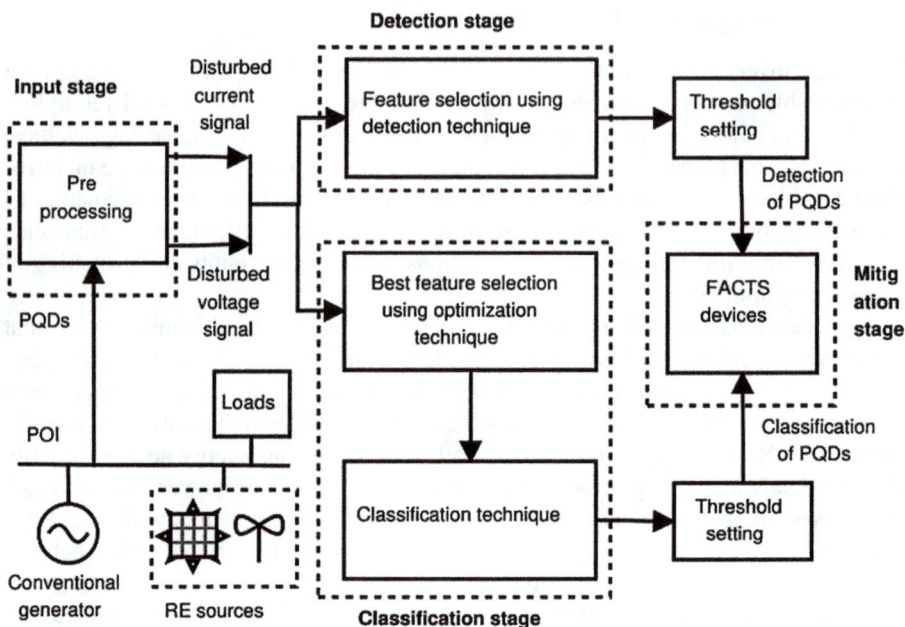

FIGURE 8.23   Various stages involved in analysing power quality.

some new signal processing-based techniques play vital role in PQDs detection. Similarly, AI-based classification techniques can also be to categorize PQD signals based on their features.

8.4.6   FAULT DIAGNOSIS IN RENEWABLE ENERGY

Fault diagnosis serves an important role in pursuing the relationship between the monitoring data and the condition of machines, which is a concerned issue in machine condition management. Intelligent Fault Diagnosis (IFD) refers to applications of ML techniques to machine fault diagnosis of the machines which are used in renewable energy systems. The ML models can adaptively learn the diagnosis knowledge of machines from the captured or collected data rather than from the experience and knowledge of engineering personnel.

The main objective of applying ML is to construct diagnosis models which can automatically connect the relationship between the data and condition of the machines. By employing AI technique, the procedure of fault diagnosis is expected to be smart enough to detect and recognize the health states of machines. Traditionally, the method of fault diagnosis is developed manually by inspecting condition of machines which increases the labour intensity and reduces the diagnosis accuracy. The advanced signal processing methods enable to identify the location where the faults happened in machines. However, these methods greatly rely on specialized knowledge, hence modern industrial applications prefer the fault diagnosis methods which can automatically recognize the health states of machines.

Data Collection	Artificial Feature Extraction		Health State Recognition
Employ sensors to collect data such/ as • Vibration • Acoustic emission • Instantaneous Speed • Current • ...	Extract some commonly used features from the collected data by using • Time-domain analysis • Frequency-domain analysis • Time-frequency-domain analysis	Select sensitive features from the extracted by • Filters such as Relief, mRMR, and DE • Wrapper such as LVM • Embedded methods such as L1 and L2 regularization	Input sensitive features to traditional machine learning based models such as • Expert system • ANN • SVM • ...

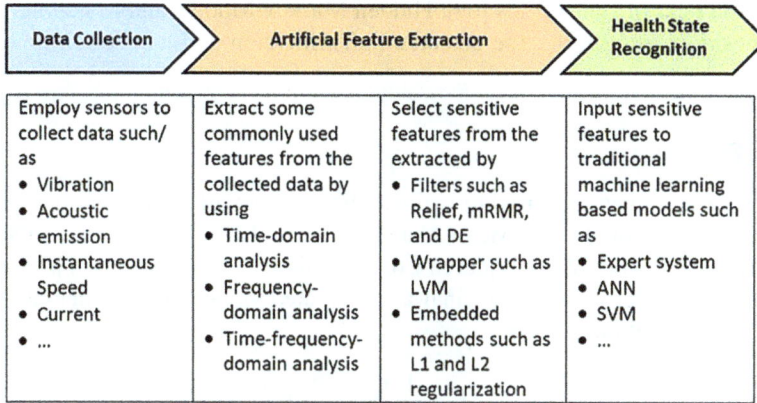

FIGURE 8.24 Fault diagnosis method in renewable energy systems using machine learning.

Some ML theories used for fault diagnosis are ANN, SVM and DNN. The procedure includes three steps: data collection, artificial feature extraction and health state recognition, which are shown in Figure 8.24.

In data collection, the sensors are mounted on machines to constantly collect data. Various sensors are used to capture data, such as vibration, acoustic emission, temperature and current transformer. In addition, the data from multi-source sensors can provide additional information, which can be used to achieve higher diagnosis accuracy.

Artificial feature extraction has two steps. First, time-domain features, frequency-domain features and time–frequency-domain features are extracted from the collected data. These features contain the health information reflecting the health states of machines. Second, feature selection methods, such as filters, wrappers and embedded methods, are used to select sensitive features to health states of machines from the extracted features.

Health state recognition uses ML-based diagnosis models to establish the relationship between the selected features and the condition of machines. The diagnosis models are first trained with labelled samples. Then, the models can recognize the condition of machines when the input samples are unlabelled [68].

Wind turbines are subject to faults which affect both their performance and their security. Gearbox and bearing failure and various sensor faults often occur, such as sensor bias fault, sensor constant gains and others. Thus, designing a reliable automated diagnosis system is essential to achieve fault detection and isolation and to reduce the maintenance cost. The ML model is defined based on the functioning of the wind turbine.

The variables considered for diagnoses in the models are wind speed, air density and the shading effect, ambient temperature and wind direction. Several classifiers, such as the Cluster Centre Fuzzy Logic, ANNs, k-NN and ANFIS, are used for diagnostic purposes in wind farms. SVM is the popular choice for the fault diagnosis issues in renewable energy systems. Besides fault detection in wind turbines, there are other RE-related applications where classification algorithms are being used. In the field of smart grids [69], Hidden Markov Models and Matching Pursuit Decomposition are used for the detection, identification and location of power system faults.

8.4.7 PERFORMANCE METRICS

In ML, several metrics are used to validate the accuracy of the prediction. They are Mean Absolute Error (MAE), Mean Absolute Percent Error (MAPE), Mean Squared Error (MSE) and Root Mean Squared Logarithmic Error (RMSLE) [70]. The MAE is the difference between the original and expected values and is calculated using the following equation:

$$MAE = \frac{1}{N} \sum (y_a - y_p) \qquad (8.1)$$

where y_a is the actual value, y_p is the predicted value and N is the number of samples.
The MSE is the square of MAE error and is calculated by the following equation:

$$MSE = \frac{1}{N} \sum (y_a - y_p)^2 \qquad (8.2)$$

The RMSLE is the logarithmic relationship between the actual data and the predicted value, and is obtained by the following equation:

$$RMSLE = \sqrt{\frac{1}{N} \sum \left(\log(y_a + 1) - \log(y_p + 1)\right)^2} \qquad (8.3)$$

The MAPE is a measure of the accuracy of a prediction, and is obtained by the following equation:

$$MAPE = \frac{1}{N} \sum \left(\frac{y_a - y_p}{y_a}\right) \times 100 \qquad (8.4)$$

8.4.8 CHALLENGES IN IMPLEMENTATION OF VARIOUS MODELS

The ML applications are growing fast owing to their ability to handle large data and high-computational power. Several research works are being carried out for applying machine to renewable energy forecasting applications. However, ML-based forecasting models have some challenges. It requires resources, skills and knowledge beyond data science to integrate ML algorithms into applications. The key challenges that are present when applying the ML are discussed in the following sections:

8.4.8.1 Problem Definition

In ML, defining the problem plays a major role in selection of techniques, data sources and resource personnel. This also determines the kind of model, data, talent and investment required.

8.4.8.1.1 Obtaining Data

One of the key challenges of machine learning is collecting and organizing the data which are required to train the ML models. The data may be fragmented and scattered across various databases, servers and networks. When collecting the data from various sources, the problems like inconsistency between database schemes, mismatching conventions, missing data, outdated data are unavoidable. Hence one of the main tasks in the ML strategy when the data is obtained from different databases is cleaning, verifying the quality and provenance and consolidating them into a dataset that can support the training and testing of the ML models.

8.4.8.2 Maintenance

The ML models are prediction machines which find patterns in data and forecast future outcomes from current observations. As the world around us changes, the data patterns and models trained will also decay. ML models have to be trained continually otherwise they will become less accurate over the time. To provide a meaningful output, ML requires constant maintenance, management and course-correction. Therefore, ensuring the infrastructure and processes to obtain a continuous stream of new data and updating models at regular interval are the key part of successful ML strategy.

8.4.8.2.1 Expert Team

The ML needs a cross-functional team which includes experts from different disciplines and backgrounds. ML needs technical support beyond data science skills. For example, technical experts are needed to verify the veracity of training data and the reliability of the model's inferences. Product managers are required to establish the objectives and desired outcomes from the ML strategy. User researchers can validate the model's performance through interviews and feedback from end-users. An ethics team is required to identify sensitive areas where the ML models might create harm. Software engineers will integrate the models into other software being used by the organization. Data engineers set up the data infrastructure and plumbing to feed the models during training and maintenance, and the IT team provides the computing, network and storage resources needed to train and serve the ML model [71].

Similarly, DL-based forecasting models also have the following challenges.

8.4.8.2.1.1 Theoretical Concern
The theoretical problems of DL are mainly due to statistics and calculation capability. To apply DL to any non-linear function, a shallow network and a deep network can be used to represent the non-linear function. Obviously, DL can represent non-linear function far better than its counterpart. However, this does not mean that DL is better for learning the non-linear function. In renewable energy forecasting, the following points should be considered. (1) Complexity of predicting samples. (2) Number of training samples required to learn the DL network. (3) Computing resources required for training the prediction samples.

8.4.8.2.1.2 Modelling the Deep Learning
In the case of a large amount of data, a complex model is required to extract the informative features. Since DL models are

more powerful, the features and other information extracted are more valuable. The DL models have more layers of hidden neurons, which enable the DL to learn renewable energy time series data easier. However, designing a hierarchical DL model with powerful feature learning and establishing appropriate prediction model for a certain forecasting dataset are some of the key issues in the application of DL. Overcoming the above-mentioned challenges can improve the accuracy of the DL prediction model [10].

8.5 CONCLUSION AND FUTURE SCOPE

This chapter gives a brief introduction to the basic knowledge about the application of ML in the renewable energy systems. It discussed the ML forecasting methods, data pre-processing, parameters required for the effective application of ML in renewable energy systems, the applications of various algorithms in renewable energy sources and various metrics to test the forecasting methods. In the present scenario, renewable energy source is integrated with conventional energy to empower electric grids, which introduces new challenges due to its interference and volatility.

Prediction of energy using ML plays a crucial role in solving these challenges. While predicting energy consumption, it is necessary to select an appropriate prediction method which will produce results and characteristics of the chosen ML model. Moreover, individual forecasting models have some limitations, hence combination of prediction models and hybrid methods based on ML can be used to overcome the limitations.

With the current developments in ML in the energy sector, it is safe to assume that ML utilization in the renewable energy sector will increase in the future. The growing customer base will help to modernize the overall renewable energy structure. The costs of infrastructure and distribution lines are prohibitively expensive. Furthermore, stringent regulations that allow monopolies to develop in the energy sector are a problem. To address these issues, the various energy sector organizations are focusing their efforts on smart grid technology. Many other countries, in addition to the United States, intend to invest in and develop smart grids with ML acting as the brain.

These smart grids will assist energy industries in allocating their energy resources in the best possible way by gathering and synthesizing massive data with the help of ML and Big Data. The dataset will then be analysed by ML to identify patterns and anomalies. As a result, they will revolutionize the energy sector's demand and supply. This advanced grid system will rely heavily on ML. Because the number of control points in grids will continue to grow rapidly from millions to billions, ML will be the only efficient way out.

As a result, smart grids come into play to meet these requirements. Smart grids are power grids that use IoT and AI to create a digital power grid that allows for two-way communication between consumers and utility companies. Smart grids are outfitted with smart metres, sensors and alerting devices that collect and display data to consumers in order for them to improve their energy consumption habits. It can also be fed into ML algorithms to predict demand, improve performance, lower costs and avoid system failures.

REFERENCES

1. Bahman Zohuri. Nuclear Fuel Cycle and Decommissioning, 2020. Albuquerque, NM: Electrical and Computer Engineering Department, University of New Mexico.
2. Ahmad Rafiee, Kaveh Rajab Khalilpour. Renewable Hybridization of Oil and Gas Supply Chains. London UK: Elsevier Inc., 2019.
3. https://en.wikipedia.org/wiki/Renewable energy (accessed on 04-11-2021).
4. http://www.greenrhinoenergy.com/solar/radiation/empiricalevidence.php (accessed on 04-11-2021).
5. Alistair G. L. Borthwick. Marine renewable energy seascape. Engineering 2016; 2:69–78. doi: 10.1016/J.ENG.2016.01.011
6. https://www.c2es.org/content/renewable-energy/ (accessed on 04-11-2021).
7. Frías-Paredes Laura, Mallor Fermín, Gastón-Romeo Martín, León Teresa. Assessing energy forecasting inaccuracy by simultaneously considering temporal and absolute errors. Energy Conversion and Management 2017; 142:533–546. doi: 10.1016/j.enconman.2017.03.056
8. Hodge Bri-Mathias, Martinez-Anido Carlo Brancucci, Wang Qin, Chartan Erol, Florita Anthony, Kiviluoma Juha. The combined value of wind and solar power forecasting improvements and electricity storage. Applied Energy 2018; 214:1–15.
9. Hao Yan, Tian Chengshi. A novel two-stage forecasting model based on error factor and ensemble method for multi-step wind power forecasting. Applied Energy 2019; 238:368–383.
10. Huaizhi, Wang, Zhenxing Lei, Xian Zhang, Bin Zhou, Jianchun Peng. A review of deep learning for renewable energy forecasting. Energy Conversion and Management 2019; 198:111799. doi: 10.1016/j.enconman.2019.111799
11. Junfei Qiu, Qihui Wu, Guoru Ding, Yuhua Xu, Shuo Feng. A survey of machine learning for big data processing. EURASIP Journal on Advances in Signal Processing 2016; 2016:67. doi: 10.1186/s13634-016-0355-x
12. https://datafloq.com/read/machine-learning-explained-understanding-learning/ (accessed on 04-11-2021).
13. https://www.analytikus.com/post/2018/09/24/machine-learning-explained-understanding-supervised-unsupervised-and-reinforcement-learni (assessed on 04-11-2021)
14. https://www.simplilearn.com/tutorials/deep-learning-tutorial/deep-learning-algorithm (accessed on 04-11-2021).
15. S. L. Brunton, J. N. Kutz. Data-Driven Science and Engineering. Cambridge, UK: Cambridge University Press, 2019.
16. MdMijanur Rahman, Mohammad Shakeri, Sieh Kiong Tiong, Fatema Khatun, Nowshad Amin, Jagadeesh Pasupuleti, Mohammad Kamrul Hasan. Prospective methodologies in hybrid renewable energy systems for energy prediction using artificial neural networks. Sustainability 2021; 13:2393. doi: 10.3390/su13042393
17. M. Sharifzadeh, A. Sikinioti-Lock, N. Shah. Machine-learning methods for integrated renewable power generation: A comparative study of artificial neural networks, support vector regression, and gaussian process regression. Renewable and Sustainable Energy Reviews 2019; 108:513–538.
18. R. Li, Y. Jin. A wind speed interval prediction system based on multi-objective optimization for machine learning method. Applied Energy 2018; 228:2207–2220.
19. Xiangyun Qing, Yugang Niu. Hourly day-ahead solar irradiance prediction using weather forecasts by LSTM. Energy 2018; 148:461–468. doi: 10.1016/j.energy.2018.01.177
20. Camila Correa-Jullian, Jose Miguel Cardemil, Enrique Lopez Droguett, Masoud Behzad. Assessment of deep learning techniques for prognosis of solar thermal systems, Renewable Energy 2020; 145:2178e2191. doi: 10.1016/j.renene.2019.07.100
21. Tayeb Brahimi. Using artificial intelligence to predict wind speed for energy application in Saudi Arabia. Energies 2019; 12:4669. doi: 10.3390/en12244669

22. Jaume Manero, Javier Béjar, Ulises Cortés. "Dust in the wind...", deep learning application to wind energy time series forecasting. Energies 2019; 12:2385. doi: 10.3390/en12122385

23. Majid Dehghani, Hossein Riahi-Madvar, Farhad Hooshyaripor, Amir Mosavi, Shahaboddin Shamshirband, Edmundas Kazimieras Zavadskas, Kwok-wing Chau. Prediction of hydropower generation using grey wolf optimization adaptive neuro-fuzzy inference system. Energies 2019: 12:289. doi: 10.3390/en12020289

24. Dan Assouline, Nahid Mohajeri, Agust Gudmundsson, Jean-Louis Scartezzini. A machine learning approach for mapping the very shallow theoretical geothermal potential. Geothermal Energy 2019; 7:19. doi: 10.1186/s40517-019-0135-6

25. E. Shayana, V. Zareb, I. Mirzaee. Hydrogen production from biomass gasification: A theoretical comparison of using different gasification agents. Energy Conversion and Management 2018; 159:30–41. doi: 10.1016/j.enconman.2017.12.096

26. Deivis Avila, G. Nicolás Marichal, Isidro Padrón, Ramón Quiza, Ángela Hernández. Forecasting of wave energy in Canary Islands based on artificial intelligence. Applied Ocean Research 2020; 101:102189. doi: 10.1016/j.apor.2020.102189

27. S. Salcedo-Sanz, L. Cornejo-Bueno, L. Prieto, D. Paredes, R. García-Herrera. Feature selection in machine learning prediction systems for renewable energy applications. Renewable and Sustainable Energy Reviews 2018; 90:728–741. doi: 10.1016/j.rser.2018.04.008

28. G. John, R. Kohavi, K. Pfleger. Irrelevant features and the subset selection problem. In: Proceedings of the 11th International Conference on Machine Learning, July 10–13. New Brunswick, NJ, 1994.

29. Cyril Voyant, Gilles Notton, Soteris Kalogirou, Marie-Laure Nivet, Christophe Paoli, Fabrice Motte, Alexis Fouilloy. Machine learning methods for solar radiation forecasting: A review, Renewable Energy 2017; 105:569–82. doi: 10.1016/j.renene.2016.12.095

30. S. Atique, S. Noureen, V. Roy, V. Subburaj, S. Bayne, J. Macfie. Forecasting of total daily solar energy generation using ARIMA: A case study. In Proceedings of the 2019 IEEE 9th Annual Computing and Communication Workshop and Conference (CCWC). Las Vegas, NV, 7–9 January 2019; pp. 114–119.

31. A.F. Zambrano, L.F. Giraldo. Solar-irradiance forecasting models without on-site training measurements. Renewable Energy 2020; 152:557–566.

32. J.F. Torres, A. Troncoso, I. Koprinska, Z. Wang, F. Martínez-Álvarez. Big data solar power forecasting based on deep learning and multiple data sources. Applied Energy 2019; 238:1312–1326.

33. S. Leva, A. Dolara, F. Grimaccia, M. Mussetta, E. Ogliari. Analysis and validation of 24 hours ahead neural 2 network forecasting of photovoltaic output power. Mathematics and Computers in Simulation. doi: 10.1016/j.matcom.2015.05.010

34. S. Monjoly, M. André, R. Calif, T. Soubdhan. Hourly forecasting of global solar radiation based on multiscale decomposition methods: A hybrid approach. Energy 2017; 119:288–298.

35. G.-Q. Lin, L.-L. Li, M.-L. Tseng, H.-M. Liu, D.-D. Yuan, R. Tan. An improved moth-flame optimization algorithm for support vector machine prediction of photovoltaic power generation. Journal of Cleaner Production 2020; 253:119966.

36. Alireza Zendehboudi, M.A. Baseer, R. Saidur. Application of support vector machine models for forecasting solar and wind energy resources: A review. Journal of Cleaner Production 2018; 199:272–285.

37. Halil Demolli, Ahmet Sakir Dokuz, Alper Ecemis, Murat Gokcek. Wind power forecasting based on daily wind speed data using machine learning algorithms. Energy Conversion and Management 2019; 198:111823. doi: 10.1016/j.enconman.2019.111823

38. A. Khosravi, L. Machado, R.O. Nunes. Times-series prediction of wind speed using machine learning algorithms: A case study Osorio wind farm, Brazil. Applied Energy 2018; 224:550–66. doi: 10.1016/j.apenergy.2018.05.043

39. Sumit Saroha, Sanjeev Kumar Aggarwal, Preeti Rana. Wind power forecasting. Forecasting in Mathematics - Recent Advances, New Perspectives and Applications, IntechOpen, 2021, 10.5772/intechopen.94550

40. Sana Mujeeb, Turki Ali Alghamdi, Sameeh Ullah, Aisha Fatima, Nadeem Javaid, Tanzila Saba. Exploiting deep learning for wind power forecasting based on big data analytics. Applied Science 2019; 9:4417. doi: 10.3390/app10175975

41. H. Demolli, A.S. Dokuz, A. Ecemis, M. Gokcek. Wind power forecasting based on daily wind speed data using machine learning algorithms. Energy Conversion and Management 2019; 198:111823.

42. A. Zameer, J. Arshad, A. Khan, M.A.Z. Raja. Intelligent and robust prediction of short-term wind power using genetic programming-based ensemble of neural networks. Energy Conversion and Management 2017; 134:361–372.

43. H. Liu, X. Mi, Y. Li. Smart deep learning-based wind speed prediction model using wavelet packet decomposition, convolutional neural network and convolutional long short term memory network. Energy Conversion and Management 2018; 166: 120–131.

44. D. Zeng, Y. Liu, J. Liu, J. Liu. Short-term wind speed forecast based on best wavelet tree decomposition and support vector machine regression. Advances in Automation and Robotics 2011; 2:373–379.

45. P.G. Nieto, E. Garcia-Gonzalo, J.P. Paredes-Sánchez, A.B. Sánchez, M.M. Fernández. Predictive modelling of the higher heating value in biomass torrefaction for the energy treatment process using machine-learning techniques. Neural Computing and Applications 2019; 31:8823–8836.

46. Xiaoli Zhang, Yong Peng, Wei Xu, Bende Wang. An optimal operation model for hydro-power stations considering inflow forecasts with different lead-times. Water Resources Management 2019; 33:173–188. doi: 10.1007/s11269-018-2095-1

47. Seyed Hashem Samadi, Barat Ghobadian, Mohsen Nosrati. Prediction of higher heating value of biomass materials based on proximate analysis using gradient boosted regression trees method. Energy Sources, Part A: Recovery, Utilization, and Environmental Effects. doi: 10.1080/15567036.2019.1630521

48. E. Shayan, V. Zare, I. Mirzaee. Hydrogen production from biomass gasification: A theoretical comparison of using different gasification agents. Energy Conversion and Management 2018; 159:30–41.

49. E.E. Ozbas, D. Aksu, A. Ongen, M.A. Aydin, H.K. Ozcan. Hydrogen production via biomass gasification, and modeling by supervised machine learning algorithms. International Journal of Hydrogen Energy 2019; 44:17260–17268.

50. F. Elmaz, O. Yücel, A.Y. Mutlu. Predictive modeling of biomass gasification with machine learning-based regression methods. Energy 2019; 191:116541.

51. P.J. García Nieto, E. García-Gonzalo, F. Sánchez Lasheras, J.P. Paredes-Sánchez, P. Riesgo Fernández. Forecast of the higher heating value in biomass torrefaction by means of machine learning techniques. Journal of Computational and Applied Mathematics 2019; 357:284–301.

52. Dripta Sarkar, Michael A. Osborne, Thomas A.A. Adcock. Prediction of tidal currents using Bayesian machine learning. Ocean Engineering 2018; 158:221–231. doi: 10.1016/j.oceaneng.2018.03.007

53. Abdollah Kavousi-Fard, Wencong Su. A combined prognostic model based on machine learning for tidal current prediction. IEEE Transactions on Geoscience and Remote Sensing 2017; 0196–2892.

54. Hamed H. H. Aly. A novel approach for harmonic tidal currents constitutions forecasting using hybrid intelligent models based on clustering methodologies. Renewable Energy 2020; 147(1):1554–1564. doi: 10.1016/j.renene.2019.09.107

55. M. Ali, R. Prasad. Significant wave height forecasting via an extreme learning machine model integrated with improved complete ensemble empirical mode decomposition. Renewable and Sustainable Energy Reviews 2019; 104:281–295.

56. A. Kavousi-Fard, W. Su. A combined prognostic model based on machine learning for tidal current prediction. IEEE Transaction on Geoscience and Remote Sensing 2017; 55:3108–3114.

57. H.H.A. Hamed. A novel approach for harmonic tidal currents constitutions forecasting using hybrid intelligent models based on clustering methodologies. Renewable Energy 2019; 147:1554–1564.

58. D. Avila, G.N. Marichal, I. Padrón, R. Quiza, Á. Hernández. Forecasting of wave energy in Canary Islands based on artificial intelligence. Applied Ocean Research 2020; 101:102189.

59. S. Shi, R.J. Patton, Y. Liu. Short-term wave forecasting using Gaussian process for optimal control of wave energy converters. IFAC Papers on Line 2018; 51:44–49.

60. Nima Safari, Aslam Ansari, Alireza Zare, C.Y. Chung. A novel decomposition based localized short term tidal current speed and direction prediction model. IEEE Power & Energy Society General Meeting 2017. doi: 10.1109/PESGM.2017.8274667

61. https://www.thinkgeoenergy.com/how-data-can-transform-the-geothermal-energy-industry/ (accessed on 04-11-2011).

62. P. Gangwani, J. Soni, H. Upadhyay, S. Joshi. A deep learning approach for modeling of geothermal energy prediction. International Journal of Computer Science and Information Security 2020; 18:62–65.

63. M.B. Diaz, K.Y. Kim, T.-H. Kang, H.-S. Shin. Drilling data from an enhanced geothermal project and its pre-processing for ROP forecasting improvement. Geothermics 2018; 72:348–357.

64. Bruno Baruque, Santiago Porras, Esteban Jove, Jose Luis Calvo-Rolle. Geothermal heat exchanger energy prediction based on time series and monitoring sensors optimization. Energy 2019; 171:49–60.

65. Doaa A. Bashawyah, Abdulhamit Subasi. Power Quality Event Detection Using FAWT and Bagging Ensemble Classifier. Genova, Italy, IEEE, 2019.

66. Gajendra Singh Chawda, Abdul Gafoor Shaik, Mahmood Shaik, P. Sanjeevikumar, Jens Bo Holm-Nielsen, Om Prakash Mahela, K. Palanisamy. Comprehensive review on detection and classification of power quality disturbances in utility grid with renewable energy penetration. IEEE Access 2020. doi: 10.1109/ACCESS.2020.3014732

67. Juan Carlos Bravo-Rodríguez, Francisco J. Torres, María D. Borrás. Hybrid machine learning models for classifying power quality disturbances: A comparative study. Energies 2020; 13:2761. doi: 10.3390/en13112761

68. Yaguo Lei, Bin Yang, Xinwei Jiang, Feng Jia, Naipeng Li, Asoke K. Nandi. Applications of machine learning to machine fault diagnosis: A review and roadmap. Mechanical Systems and Signal Processing 2020; 138:106587. doi: 10.1016/j.ymssp.2019.106587

69. H. Jiang, J. Zhang, W. Gao, Z. Wu. Fault detection, identification, and location in smart grid based on data-driven computational methods. IEEE Transactions on Smart Grid 2014; 5:2947–2956.

70. Prince Waqas Khan, Yung-Cheol Byun, Sang-Joon Lee, Dong-Ho Kang, Jin-Young Kang, Hae-Su Park. Machine learning-based approach to predict energy consumption of renewable and non-renewable power sources. Energies 2020; 13:4870. doi: 10.3390/en13184870

71. https://venturebeat.com/2021/04/22/challenges-of-applied-machine-learning/ (accessed on 04-11-2021).

APPENDIX A

TABLE 8.A
Machine Learning Models for Solar Energy Prediction

Models	Techniques
AI	Gated recurrent units; RF, SVR, RF; RF; RF, gradient-boosted regression, extreme GB ; Linear regression, decision trees, SVM, ANN; ANN, SVM, GB, RF; ANN; CNN; DNN; DNN, RNN, LSTM; LSTM, auto-LSTM, gate recurrent unit (GRU), machine learning and statistical hybrid model; LSTM; LSTM, GRU; Copula-based non-linear quantile regression; Multi-method; Smart persistence; K-nearest-neighbours, GB; K-nearest-neighbours, SVM; Angstrom-Prescott; Multi-layer feed-forward neural network; Support vector classification; GPR; Regime-dependent ANN; ELM; Adaptive forward–backward greedy algorithm, leapForward, spikeslab, Cubist and bagEarthGCV; Static and dynamic ensembles
Statistical	ARIMA
Hybrid	Wavelet decomposition-hybrid; Improve moth-flame optimization algorithm-SVM; SVM-PSO; Cluster-based approach, ANN, SVM; SVM, Horizon, General; ANN, Principle component analysis; Auto regressive mobile average, MLP, Regression trees; Ensemble EMD-least square SVR; Least absolute shrinkage and selection operator, LSTM; RF, SVR, ARIMA, k-nearest neighbours; Mycielski-Markov; VMD-deep CNN; PSO-ELM; Multi-objective PSO; Artificial bee colony-empirical models; Gated recurrent unit and attention mechanism

APPENDIX B

TABLE 8.B
Machine Learning Models for Wind Energy Prediction

Models	Techniques
AI	Gaussian process regression (GPR), Support vector regression (SVR), Artificial neural networks (ANN); xGBoost regression, SVR, Random forest (RF); Least squares support vector machine (SVM); SVR, ANN, Gradient boosting (GB), RF; RF; GB trees; Multi-layer perceptron (MLP); Deep neural network (DNN)-principal component analysis; Feedforward ANN; Efficient deep convolution neural network; Linear regression, neural networks, SVR; Convolutional neural networks (CNN); DNN; Efficient deep CNN; Stacked autoencoders, backpropagation; Predictive deep CNN; Improved radial basis function neural network-based model with an error feedback scheme; ANN and genetic programming; Long short-term memory (LSTM); Improved LSTM-enhanced forget-gate network; LSTM-ANN; Auto-LSTM; Shared weight LSTM network; Ensemble-LSTM; Instance-based transfer-GB decision trees; Extreme learning machine (ELM); Empirical mode decomposition (EMD)-stacked autoencoders-ELM; Pattern sequence-based forecasting

(Continued)

TABLE 8.B *(Continued)*
Machine Learning Models for Wind Energy Prediction

Statistical	Physics-informed statistical
Hybrid	Adaptive neuro-fuzzy inference system, Particle swarm optimization (PSO), Genetic algorithm; Improved dragonfly algorithm-SVM; Deep belief network with genetic algorithms; Type-2 fuzzy neural network-PSO; Multi-objective ant Lion algorithm-Least squares SVM; Complete ensemble EMD-multi-Objective grey wolf optimization-ELM; Variational Mode decomposition (VMD)-Backtracking Search-regularized ELM; Coral reefs optimization algorithm with substrate layer, ELM; Stacked extreme-learning machine; VMD-singular spectrum analysis-LSTM-ELM; Ensemble EMD-deep Boltzmann machine; ELM-Improved complementary ensemble EMD with Adaptive noise-Autoregressive integrated moving average (ARIMA); Bayesian model averaging and Ensemble learning; Sparse Bayesian-based robust functional regression; Kernel principal component analysis-Core vector regression-Competition over resource; Wavelet packet decomposition-LSTM; Empirical wavelet transformation, Recurrent neural network (RNN)

APPENDIX C

TABLE 8.C
Machine Learning Models for Hydropower, Biomass and Tidal Energy Prediction

Sources of Energy	Models	Techniques
Hydropower	AI	Bayesian linear regression
	Hybrid	Grey wolf optimization-adaptive neuro-Fuzzy inference system; Long-medium and short-term, Bayesian stochastic dynamic programming
Biomass	AI	Linear regression, k-nearest neighbours' regression, SVM, Decision tree regression; Gradient boosted regression trees; Decision tree regression, MLP
	Hybrid	SVM-Simulated annealing; PSO-SVM
Wave	AI	Fuzzy inference systems, ANN; GPR; Interval type-2 fuzzy inference system
	Hybrid	Fuzzy inference systems, ANN; GPR; Interval type-2 fuzzy inference system
Tidal	AI	Wavelet-SVR
	Hybrid	Wavelet and ANN, Fourier series based on least square method; GPR-Bayesian; Ensemble EMD-Least squares SVM; Modified harmony search
Geothermal	AI	LSTM; Multiple regression-ANN; RF
	Hybrid	Time dependent neural networks

9 Effective Contribution of Green Human Resources Practices on Environmental Sustainability in the Era of Industry 4.0

Evidence from India

V. Sathya Moorthi
Department of Business Administration,
Kalasalingam Academy of Research and
Education, Virudhunagar, Tamil Nadu, India

CONTENTS

DOI: 10.1201/9781003369554-9

9.1   INTRODUCTION

In most recent times, the world has become more aware about climate change, environmental concerns, and sustainable development [1]. The increasing concern about environment has led companies all over the world to start practicing various pro environmental activities that ensure both the competitive edge and green organizational identity [2]. These became one of the managerial agendas for various companies all around the world [3]. As a result, numerous green revolution has happened in different disciplines, such as human resource management, supply chain management, etc. [4]. The corporate companies from all the countries, whether it is developed or developing, have thought to incorporate green practices in all business processes. The fact is that capturing international market becomes tougher in the competitive situation. To overcome this, companies gain benefits through ecofriendly practices [5].

These green practices require environmental concern behavior among the employees, which can be achieved through Green Human Resource Management (GHRM) [6]. The GHRM has a unique and significant role in encouraging ecofriendly practices and behaviors within organizations [7]. The various studies taught us the importance of GHRM to the enterprises. GHRM is a stimulating factor for creating pro-environmental behaviors among the employees and influence employees' commitment toward environment [8]. GHRM also enhances green performance of the organization [9]. It demonstrates the responsibilities of organizations and employees toward environment [10]. Each component of GHRM ensures the sustainable development of the organizations and organizational reputation [11]. GHRM facilitates proactive measures, through which a company can achieve green organizational goal and competitive edge in the industry [12].

Green Environmental Behavior (GEB) refers to actions that do not harm that environment and possible ways of preventing from environmental harm [13]. Different literature suggests that GHRM is the prominent predictor of GEBs of employees. However, these studies are confined to green behavior in workplace alone. Moreover, very few studies suggest the role of management in creating green behavior among employees.

Hence this study intends to explore the adaptation of GHRM practices in ensuring a sustainable ecosystem and the significance of GTL in creating such an environment in this era of Industry 4.0. This has been addressed by first studying the direct effect of GTL on employees' GEB. Secondly, by investigating indirect effects of GTL on GEB of the employees. By the way, this study aims to explore the mediating role of GHRM in the relationship between GTL and GEB.

Green Transformational Leadership (GTL) refers to "Behaviors of the leaders who always motivating and inspire the followers towards to achieve green environmental sustainability". Various literatures demonstrated that GTL plays a significant role in creating green culture in the organization. GTL promotes green creativity among the employees and has power to change the mindset of them [14]. Behavioral changes are highly dependent on leaders and their nature [15]. It also plays an important role in enhancing performance [16]. In this way, the present study stresses the significance of GTL, in order to make sustainable behavior and enhance green performance.

To investigate this, India was taken as evidence, since India is in the threshold of fourth industrial revolution. Still many manufacturing companies, especially small- and medium-level companies, struggle to implement Industry 4.0, which consists of enabling technologies including the Internet of Things (IoT), cloud computing, machine learning and deep learning into the production process. This is because of various encounters like huge investment, lack of required skill and information, and data security [17].

Sustainable Development Goals (SDGs) are designed to develop performance of the economy without affecting the environment. Technology-backed Industry 4.0 concept aids to achieve sustainable development targets [18]. In order to achieve SDGs, companies move forward to implement technology-backed Industry 4.0. The preparedness of leaders and employees is crucial to take forward the idea of Industry 4.0. The leaders are important because they are going to spend huge investments, both money and knowledge, on the technologies. Similarly, the role of employees is also essential because they are going to use and execute ideas into practice.

This study focuses mainly on how the thought of green leadership and green HR policies together influence the behavior of the employees in order to protect the environment and to achieve sustainable development in the era of Industry 4.0 and evidence produced from India.

9.1.1 OBJECTIVES OF THE CHAPTER

The objectives of this chapter are as follows:

- To study the green HRM practices as an emerging trend for Industry 4.0 and adoption of green HRM practices for sustainable ecosystem;
- To study the significant role of GTL in promoting green behavior to create green culture;
- To examine the relationship between the GTL of the organizational leaders and the GEB of the employees;
- And, to analyze the mediating role of GHRM policies on the relationship between the GTL and GEB.

9.1.2 ORGANIZATION OF THE CHAPTER

This chapter is organized as follows: Section 9.2 elaborates literature review toward the global trend in sustainability, GHRM, GTL, and GEB. Section 9.3 gives hypothesis development. Section 9.4 covers conceptual framework. Section 9.5 overviews methods. Sections 9.6–9.8 highlight results and interpretation, discussion and limitations of the study. Section 9.9 concludes the chapter with future scope.

9.2 LITERATURE REVIEW

9.2.1 GLOBAL TREND IN SUSTAINABILITY

In recent years, the world has been discussing more about creating and maintaining sustainability. The downright trend of developing countries has exposed its devastating impact on our societies. Hence all societies are striving toward creating a sustainable ecosystem and its development [19]. The historical Glasgow agreement stated that behavioral changes are required to combat mindset that focuses only on the economic well-being of the country. It insisted that all sections of the societies must join together to create green concerned behavior [20]. This shows us that the world recognizes that technology alone will not bring any solutions, and behavioral changes are needed to create a sustainable ecosystem [21].

Among the emerging nations, India plays a pivotal role in adaptation and implementation measures to attain sustainability. At the same time, India has more number of challenges because of extreme socioeconomic inequalities [22]. This needs to be addressed with appropriate behavioral measures to ensure a sustainable ecosystem [23]. This study advocates that those behavioral changes are achieved through policy changes and emergence of new transformational leadership.

9.2.2 GREEN HUMAN RESOURCE MANAGEMENT

GHRM plays a vital role in developing a green culture in society. It examines people's behavior in each of their activities aligned with the green environmental goal of the particular society [24]. Previous studies explain the categories of GHRM policies and practices. First, GHRM enables green ability and creativity through green recruitment policies, selection procedures, and training to develop green activities. Next, it motivates the employees by providing rewards to ensure green performance. Finally, it stimulates them to involve in green behavior and creates culture among them [6].

9.2.3 GREEN TRANSFORMATIONAL LEADERSHIP

Transformational leadership is an essential part of management, and their roles are inevitable in shaping the behavior of the employees. Transformational leadership always helps employees to achieve a high level of performance and allows to think about each activity in a creative manner [25]. Transformational leadership creates a platform where employees think and apply new ideas and motivates them to achieve organizational and individual goals. Moreover, it stimulates intellectual ideas to perform the work effectively [26]. Creative behavior has been enabled with the help of transformational leadership in the workplace [27]. Bass defined that GTL helps the employees to think and act to create a way of performing the work, thereby protect the environment from a manmade hazard.

9.2.4 GREEN ENVIRONMENTAL BEHAVIOR

In recent days, organizations are framing objectives with green concern concepts. These objectives are more concerned with environmental protection and sustainability. Organizations also change business operations and activities to ensure environmental protection. One way to attain environmental sustainability is creating a green culture in a workplace where employees must possess GEB [28]. GEB refers to a behavior that helps achieve environmental sustainability and perform pro-environmental activity in a workplace [29]. The role of leaders and management is key for creating such culture and initiating many levels of pro-environmental activities to ensure GEB in the organizations [30]. This study extends the social implication of the GEB.

9.3   HYPOTHESIS DEVELOPMENT

H1: Green Transformational Leadership (GTL) influences Green Environmental Behavior (GEB) of the employees.

H2: Green Transformational Leadership (GTL) leads to framing Green Human Resource Management (GHRM) policies.

H3: Green Human Resource Management (GHRM) influences Green Environmental Behavior (GEB) of the employees.

The first three hypotheses (H1–H3) are proposed to test the association between GTL and GEB, GTL and GHRM, and GHRM and GEB. Following these hypotheses, the indirect effect is planned to test the association between GTL and GEB through mediating element GHRM (GEB).

H4: Green Human Resource Management (GHRM) mediating the association between Green Transformational Leadership (GTL) and Green Environmental Behavior (GEB).

9.4   CONCEPTUAL FRAMEWORK

Conceptual framework has been designed based on the earlier discussion. The proposed model is analyzed in two parts. The first part is about testing the total effects between GTL and GEB.

The next part is about examining the mediating effect of the model. To do this, GHRM is added as the mediating element between GTL and GEB (Figure 9.1). In this way, this chapter examines the direct effect and indirect effect of the model.

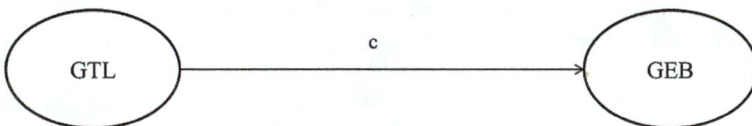

FIGURE 9.1   Total effects between GTL and GEB.

Note: Where c is the total effect and c = Direct effect + Indirect effect; c = ab + c'.

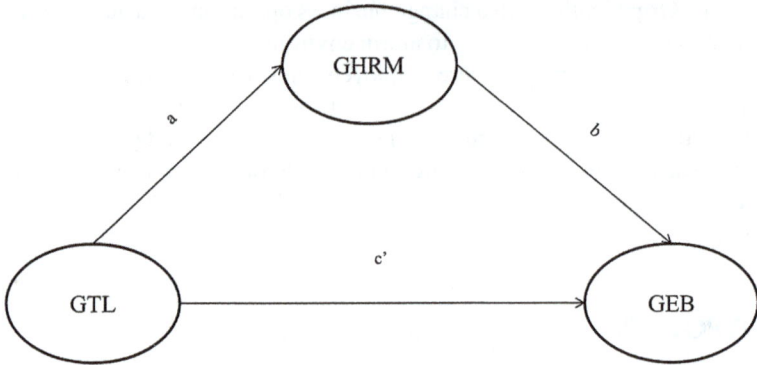

FIGURE 9.2 Mediating effect of GHRM between GTL and GEB.

Note: c' is a Direct effect and ab is an Indirect effect.

Figure 9.2 shows the mediating effect of GHRM.

9.5 METHODS

9.5.1 SAMPLE AND PROCEDURE

Data was gathered for the analysis from diverse manufacturing industries (Chemical, Cement, and Medicine) in India. Each company's HR managers are contacted directly and educated about the practices followed by the companies to protect the environment and study the procedure for practicing GHRM.

9.5.2 MEASURES OF MODEL

The study consisting of three core variables, namely, GTL, GHRM, and GEB, were observed through five-point Likert scale. A three-items scale developed by Tang et al. [31] is used to measure GHRM. Originally six items were developed by them. But context of present study has been reduced into three items, namely, Green training, Green motivation and performance management, and Green involvement of the employees. Items such as "My organization relates employees' eco friendly behavior to compensation and rewards" were used. GEB was a measured scale developed by [32] consisting of seven items. But for context of the current study, these were reduced into three. Items include "I turn lights off when not in use" and "I always prefer to buy recycle products for my personal and family use". GTL was a self-developed scale based on the idea of which contained two items: "My leader always motivates to perform green sustainable activity" and "My leader always inspires me by performing green sustainable activity".

9.5.3 ANALYSIS AND MODEL EVALUATION

The present study adopted Path Analysis – Partial Least Square (PLS) path modeling of SMART PLS 3.3 software to test the proposed hypothesis. Path Analysis (Structural Equation Model [SEM]) has been used to establish the organized relationship between different variables and has helped to validate the functional and analytical hypotheses. SMART PLS 3.3 software has been used for analyzing the SEM, which consists of two steps of assessment: First, assessment of measurement model, which helps to examine the relationship between measurement items and latent constructs. Second, examining the structural model, which helps to establish the relation between different constructs [33].

9.5.4 MEASUREMENT MODEL VALIDATION

In the process of assessment of the measurement model, number of items can be removed from the analysis when the value of factor loadings is low (<0.6) [34]. In this analysis, all the value of factor loadings is higher than the cutoff value of 0.6. Hence it is not necessary to eliminate any observed items from the analysis and all are used for further process of analysis. To test the reliability of the latent variables with observed variables, Cronbach's alpha and Composite Reliability (CR) were used in this analysis. Composite reliability can be achieved when each CR value is larger than the threshold value of 0.7 [35]. In this analysis, all CRs were higher than the recommended value of 0.7. Similarly, the value of Cronbach's alpha of each latent construct was higher than 0.7 threshold. Next, we analyzed the validity of constructs in the process of measurement model. Convergent validity and discriminant validity were checked for the constructs. Convergent validity was checked using Average Variance Extracted (AVE), which has an acceptable level of higher than 0.5 for each construct. The outcome values of reliability and validity with factor loadings are shown in Table 9.1.

Then the discriminant validity is checked using the Fornell-Larcker criterion and heterotrait-monotrait (HTMT) ratio. As per the Fornell-Larcker criterion, the value

TABLE 9.1
Factor Loadings, Reliability, and Convergent Validity

	Loadings	Cronbach's Alpha (CA)	Composite Reliability (CR)	Average Variance Extracted (AVE)
GTL		**0.728**	**0.88**	**0.785**
GTL1	0.868			
GTL2	0.904			
GHRM		**0.839**	**0.893**	**0.676**
GHRM1	0.858			
GHRM2	0.838			
GHRM3	0.858			
GHRM4	0.728			
GEB		**0.705**	**0.88**	**0.63**
GEB1	0.804			
GEB2	0.838			
GEB3	0.736			

Note: CA, CR, and AVE of respective constructs are represented in bold with italic words.

TABLE 9.2

Fornell-Larcker Criterion

	GEB	GHRM	GTL
GEB	*0.794*		
GHRM	0.453	*0.822*	
GTL	−0.332	−0.488	*0.886*

Note: The square root of AVE values is represented in bold with italic.

derived from the square root of AVE for the latent construct must be greater than the inter-latent construct correlations [36]. Table 9.2 shows that the square root of AVE value is greater than the correlation with other constructs.

The discriminant validity was also checked by using HTMT ratio of correlations, which have values of less than 0.90 [36]. The values from Figure 9.3 and Table 9.3 clearly indicate that the values of HTMT ratios were below the standard threshold 0.90. Hence discriminant validity is achieved.

9.5.5 STRUCTURAL MODEL VALIDATION

Once the validation of reliability and validity of the model was made, the next step in Path Analysis – Partial Least Square (PLS) – is the assessment of structural model to examine the relationship among the latent constructs, based on the R^2 value,

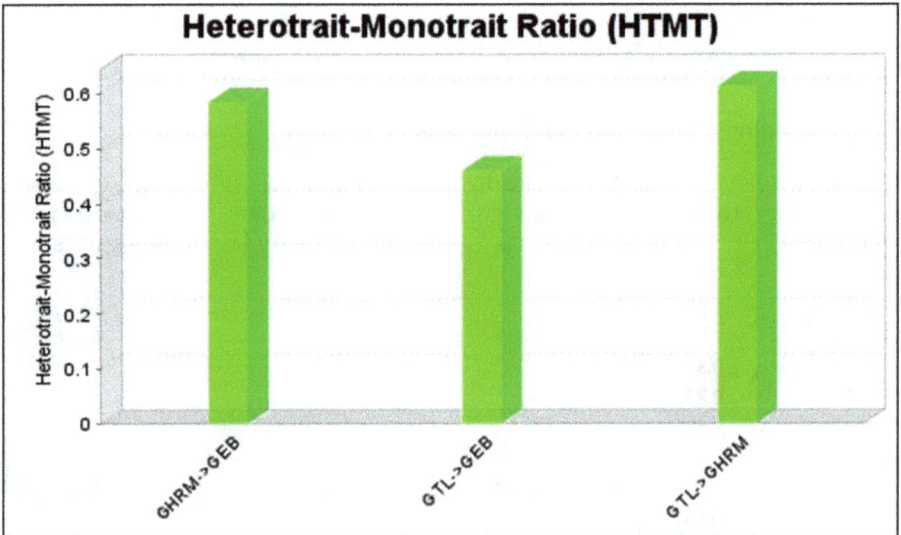

FIGURE 9.3   Heterotrait-Monotrait ratio of endogenous variable.

Note: HTMT ratio value should be less than 0.90.

TABLE 9.3

Heterotrait-Monotrait (HTMT) Ratio of Correlations

	GEB	GHRM	GTL
GEB			
GHRM	0.586		
GTL	0.461	0.615	

Q^2 value, and significance of structured path and its significance [33]. This study used a bootstrap re-sampling technique to assess the path coefficients significance with 5000 replicates and 300 cases. R^2 helps to decide the strength of each structural path of the proposed model and predictive accuracy of the model. R^2 is also used to explain the variance occurred in the dependent variable by independent variables [33]. Values of R^2 greater than or equal to 0.10 is desirable [37]. In this study, the value of R^2 was over the desirable value of 0.10 and its significance. The predictive accuracy is explained as substantial, moderate, and weak when the value of R^2 is over 0.75, 0.50, and 0.25, respectively [38]. In this study, independent variables GTL and GHRM described 27.1% of variance in the dependent variable GEB. This means that independent variables GTL and GHRM have moderate impact on the dependent variable GEB. The predictive relevance of endogenous constructs was determined by cross-validated redundancy index (Q^2) of the projected model [39]. The value of Q^2 higher than 0 infers that the respective model has predictive relevance and if it is less than 0 conveys that the model has a lack of predictive relevance [40]. In this study, cross-validated redundancy index (Q^2) is greater than 0 for each endogenous construct. This conveys that our endogenous constructs of GHRM and GEB have predictive relevance. In addition, the model fit was assessed by using Standardized Root Mean Square Residual (SRMR). The value of SRMR should be less than the required value of 0.10 [41]. In this study, the model was treated to be an acceptable level of fit, since the value of SRMR was 0.075, which was below the required level of 0.10. The resultant value of R^2 and Q^2 are shown in Table 9.4.

TABLE 9.4

R^2 and Q^2 Values of Variables

	R^2	Q^2
GEB	0.271	0.136
GHRM	0.290	0.157
GTL		

9.6   RESULTS AND INTERPRETATION

To test the hypothesis mentioned in this study, the amount of variance of path coefficients in the dependent variable for each independent variable was measured [41]. As shown in Table 9.5 and Figure 9.4, GTL has a significant and positive effect on GEB ($\beta = 0.145$, t = 2.741, p < 0.05). This is for testing our first hypothesis, and it established the association between GTL and GEB. Likewise, we can see the significant and positive leads of GTL toward framing the GHRM policies ($\beta = 0.488$, t = 14.205, p < 0.05); this confirmed our second hypothesis. Similarly, GHRM policies also have a significant and positive impact on GEB ($\beta = 0.382$, t = 7.433, p < 0.05). This confirmed our third hypothesis, which is used to establish the association between GHRM and GEB of the employees.

9.6.1   MEDIATION ANALYSIS

In this part, mediation analysis was used to measure the role of GHRM as a mediator on the association between GTL and GEB. For achieving this purpose, the study

TABLE 9.5
Hypothesis Testing by Bootstrapping Procedure

	Path Coefficient	Standard Deviation (STDEV)	T Statistics (|O/STDEV|)	p Values
GTL -> GEB	0.145	0.053	2.741	0.006
GTL -> GHRM	0.488	0.034	14.205	0.000
GHRM -> GEB	0.382	0.051	7.433	0.000

Note: p value should be less than 0.05 to accept the hypothesis.

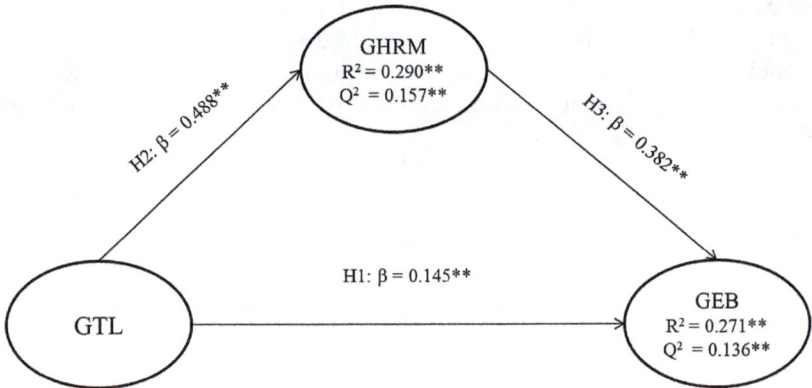

FIGURE 9.4   Hypothesis testing by bootstrapping procedure.

Note: **p < 0.05.

TABLE 9.6
Mediation Analysis

Total Effect (GTL → GEB)		Direct Effect (GTL → GEB)		Indirect Effect (GTL → GEB)					
Coefficient	p value	Coefficient	p value		Coefficient	SD	T Value	p Value	BI (2.5%–97.5%)
0.332	0.000	0.145	0.006	GTL → GHRM → GEB	0.187	0.032	5.833	0.000	0.121–0.247

employed bootstrapping function of SMART PLS [37]. In this way, we need to address the answer for the following three questions [41]:

1. Is there a significant total effect exist between GTL and GEB when the mediator variable GHRM is excluded from the model?
2. Are there any significant changes in the direct effect, from the total effect, that exist between GTL and GEB when the mediator variable GHRM is included into the model?
3. Is there a significant indirect effect of mediating variable GHRM exists between the GTL and GEB?

To address the answer to the first question, we excluded the GHRM from the model and ran the model with GTL and GEB. The outcome exposed that the total influence of GTL on GEB was significant ($\beta = 0.332$, t = 7.894, p < 0.05). To answer the second question, we analyzed the insertion of GHRM as a mediating variable; the level of effect of GTL on GEB has come down but it was significant ($\beta = 0.145$, t = 2.780, p < 0.05). It means that the direct effect was significant in the presence of mediating variable. When attending to answer the third question, the indirect effect of GHRM on the relationship between GTL and GEB was analyzed. The indirect effect of GTL on GEB through GHRM was significant ($\beta = 0.187$, t = 5.833, p < 0.05). Thus, in this way, this study demonstrates that the relationship between GTL and GEB is partially mediated by GHRM. The values are shown in Table 9.6.

9.7   DISCUSSION

Global warming and climate change are the serious noticeable issues and posing various challenges to businesses. If we work toward protecting environment alone, it may affect the process of businesses. Thus, there must be solutions that address the problem of global warming and help to manage the challenges of business. Furthermore, irrational solutions may lead to some other issues to the living organisms. Hence it is important to find the scientific solution to those issues [42].

Many researchers are focusing majorly on scientific methods to eliminate the problem of global warming and to ensure sustainable development throughout the world [6]. Technology-based solutions will be beneficial for businesses and support environmental activities. However, technology alone cannot change any-thing. Behavioral changes are also required to harvest the complete benefit of these researches. Behavioral changes must begin from two places. One is from education and the other is from the workplace. The changes from these two places will be reflected completely in the society and will add social value [43]. Organizational poli-cies, norms, and regulations are among the various sources of employees' behavioral change [44]. Leadership is one of the influential factors in framing policies and norms for the organization [45]. This study develops the conceptual idea of connecting trans-formational leadership with employees' behavior to achieve the proposed outcome.

Various literatures are focusing mainly on the way to frame and execute Green Human Resource policies. Very few studies shed light on the outcomes of GHRM. This study adds another concept of Green intentional leadership and their efforts into GHRM to change people's behaviors toward creating green environment. The findings of this study stressed the significant role of GTL in promoting green sus-tainable and environmentally concerned behavior through green policies. This key finding was backed up by the conceptual model of this study. First finding of current study is that the GTL has significant and positive impact on GEB of the employees. Prior studies suggest that the transformational leadership acts as a tool to change the behavior of the employees. Leaders are supporting the employees to be empowered and to create pro-innovative organizations [25]. Furthermore, GTL encourages the employees to take up the green activities in their work to ensure green performance of the organizations [46]. This finding is unvarying from the prior studies stating that GTL influences the GEB of the employees. GTL has been an influential factor for modeling employee behavior toward green activities in the process of organiza-tion. If the organizational leadership change toward the green activities, that will be reflected in the process of policy framing and that will lead to changes in the behav-ior of the employees. Thus, the present study advocates how leaders change their idea to change employees' behavior with the help of policies.

Another finding of the study endorses with prior studies that GHRM mediates the association between GTL and GEB. Generally, Human Resource (HR) policies are aided in changing behavior and even any leader trying to control the behavior of employees through HR policies. In such a way, GHRM always plays an inevitable role in shaping employees' behavior. It has a positive impact on the pro-environmental behavior of the employees [28]. Past studies exposed the mediating role of GHRM between the various associations that led to pro-environmental behaviors [47]. The finding is consistent with preceding studies that GHRM has a positive and significant influence over GEB of the employees. The true efforts inculcated by employers in selecting, training, and motivating the employees toward green activities and practices will deliver fruitful outcomes to the organization. Moreover, these efforts put forth by the leaders will have a unique effect and will add social value to the organizations.

Leadership plays a crucial role in framing HR policies and it will have a positive impact on the behavior and performance of the people [48]. The findings revealed that there is a positive impact of GTL on framing Green Human Resource Policies. When thought of the leaders leads a green organization and the same thought will blend with GHRM policies, it will yield the pro-environmental concerned behavior among the people.

9.7.1 THEORETICAL IMPLICATIONS

This study contributes several theoretical aspects to the concept of sustainable development in the era of Industry 4.0. Firstly, this study provides sound theoretical contributions to the concept of GTL. It indicates the strong role of transformational leadership in framing policy decision toward sustainability in organizations. In addition, this study creates essential relation between GTL and GHRM and GEB of the employees. Thus, this research enriches and enlarges the current knowledge about the concept of GTL and its significant outcome. This also strengthens the theoretical building by adding GHRM and GEB into the GTL.

Secondly, the current research ensures the significant role of GHRM in changing behavior of the employees. It explores the importance of leadership in framing an HRM system that makes sure the sustainable activity in the organization. This study confirms that GTL is one of the precedent factor for GHRM, and GEB is the outcome of GHRM.

Finally, this research established the mediating role of GHRM between GTL and GEB. It adds existing theoretical knowledge on the concept of environmental behavior and confirms the influential factors on environmental behavior. Hence, the present study addresses a foremost research gap by examining the relationship among GTL, GHRM, and GEB and the role of GTL and the mediating effect of GHRM on the GEB of the employees.

9.7.2 PRACTICAL IMPLICATIONS

This study will have some real-world implications as well for society and its development toward GEB. Among the various severe causes of global warming, business process is one of the causes of higher emission of CO_2, depletion of ozone layer, water wastage, and higher utilization and deforestation. This study provides a behavioral solution to environmental issues and ensures sustainable development in the business process.

GTL will facilitate business processes to achieve sustainable development by designing green policies. This study stressed that the thought process to carry out green activities must begin with top-level management. Moreover, this study proves that sincere efforts taken by leaders will reflect in the form of behavioral changes among employees. They should initiate green activities and ensure that they reach all levels of management. This GTL will add a benefit for the leaders to achieve competitive advantage for their company in the market.

Generally, behavioral changes, among the employees in a workplace, have been achieved through HR policies. The intent of every leader will reflect in the form of policies. This study proves that there is a positive and significant impact of GTL on GHRM policies. Adoption of GHRM policies and procedures has become essential for developing countries like India to ensure sustainable development. Enactment of GHRM can enable employees to carry out activities that protect the environment. Any firm, which executes GHRM practices, including green performance training, green performance rewards, etc., makes considerable changes in behavior of the employees toward the environment. For example, green training will guide the employees on how to perform a particular task without affecting the environment with full commitment and concern toward protecting the environment. Similarly, green rewards will motivate employees to engage in activities to protect and preserve the environment. These behavioral changes will explicit employees' behavior in every activity outside the workplace as well, including buying recycled products, reducing energy consumptions, educating their kids toward using recycling products, etc. When Green leadership delivers their idea through HR policies, any firm can easily achieve sustainable development.

9.8   LIMITATIONS

The outcomes of the present study will help to enhance concepts of GHRM, GTL, and GEB, both theoretically and practically. Still some part of this study needs to focus on future research. First, this study used data collected from chemical, cement, and medicine manufacturing companies, and the outcome of the study is limited to these companies and companies which are producing similar nature of products. The validity of outcome of this study to other types of manufacturing companies is a concern.

Second, this study examined the mediating effect of GHRM on the association between GTL and GEB and neglected the moderator effects on the associations among the GTL, GEB, and GHRM.

Finally, this study examines the role of GTL and GHRM applicability in the context of India alone on the focus of the emerging economy.

9.9   CONCLUSION AND FUTURE SCOPE

Green energy and sustainable development are not only technical practices but also behavioral practices. The companies from various countries give immense attention to the sustainable development of business through adopting green leadership, green policies and practices, and green behavior in the companies all over the world. This will be supporting the changes required to implement the concept of Industry 4.0. It is very important to know about GHRM practices and their benefits to achieve the goal of environmental behavior. Also, green behavioral changes will occur when thought of green leadership hands together with GHRM policies.

There is more scope in the concept of GTL, GHRM, and GEB in the future research. The researchers use the outcome of this study as a reference for the concept of GTL, GHRM, and GEB. Future research must examine the same proposed model in other types of manufacturing companies and test the consistency of the result of this study across different manufacturing companies. Future studies can be carried out with the aim of establishing relationship with moderators in the existing relationship among GHRM, GTL, and GEB, which will provide additional useful results. Furthermore, the outcome of the present study must expand into other developing economies and some under-developed countries to have wider applicability in future.

REFERENCES

1. H. V. Ford *et al.*, "The fundamental links between climate change and marine plastic pollution," *Sci. Total Environ.*, vol. 806, pp. 1–11, 2022, doi: 10.1016/j.scitotenv. 2021.150392.

2. B. Afsar, and W. A. Umrani, "Corporate social responsibility and pro-environmental behavior at workplace: The role of moral reflectiveness, coworker advocacy, and environmental commitment," *Corp. Soc. Responsib. Environ. Manag.*, vol. 27, no. 1, pp. 109–125, 2019, doi: 10.1002/csr.1777.

3. J. González-Benito, and Ó. González-Benito, "A review of determinant factors of environmental proactivity," *Bus. Strateg. Environ.*, vol. 15, no. 2, pp. 87–102, 2006, doi: 10.1002/bse.450.

4. A. A. Marcus, and A. R. Fremeth, "Green management matters," *Acad. Manag.*, vol. 23, no. 3, pp. 17–26, 2016.

5. C. J. C. Jabbour, A. B. L. Jabbour, A. A. Teixeira, and W. R. S. Freitas, "Environmental development in Brazilian companies: The role of human resource management," *Environ. Dev.*, vol. 3, no. 1, pp. 137–147, 2012, doi: 10.1016/j.envdev.2012.05.004.

6. D. W. S. Renwick, C. J. C. Jabbour, M. Muller-Camen, T. Redman, and A. Wilkinson, "Contemporary developments in green (environmental) HRM scholarship," *Int. J. Hum. Resour. Manag.*, vol. 27, no. 2, pp. 114–128, 2016, doi: 10.1080/09585192.2015.1105844.

7. E. Bombiak, "Green human resource management – The latest trend or strategic necessity?" *Entrep. Sustain. Issues*, vol. 6, no. 4, pp. 1647–1662, 2019, doi: 10.9770/jesi.2019.6.4(7).

8. N. Y. Ansari, M. Farrukh, and A. Raza, "Green human resource management and employees pro-environmental behaviours: Examining the underlying mechanism," *Corp. Soc. Responsib. Environ. Manag.*, vol. 28, no. 1, pp. 229–238, 2021, doi: 10.1002/csr.2044.

9. Y. J. Kim, W. G. Kim, H. M. Choi, and K. Phetvaroon, "The effect of green human resource management on hotel employees' eco-friendly behavior and environmental performance," *Int. J. Hosp. Manag.*, vol. 76, no. March 2018, pp. 83–93, 2019, doi: 10.1016/j.ijhm.2018.04.007.

10. B. Zhang, "Environmental protection responsibility of enterprises: Green human Resource management," *IOP Conf. Ser. Mater. Sci. Eng.*, vol. 612, no. 5, 2019, doi: 10.1088/1757-899X/612/5/052022.

11. E. Bombiak, and A. Marciniuk-Kluska, "Green human resource management as a tool for the sustainable development of enterprises: Polish Young company experience," *Sustainability*, vol. 10, no. 6, 2018, doi: 10.3390/su10061739.

12. J. Zhao, H. Liu, and W. Sun, "How proactive environmental strategy facilitates environmental reputation: Roles of green human resource management and discretionary slack," *Sustainability*, vol. 12, no. 3, 2020, doi: 10.3390/su12030763.

13. L. Andersson, S. E. Jackson, and S. V. Russell, "Greening organizational behavior: An introduction to the special issue," *J. Organ. Behav.*, vol. 34, no. 2, pp. 151–155, 2013, doi: 10.1002/job.1854.

14. S. Mittal, and R. L. Dhar, "Effect of green transformational leadership on green creativity : A study of tourist hotels," *Tour. Manag.*, vol. 57, pp. 118–127, 2016, doi: 10.1016/j.tourman.2016.05.007.

15. J. R. B. Halbesleben, M. M. Novicevic, M. G. Harvey, and M. R. Buckley Ronald, "Awareness of temporal complexity in leadership of creativity and innovation: A competency-based model," *Leadersh. Q.*, vol. 14, no. 4–5, pp. 433–454, 2003, doi: 10.1016/S1048-9843(03)00046-8.

16. A. Constantine, "Determinants of organisational creativity: A literature review," *Manag. Decis.*, vol. 39, no. 10, pp. 834–841, 2001.

17. N. Singhal, "An empirical investigation of industry 4.0 preparedness in India," *Vision*, no. 2017, 2020, doi: 10.1177/0972262920950066.
18. M. A. Berawi, "The role of industry 4.0 in achieving sustainable development goals," *Int. J. Technol.*, vol. 10, no. 4, pp. 644–647, 2019, doi: 10.14716/ijtech.v10i4.3341.
19. B. Duvnjak, and A. Kohont, "The role of sustainable HRM in sustainable development," *Sustainability*, vol. 13, no. 19, 2021, doi: 10.3390/su131910668.
20. unfccc. (2021). *Report of the Conference of the Parties serving as the meeting of the Parties to the Paris Agreement on its third session, held in Glasgow from 31 October to 13 November 2021* (Vol. 03277, Issue March). https://www.ipcc.ch/report/ar6/wg1/
21. S. Bandyopadhyay, D. C. Y. Foo, and R. R. Tan, "Sustainability trends, 2021 Best paper, and plans for 2022," *Process Integr. Optim. Sustain.*, vol. 6, no. 1-2, pp. 2–3, 2022, doi: 10.1007/s41660-021-00218-y.
22. H. Kharas, J. W. Mcarthur, and K. Rasmussen, "How many people will the world leave behind? Assessing current trajectories on the Sustainable Development Goals," no. September, pp. 1–62, 2018 [Online]. Available: https://www.brookings.edu/wp-content/uploads/2018/09/HowManyLeftBehind.pdf.
23. A. M. Khalid, S. Sharma, and A. K. Dubey, "Concerns of developing countries and the sustainable development goals: Case for India," *Int. J. Sustain. Dev. World Ecol.*, vol. 28, no. 4, pp. 303–315, 2021, doi: 10.1080/13504509.2020.1795744.
24. B. Afsar, B. Al-Ghazali, and W. Umrani, "Corporate social responsibility, work meaningfulness, and employee engagement: The joint moderating effects of incremental moral belief and moral identity centrality," *Corp. Soc. Responsib. Environ. Manag.*, vol. 27, no. 3, pp. 1264–1278, 2020, doi: 10.1002/csr.1882.
25. D. I. Jung, C. Chow, and A. Wu, "The role of transformational leadership in enhancing organizational innovation: Hypotheses and some preliminary findings," *Leadersh. Q.*, vol. 14, no. 4–5, pp. 525–544, 2003, doi: 10.1016/S1048-9843(03)00050-X.
26. M. D. Mumford, "Managing creative people: Strategies and tactics for innovation," *Hum. Resour. Manag. Rev*, vol. 10, no. 3, pp. 313–351, 2000, doi: 10.1016/S1053-4822(99)00043-1.
27. R. T. Keller, "Transformational leadership, initiating structure, and substitutes for leadership: A longitudinal study of research and development project team performance," *J. Appl. Psychol.*, vol. 91, no. 1, pp. 202–210, 2006, doi: 10.1037/0021-9010.91.1.202.
28. B. Bin Saeed, B. Afsar, S. Hafeez, I. Khan, M. Tahir, and M. A. Afridi, "Promoting employee's pro-environmental behavior through green human resource management practices," *Corp. Soc. Responsib. Environ. Manag.*, vol. 26, no. 2, pp. 424–438, 2019, doi: 10.1002/csr.1694.
29. R. Wesselink, V. Blok, and J. Ringersma, "Pro-environmental behaviour in the workplace and the role of managers and organisation," *J. Clean. Prod.*, vol. 168, pp. 1679–1687, 2017, doi: 10.1016/j.jclepro.2017.08.214.
30. C. C. Baughn, N. L. Bodie, and J. C. McIntosh, "Corporate social and environmental responsibility in Asian countries and other geographical regions," *Corp. Soc. Responsib. Environ. Manag.*, vol. 14, no. 4, pp. 189–205, 2007, doi: 10.1002/csr.160.
31. G. Tang, Y. Chen, Y. Jiang, P. Paillé, and J. Jia, "Green human resource management practices: Scale development and validity," *Asia Pacific J. Hum. Resour.*, vol. 56, no. 1, pp. 31–55, 2018, doi: 10.1111/1744-7941.12147.
32. J. L. Robertson, and J. Barling, "Greening organizations through leaders' influence on employees' pro-environmental behaviors," *J. Organ. Behav.*, vol. 34, no. 2, pp. 176–194, 2013, doi: 10.1002/job.1820.
33. J. F. Hair, M. Sarstedt, L. Hopkins, and V. G. Kuppelwieser, "Partial least squares structural equation modeling (PLS-SEM): An emerging tool in business research," *Eur. Bus. Rev.*, vol. 26, no. 2, pp. 106–121, 2014, doi: 10.1108/EBR-10-2013-0128.

34. D. Gefen, and D. Straub, "A practical guide to factorial validity using PLS-graph: Tutorial and annotated example," *Commun. Assoc. Inf. Syst.*, vol. 16, pp. 91–109, 2005, doi: 10.17705/1cais.01605.

35. S. Faraj, "Why should is here ? Examining social," *Soc. Cap. Knowl. Contrib. Spec.*, vol. 29, no. 1, pp. 35–57, 2005.

36. J. Henseler, C. M. Ringle, and M. Sarstedt, "A new criterion for assessing discriminant validity in variance-based structural equation modeling," *J. Acad. Mark. Sci.*, vol. 43, no. 1, pp. 115–135, 2015, doi: 10.1007/s11747-014-0403-8.

37. R. F. Falk, and N. B. Miller, A Primer for Soft Modeling. In *The University of Akron Press* (Issue April 1992).

38. J. F. Hair, C. M. Ringle, and M. Sarstedt, "PLS-SEM: Indeed a silver bullet," *J. Mark. Theory Pract.*, vol. 19, no. 2, pp. 139–152, 2011, doi: 10.2753/MTP1069-6679190202.

39. J. L. Roldán, and M. J. Sánchez-Franco, Variance-based structural equation modeling: Guidelines for using partial least squares in information systems research. In *Research Methodologies, Innovations and Philosophies in Software Systems Engineering and Information Systems* (Issue January 2012, pp. 193 -221). https://doi.org/10.4018/978-1-4666-0179-6.ch010..

40. I. Castro, and J. L. Roldán, "A mediation model between dimensions of social capital," *Int. Bus. Rev.*, vol. 22, no. 6, pp. 1034–1050, 2013, doi: 10.1016/j.ibusrev.2013.02.004.

41. J. F. Hair, G. T. M. Hult, C. M. Ringle, and M. Sarstedt, *A primer on partial least squares structural equation modeling (PLS-SEM)*. California: Thousand Oaks. 2017.

42. S. E. Jackson, D. S. Ones, and S. Dilchert, *Managing Human Resources for Environmental Sustainability* (A. I. Kraut (ed.); Issue January), 2016. Jossey-Bass, San Francisco. www.josseybass.com.

43. S. H. Appelbaum, G. D. Iaconi, and A. Matousek, "Positive and negative deviant workplace behaviors: Causes, impacts, and solutions," *Corp. Gov.*, vol. 7, no. 5, pp. 586–598, 2007, doi: 10.1108/14720700710827176.

44. J. Burchell, "Anticipating and managing resistance in organizational information technology (IT) change initiatives," *Int. J. Acad. Bus. World*, vol. 5, no. 1, pp. 19–28, 2011.

45. K. A. Graham, J. C. Ziegert, and J. Capitano, "The effect of leadership style, framing, and promotion regulatory focus on unethical pro-organizational behavior," *J. Bus. Ethics*, vol. 126, no. 3, pp. 423–436, 2015, doi: 10.1007/s10551-013-1952-3.

46. X. Wang, K. Zhou, and W. Liu, "Value congruence: A study of green transformational leadership and employee green behavior," *Front. Psychol.*, vol. 9, no. Oct, pp. 1–8, 2018, doi: 10.3389/fpsyg.2018.01946.

47. M. Úbeda-García, E. Claver-Cortés, B. Marco-Lajara, and P. Zaragoza-Sáez, "Corporate social responsibility and firm performance in the hotel industry. The mediating role of green human resource management and environmental outcomes," *J. Bus. Res.*, vol. 123, no. June 2020, pp. 57–69, 2021, doi: 10.1016/j.jbusres.2020.09.055.

48. S. K. Singh, M. Del Giudice, R. Chierici, and D. Graziano, "Green innovation and environmental performance: The role of green transformational leadership and green human resource management," *Technol. Forecast. Soc. Change*, vol. 150, no. October 2019, p. 119762, 2020, doi: 10.1016/j.techfore.2019.119762.

10 Recent Developments in Waste Valorization

An Overview of Indian and Worldwide Perspectives

Gajendra Singh Vishwakarma[1],
Dolly Vadaviya[1], Preeti Kashyap[1],
Raviprakash Chandra[2], and Kunal Shah[3]
[1]Department of Biological Science and Biotechnology,
Institute of Advanced Research, Gandhinagar, India
[2]Department of Engineering and Physical Sciences,
Institute of Advanced Research, Gandhinagar, India
[3]Department of Anesthesiology, University of Texas MD
Anderson Cancer Center, Houston, Texas, United States

CONTENTS

10.1   INTRODUCTION

Due to the exponential population growth and development of various industrial clusters and townships, the waste generation rate in the urban areas is sharply accelerated. In addition to that in rural areas, especially in the case of agricultural sectors, the farm mechanization and upgraded technologies of irrigation and crop protection also contributed to the agricultural waste generation. The change in the lifestyle, food habits and diverse culture also triggered the rate of waste generation in India. As per the report of the Central Pollution Control Board of India, the per capita waste generation has increased at an exponential rate (0.26–0.85 kg/day) [1]. Out of that, about 80%–90% of the municipal waste is disposed of in landfills without proper management practices and open burning, leading to air, water and soil pollution [2]. Therefore, the current research and development-related centers, institutes as well as the industrial sectors are working on the alternative strategies of waste minimization and management. For achieving zero waste and enhanced recovery, waste valorization is one of the suitable techniques. From the last 10–12 years of research, the database suggested that a lot of work has been done in this field. As a sign of this, some of the industries and research organizations commercialized some of the valuable products and technologies based on the process of waste valorization. Even some of the municipal corporations and townships are also adopting concepts of the waste valorization in the management of domestic solid waste. But due to the lack of information and data availability, the combined effort is still very less; therefore, the outcome of the whole scenario is still hidden.

The objectives of the chapter are as follows:

- To discuss the concept of waste valorization and understand how this practice adds economic value to waste;
- To highlight different approaches of waste valorization integrated with the food, pharmaceutical and fishing industries;
- And, to examine the current state of affairs and recent developments in waste valorization on a local and global scale.

10.1.1   ORGANIZATION OF THE CHAPTER

This chapter is divided into five sections: Section 10.2 is about the basic concepts and fundamentals of waste valorization. Section 10.3 discusses the recent developments in waste valorization, that is, kitchen waste and food waste valorization process, the process of biorefinery and food industry, dairy and pharmaceutical industry waste valorization in addition to valorization strategies of some other industries like fisheries, brewery, slaughterhouses, forestry and paper industry. Sections 10.4 and 10.5 elaborate the status of waste valorization in India and at the global scale. Section 10.6 concludes the chapter with future scope.

10.2   FUNDAMENTALS OF WASTE VALORIZATION

Waste valorization is the process of reusing, recycling or composting waste materials and converting them into more useful products including materials, chemicals, fuels or other sources of energy. The concept of waste valorization is standing side by side with recycling and reusing technologies, namely "enhancing the value" [3].

Recently, there has been an increased push on the industrial manufacturers and the environmentalists toward higher sustainability to improve cost effectiveness and meet customers' demand for renewable chemicals and fuels [4, 5]. Valorization of waste has been one of the important research areas since past few years. It has concerned a great deal of importance as a potential alternative to the conventional solid waste disposal in landfill sites. The increasing development of environmental strategies to process such solid waste is an interesting area of research in our current society. The recent research has focused on producing energy from the waste instead of disposing and decomposing (e.g., bioethanol and biodiesel production) [6–8]. Meanwhile, useful organic chemicals can be generated from organic waste via biorefinery or white biotechnology (e.g., bioplastics and/or succinic acid) as well as by developing sustainable green production strategies [9, 10].

Waste valorization is mostly related to waste management for a long time. Fast depletion of fuel has bought this concept back to our society with renewed interest, natural and primary resources. Recently, the increased waste generation and land filling worldwide stressed on the need for more sustainable and cost-efficient waste management protocols. Different valorization techniques are currently showing great promise in meeting future industrial demands of sustainable future.

Numerous methods are utilized for valorization, it may be the valorization of organic waste through decomposition, conversion of agricultural solid waste as animal feeds and modification of biochar for the removal of heavy metals and organic pollutants, utilization of agricultural waste as nanofoods, solid-state fermentation of organic waste and municipal/domestic solid waste valorization for the production of different enzymes like glucoamylase [11, 12]. Apart from these, the concepts of biorefinery, production of fertilizer, fuel production in forms of bioethanol, biodiesel, biohydrogen and chemical extraction are also the part of waste valorization. In this regards Table 10.1 enlists some of the products, their raw materials and the process of valorization.

10.3   RECENT DEVELOPMENTS IN WASTE VALORIZATION

The current research on waste valorization is directing toward the three sustainable goals: first is the generation of alternative energy resources to replace the fossil fuels, second is the generation of high-value chemicals and third is the production of feed supplements and raw materials. In order to achieve the aforementioned goals, the combination of various traditional waste management methodologies has been modified [4]. The most commonly used strategies in waste valorization are the flow chemical technique, pyrolysis of waste for the synthesis of fuels and materials like carbon nanotubes and graphene, and biodegradation or bioconversion methods to degrade complex wastes into similar forms [4].

10.3.1   KITCHEN WASTE AND FOOD WASTE VALORIZATION

The major fraction of the kitchen waste belongs to the organic material. Therefore, the bioconversion or biodegradation methods are highly suitable for the food waste

TABLE 10.1

Food Waste Valorization through Different Treatment Procedures

Sr. No.	Product Name	Raw Materials	Process	References
1.	Nutrient-enriched biochar and biofuels	Agricultural and dairy waste	Integrated thermo-chemical conversion	[31]
2.	Syngas, char and pyrolytic oil	Carbon-based solid wastes involving CO_2 gas	Thermo-chemical processes (pyrolysis and gasification)	[32]
3.	D-limonene (terpenic compound)	Citrus waste (CW) consists mainly of peels and pressed pulp (seeds and segment membranes)	Hydro distillation, cold pressing and solvent extraction	[33]
4.	Cosmetic and pharmaceutical products	Grape stem extracts	High performance liquid chromatography	[34]
5.	Succinic acid	Organic waste with simultaneous biogas upgradation	Enzymatic hydrolysis	[35]
6.	Lignocellulosic nanofibrils (cellulose nonmaterial)	Agricultural waste (wheat straw and its pulping solid residue)	Depolymerization and degradation	[36]
7.	NPK-enriched organic fertilizer	Wet blue leather, poultry bone, water hyacinth	Bidirectional base hydrolysis followed by acid hydrolysis	[37]
8.	Biopolymer	Rapeseed meal	Green solvent extraction using *Pseudomonas putida*	[38]
9.	Silver nanoparticles	Silver-rich low temperature co-fired ceramic	Chemical metallurgy followed by leaching optimization and then, silver was selectively recovered through precipitation	[39]
10.	Lactose-derived nutraceutical	Whey	Trans-galactosylation, isomerization and hydrogenolysis	[40]
11.	Lactic acid and plasticizer	Food waste	Fungal hydrolysis, microalgae cultivation and biomass processing	[41]
12.	Feedstock	Keratinous materials	Hydrothermal methods, enzymatic hydrolysis, bioconversion	[42]

valorization [13]. The waste valorization based on microbial process mainly deals with either production of biomass-based fuel and energy resources or in the generation of food supplements, alternatives of animal feed material. Apart from this, the traditional composting or vermicomposting methods are also frequently used in the food waste valorization. The food waste generally rich in protein and fat has high energy content, and the good amount of fiber is ideally suitable for the generation of animal feed material. In this affection, the waste generated from oil extraction industries could be used in the form of oil cakes for animal feed as well as fertilizer due to its high mineral content [14, 15]. Industrial byproducts like malted barley, milling byproducts like wheat and rice, the wastes from potato industries and silage of the sugarcane top and molasses are some of the major industrial wastes utilized for the production of animal feed currently [5]. For the production of pathogen-free biofertilizers and soil conditioners, aerobic degradation of food wastes or agro-industrial wastes via microorganisms has shown to be a feasible method. Composting has not only contributed to the reduction in the burden of handling and disposal of solid wastes but has also helped in the removal of secondary pollution and led to good environmental benefits [16]. In addition, it has reduced the dependence on chemical fertilizers that leads to soil and water pollution. This valorization process is dependent on factors like temperature, humidity content, pH, aeration, degree of compaction, particle size, etc. It has been noted that aeration has been the most important factor affecting the quality of compost, whereas the C/N ratio affects the maturity of compost. Various research studies are focusing on the optimization of parameters to improve the yield and quality of compost. Greenhouse gas emissions that are generated as a result of organic matter degradation affect the sustainability of the composting process.

10.3.2 BIOREFINERY

Biorefinery is a process to valorize waste as a renewable feedstock to recover bio-based materials through sustainable biotechnology, which integrates remediation and resource recovery. Bioprocesses that utilize waste as a substrate for producing bio-based products require appropriate technologies including acidogenesis, bioelectrogenesis, photosynthesis, photo fermentation, etc. (Figure 10.1) [17]. The fossil-based economy worldwide has led to rapid economic expansion and detrimental climate change, and mismanaged disposal of waste in natural habitat. Waste remediation process like biorefinery is intended to reduce/remove pollutants to safeguard environment [18]. Solid, liquid, gaseous or any kind of waste has an inherent net positive energy that can be recovered by biorefineries to produce bio-based products and biofuels, which could pave the way to achieve a circular closed loop and low-carbon bioeconomy [6].

This concept is based on the separation of building blocks of biomass units such as proteins, fatty acids, carbohydrates and other compounds for the production of chemicals, biofuels and other products of higher market values. The feedstock for biorefineries is categorized mainly as carbohydrates and lignin, triglycerides and organic residues that are obtained from agricultural industry, forestry, household residues and aquaculture [19]. Availability of biomass is the key factor for the sustainability of biorefineries. Corn, sugarcane, wheat, and other grains plus rapeseed and

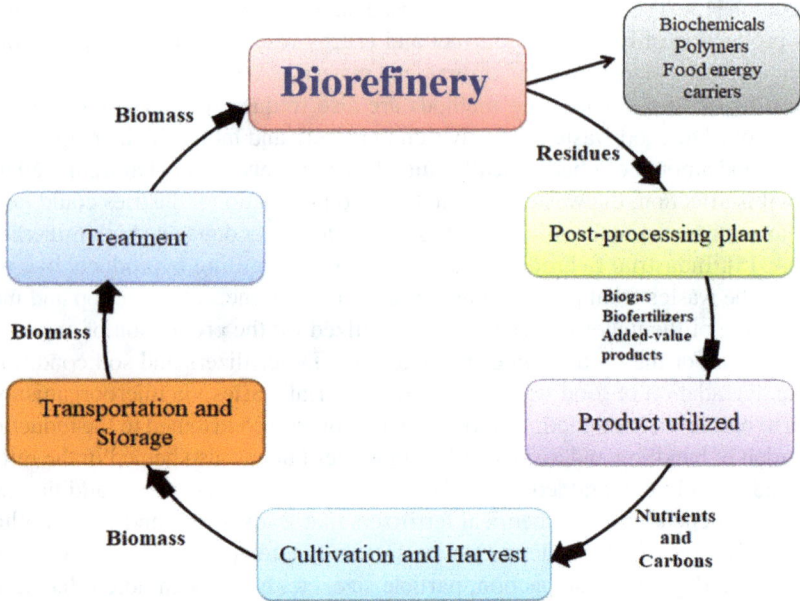

FIGURE 10.1 Working mechanism of biorefinery.

palm oil are categorized as first-generation feedstock's for bio-refineries. Second-generation biomass includes wastes like residues from the food industry, forestry and nonfood crops; agriculture is becoming a pioneering subject for the research field [20]. This can also help in aiding availability of feedstock that is independent from soil fertility, its yields making it independent from factors like climate, difficulties in harvesting and overcoming high production cost, helps easy transportation and storage and also having reduction in GHG emission. The third generation (algae) and fourth generation (vegetable oil and biodiesel) are also used as feedstock to attempt to obtain renewable biofuels in biorefinery. Food wastes are traditionally used as feedstock for valorization strategies like separation, extraction and fermentation.

A systemized interlinking of bioprocesses is the key to constructing a waste biorefinery. Pourbafrani and co-workers applied a dilute-acid process to hydrolyze citrus waste as pretreatment and obtained firstly limonene. After centrifugation of slurry, the liquid part was utilized in the fermentation and distillation steps to carry out the production of ethanol. Also, pectin was separated after the hydrolysis step by applying a precipitation process. The spillage from the distillation step was used for biogas production [5].

10.3.3 Food Industry, Dairy Waste, and Pharmaceutical Industry Valorization

Food-processing industries produce enormous amounts of waste and by-products and form the second largest generators of the waste after household sewage wastes. Moreover, with the increasing industrialization and urbanization, it is essential to examine that the waste generated from food industries and its utilization and by-products from food processing do play a major role in the upgradation of economic performance, ensuring environmental sustainability in the upcoming years [21]. The compounds recovered from food wastes could have the potential to be employed in different food and biotechnological applications along with the restoration of natural resources. Apart from the conventional methods such as liquid–liquid extraction and mechanical shaking, the recent development of green extraction techniques such as microwave-assisted extraction (MAE), ultrasound-assisted extraction (UAE), enzyme-assisted extraction (EAE) and pulsed electric field (PEF) are seen as significant steps in recovering by-products from food and vegetal wastes [22].

Many researchers have found this emerging technology effective in the extraction process with the combination of other techniques such as osmotic shock and mechanical press. Furthermore, the combination of PEF and solid/liquid extraction produces a lesser amount of food waste compared to existing transformation technologies. Other than this, Electroporator (the electrical system), for the extraction process, which is similar to any other than PEF system is also used for food applications with different specifications. It is composed of a treatment chamber, a pulsed power modulator and a control unit [22].

Industrial fermentation has been developed for valorizing vegetable and fruit processing byproducts and to improve their nutritional and consumption value or to produce biologically active compounds. Fermentation of a wide variety of products including *barley, soya, rice, citrus and milling* by-products has been reported [23]. Different applications reported have been focused on increasing the nutritional value of vegetable by-products, while several *Lactobacillus* and *Penicillium* species have been used to provide high-purity lactic acid. Bacteria and fungi like *Bacillus subtilis, Rhizopus oligosporus* or *Fusarium flocciferum* may be used to efficiently produce protein extracts with high biological value, and a wide variety of functional carbohydrates and glycosidases have been produced employing *Aspergillus, Yarrowia* and *Trichoderma* species [3].

In addition, vegetal waste and by-products can be a valuable and promising source of natural pigments. Various advances, innovations and challenges in the extraction and application of these natural pigments to harvest their potential in functional bioactive have been developed, which are safe, biodegradable and are pigment alternatives to the currently used artificial coloration paradigm extracted from the renewable waste resources [24].

Dairy waste industries, on the other hand, accounts for the large amount of whey that is produced during cheese- and milk-related production processes, which further points out toward the increasing demand by society for more sustainable processes, in order to support Sustainable Development Goal's (SDG's) concept [25].

The main organic residual from the dairy industry consists of chemically modified liquids with a high organic load, considerable variations in pH and a relatively large load of suspended solids. Basically, there are two types of whey: sweet and acid. Sweet whey is a by-product from white hard cheese production, for example, Cheddar or Swiss cheese and is the most common type, while the acid whey (or sour whey) is a by-product from cottage/cream cheese, Skyr yoghurt or Greek yogurt production. Compared to sweet whey, acid whey has less protein, is more acidic and has

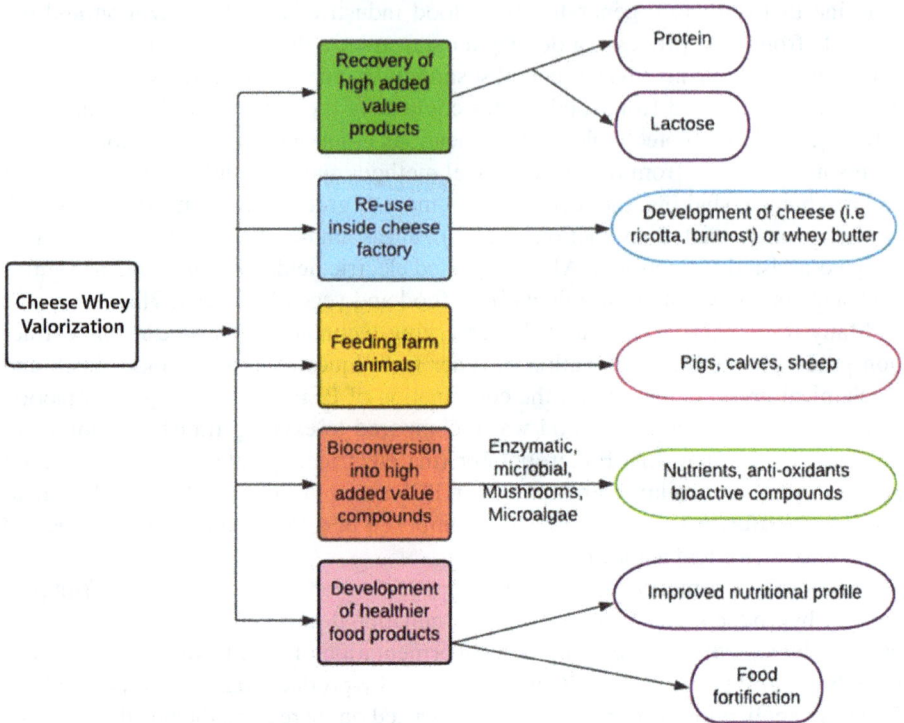

FIGURE 10.2   Process for the valorization of cheese whey.

a more distinct (sour) taste, making it more difficult to valorize [26]. Casein whey is derived from the production of calcium caseinate and has properties similar to acid whey, with a pH between that of acid whey and sweet whey. Though in general, sweet whey holds more importance (Figure 10.2) [27].

Thus, valorization of dairy residuals accounts for its ability to separate solids from whey. A number of approaches have been developed for storing whey before processing. Like, wastewater can be refrigerated and replenished when signs of acidification are observed. Alternatively, anaerobic seed sludge used for fermentation and fresh raw cheese whey is kept at 4°C until used. Finally, the wastewater can be stored at 20°C and thawed before use [27].

Cheese whey (CW), as discussed earlier, is the liquid resulting from the precipitation and removal of milk casein during cheese-making, while there is another CW, called the second cheese whey (SCW), derived from the production of cottage and ricotta cheeses [28]. They are the main by-products of dairy industry, and the major constituent of CW and SCW is lactose, contributing to the high biological oxygen demand (BOD) and chemical oxygen demand (COD) content. Because of this reason, CW and SCW are considered as high-polluting agents, making their disposal a problem for the dairy sector [29]. CW and SCW, however, also consist of lipids, proteins and minerals, making them useful for the production of various compounds. Thus, various microbial processes, useful to promote the bioremediation of CW and

SCW, whey-derived products and the production of health-promoting whey drinks, vinegar and biopolymers, which may be exploited as value-added products in different segments of food and pharmaceutical industries have been developed recently [30].

10.3.4 PHARMACEUTICAL INDUSTRIES

India, as we know, holds an important designation in the global pharmaceuticals sector. The Indian pharmaceutical is the 3rd largest in terms of volume and 13th largest in terms of value. Moreover, India is the largest producer of generic drugs, which includes biopharmaceuticals, bio-services, bio-agriculture, bio-industry comprising vaccines, therapeutics and diagnostics along with biomedical wastes such as used gloves, syringes, PPE kits, etc. However, when we talk about waste valorization in such industries, things get a little bit complicated as there is a huge production of such waste which is not easy to discard but natural sources and plant-based sources make it easier, and thus, with the help of biotechnological applications, the various new innovative high-value-added products are being developed [43].

For example, ergosterol, a vital component of the fungal and protozoan cell membranes, is known to play a fundamental role in membrane fluidity and integrity while acting as a drug target of several antifungal agents. Moreover, ergosterol and its derivatives isolated from mushrooms and other natural sources have been reported to have several therapeutic properties and importance of vitamin D (ergocalciferol and cholecalciferol) in anticancer research [44].

Another example is of the subgenus Bryophyllum, which includes about 25 plant species native to Madagascar and is widely used in traditional medicine worldwide. Different formulations from Bryophyllum have been employed for the treatment of several ailments, including infections, gynecological disorders and chronic diseases such as diabetes, neurological and neoplastic diseases. Two major families of secondary metabolites have been reported responsible for these bioactivities: phenolic compounds and bufadienolide, though they are found in less amount and are biosynthesized in response to different biotic and abiotic stresses, but with the help of novel approaches to valorization of Bryophyllum, allowing a sustainable production of the same. This prevents a massive exploitation of wild plant resources, which are now considered as a promising source of plant bioactive compounds, with enormous antioxidant and anticancer potential, which could be used for their large-scale biotechnological exploitation in cosmetic, food and, specifically, in pharmaceutical industries. [45].

10.3.5 OTHER INDUSTRIES

10.3.5.1 Fisheries

The use of fisheries, its by-products and the processing units can represent an interesting source of highly added value bioactive compounds, such as proteins, carbohydrates, collagen, polyunsaturated fatty acids, poly-phenolic constituents, vitamins, alkaloids, toxins, chitin and carotenoids; nevertheless, their biotechnological potential is still largely underutilized [46].

Fish internal organs are known to possess a rich and nutritional source of bio-active compounds; the fish gut microbiota biosynthesizes essential or short-chain fatty acids, vitamins, minerals or enzymes and is a source of probiotics, producing bioactive compounds with antibiotic and biosurfactant/bioemulsifier activities. Thus, valorization technologies for such raw materials play a great role in providing additional value for companies involved in various fields.

An interesting alternative to the methods used for valorization has recently been suggested, which involves the use of innovative green technologies such as ultrasound-assisted extraction and supercritical fluid extraction (SFC) of seafoods and their by-products, simultaneously preserving and enhancing the quality and the extraction efficiency, as well as minimizing functional properties' losses of the bioactive compounds extracted from marine by-products. The SCF extraction has been largely applied for commercial products in the food industry; however, in order to obtain high-purity products with good added value, the method is more specialized in the nutraceuticals and pharmaceutical applications; but in the biorefinery scenario, SCF plays a crucial role, since it does not alter the biomass in the perspective of further extraction [47].

10.3.5.2 Brewery

In the brewery sector, the form of organic residual which is produced in most abundance is the brewer's spent grain (BSG). It consists primarily of lignocellulosic material (such as cellulose, lignin and hemicelluloses) consisting of fibers, ashes, proteins and others and is nearly 80% carbohydrate. Mostly used for animal feed but also serves as a feedstock for biogas (biomethane) and biofertilizer production [48]. Apart from this, BSG is used for lignocelluloses yeast carriers for beer fermentation, has a potential as an inexpensive bioabsorbent and is used to remove industrial dyes. Due to its high fiber and low ash content, it is suitable for producing bricks too. However, in context to valorization, the recent research has focused mainly on processing technologies, specifically, on the implementation of feedstock and pretreatment processes for different potential applications in order to expand the scope of possible end products. Attention is placed on upgrading BSG and similar residuals to higher-value products but pathways are still predominantly "technology push". Henceforth, the BSG valorization value chain is characterized as low complexity, high efficiency and low supplier capability [49].

10.3.5.3 Slaughter Houses

Industries related to these houses have by-products that are the residuals from slaughtering and processing animals including organs, blood, bones and horns, skins and other fatty tissues, wastewater, and fish remnants in various forms. The meat industry is known to be a high-waste-generating industry, but here, the concept of valorization of animal by-products is demand-driven and similar to dairy industries. The slaughterhouses because of the diversity of animal by-products and the complexity of processing of residuals make them weaker valorization value chain products as compared to dairy industries. Thus, these products go to incineration or landfilling in some parts of the world. Therefore, in order to promote a more modular value chain of valorization in such industries, moving toward a symbiotic-shared site relationship between the slaughterhouses and their rendering operations can help in the advancement and improvement of technologies in this arena [50].

10.3.5.4 Forestry and Paper Industry

Forest, in general, represents the use of enormous forms of residual products along with cascading uses of wood, timber, pulp, paper, biomass and biofuels produced from biorefineries involving energy utilization. Forestry residuals are considered as the largest and most important wood-based biomass source for biofuel production in the future. Though, research in the valorization of forestry by-products is highly dependent on case studies and surveys but still a lot needs to be explored in regard to future prospects and the need for sustainable environment [27].

Forest biomass is basically residues that are composed primarily of cellulose, hemicellulose and lignin in varying proportions depending upon the variability of species. Residues from forest operations are known to have heterogeneous compositions and varied physical properties due to the presence of branches, foliage, tree tops and bark, compared to those derived from wood-manufacturing industries. Several technological approaches have been developed to add value to forest biomass residues through their conversion to biomaterials such as wood-based composite panels, wood-plastic composites, wood pellets and biofuels, such as biochar, bio-oil, syngas (thermochemical approach) and biogas biochemical approach [51].

Research is still required for their conversion into value-added products because their complex structural and chemical mechanisms resist decomposition in developing high-quality and economically viable biofuels and biomaterials. In contrast, wood-based panels, composites, pellets and biofuels produced by the wood-manufacturing industries exhibit superior properties and characteristics for commercialization. Recent studies and advances regarding valorization of forest biomass residues are a welcome recognition of the need to transition to a sustainable economy, and a definitive strategy for achieving objectives that have been set for reducing greenhouse gas emissions [52].

10.4 STATUS OF WASTE VALORIZATION RESEARCH IN INDIA

In India, with the initiation of the Swachh Bharat Abhiyan campaign in 2014 throughout the country, the massive problem of management of collected waste emerged. In this regard, various research institutes like The Indian Council of Agricultural Research, New Delhi, Council of Scientific and Industrial Research labs and other states and central universities as well as different industries came forward with various innovative ways to generate some valuable products from the waste. In this regard, the Indian Agriculture research focused on the management of waste generated during agricultural practices. The Council aggressively worked on the management of agricultural waste together with converting it into some useful products, thereby generating wealth from waste [53]. Consequently, from all over India, various researchers reported various setup and strategies of waste valorization as shown in Table 10.2.

TABLE 10.2
Various Setup and Strategies of Waste Valorization Done by Various ICAR Units in India [53]

Sr. No.	Product Name	Raw Materials	Features	Uses	Advantages	Developers
1.	Lac dye	Effluent of stick lac washing	Anthraquinone derivatives, Avg. recovery yield 0.25% by weight	Mordant dye for dyeing wool and silk, Rs. 3000–4000/kg	Natural and nontoxic, Used as food dye for coloring ham and sausages	S.K. Pandey, ICAR-IINRG, Ranchi
2.	Mushroom production	Oil palm bunch refuse, oil palm mesocarp	High-quality protein rich mushroom, can be setup as a small-scale industry	For the production of paddy-straw mushroom, oyster mushroom, summer–white milky and button mushroom	Regular income to rural population, Eco-friendly disposal of oil palm factory waste	M. Kochu Babu, ICAR-IIOPR, Pedavegi
3.	Cellulose and micro-crystalline cellulose	Banana sheath fibers	Mechanical extraction via improved fiber extractor, acid hydrolysis and alkali treatment used to obtain pure cellulose	Used in handicrafts, paper cards and high-quality fabric materials, ropes, mats, etc.	Cost ratio 2.27, cellulose fibers used in textile industry, as adsorbent, chemical fibers, as reinforcement biocomposites, etc.	K.J. Jeyabaskaran, M.M Mustafa, R. Pitchaimuthu, T. Sekar, ICAR-NRCB, Tamil Nadu
4.	Yogurt with antioxidant	Wine lees	Has antibacterial and antioxidant properties, none caloric thickness, flavor enhancers, food additives such as beta-glucans, source of natural color	Beneficial for health as rich in antioxidants	Beneficial to health, helpful in proper disposal of winery wastes, cost ratio 1.5:1	Ajay Kumar Sharma, ICAR-NRCG, Pune

(Continued)

TABLE 10.2 (Continued)

Various Setup and Strategies of Waste Valorization Done by Various ICAR Units in India [53]

Sr. No.	Product Name	Raw Materials	Features	Uses	Advantages	Developers
5.	Technology of extraction of high-value polysaccharide	Ripened Lasora (*Cordia myxa*)	Polysaccharide powder obtained by heating, mechanical stirring of mucilage, precipitation with dil. acid, purification and drying	Powder can be used as an emulsifying and thickening agent in foods, Good film forming properties, hence, used in coating application	Underutilized ripened fruit can be exploited for polysaccharide extraction	Saurab Swami, P.R. Meghwal, Akath Singh, Om Prakash, ICAR-CAZIR, Jodhpur
6.	CIPHET GLOW U (face pack and toner)	Kinnow peel and its extracts	Prepared using indigenous technology, low-cost development	Gives supple, soft skin upon regular use, possess disinfectant and antiseptic properties	Nourishing, healing, cleansing, source of vitamin C and antioxidants	D.S. Uppal, H.S. Oberoi, R.T. Patil, ICAR-CIPHET, Punjab
7.	Agarbatti	"Patchouli Spent Charge"	2/3 of wood powder is substituted with Patchouli Spent Charge powder	Used to freshen up the scent of indoor areas, for spiritual purpose, etc.	Could substitute 10% of originally required powder	H.G. Ramaya, Vasaundhara, V. Palanimuthu, Dayanand Kumar, JAU, Junagdh
8.	Natural fiber	Pineapple leaf	Fiber extracted from entire length of pineapple leaves, extractor's output capacity is 30 kg green leaves per hour	Used as fabric for textile materials and in the manufacture of yarns and handicrafts	Helpful in development of diversified value-added products, provides additional income to farmers, creates rural employment	Laxmikanta Nayak, ICAR-NINEFT, Kolkata

(*Continued*)

TABLE 10.2 (Continued)
Various Setup and Strategies of Waste Valorization Done by Various ICAR Units in India [53]

Sr. No.	Product Name	Raw Materials	Features	Uses	Advantages	Developers
9.	Virgin pomegranate seed oil	Pomegranate seeds	Oil is extracted via hydraulic cold press from leftover portion of arils after juice extraction	Exponentially potent oxidant, natural anti-inflammatory agents, improves heart health, protects against cancer, atherosclerosis and used in cosmetic industries	18.8%–19.20% recovery of high-quality virgin, holds potential for employment generation	Nilesh N. Gaikwad, R.K. Pal, ICAR-NRC, Solapur
10.	Chitin and chitosan	Prawn shell waste	Demineralization and deproteinization of crustacean shell using acid and alkali. Deacetylation of chitin yields chitosan	For developing bio-degradable and anti-microbial packaging films for food, beads for removal of lead and fluoride from water	High-value products of industrial, pharma and nutraceutical importance, also supplements fishers/ aqua farmers income	P. Madhavan, P.T. Mathew, K.G. Ramachanan Nair, A.A. Zyundheen, ICAR-CIFT, Kochi
11.	Chitopro-bone health supplement	Fish and shellfish waste	Contains chitin derivatives from shrimp shell and collagen derivatives from fish scale	Use of supplements helps in improving bone health, specially the joints	100% of biological origin, helps fight inflammation and arthritis	K. Elavarsan, C.S. Tejpal, A.A. Zyundheen, ICAR-CIFT, Kochi
12.	Foliar spray	Fish waste	Applicable on wide variety of plants, can be fortified with the deficient components, if required.	An effective method of treating certain nutrient deficiencies perhaps boosting plant growth	Enhances productivity, pest-repellant properties and provides gainful employment with a high return of margin	A.A. Zyudheen, Binsi Pillai, ICAR-CIFT, Kochi

(Continued)

TABLE 10.2 *(Continued)*

Various Setup and Strategies of Waste Valorization Done by Various ICAR Units in India [53]

Sr. No.	Product Name	Raw Materials	Features	Uses	Advantages	Developers
13.	Marine melanin	Cephalopod ink	Technically simple, environment friendly, economically feasible with a greater margin of return	Powerful antioxidant, potent anti-inflammatory and anti-agent, strong photo-protective agent, skin and hair nourishing agent	International market pricing at $100–350/g	P.K. Binsi, A.A. Zyudheen, Ashok Kumar K., ICAR-CIFT, Kochi
14.	Soy-based food product	Okra (by-product from tofu and soy milk production)	3.5%–4% protein, 12%–14.5% crude fiber and 24% protein	Soy-based food product like biscuits, cakes and muffins; also used as animal feed and nitrogen fertilizer	Prepare bakery-baked soy products, fermented products rich in proteins, can be added to compost to add organic nutrients and nitrogen	L.K. Sinha, Sumedha Deshpande, ICAR-CIAE, Bhopal
15.	Baskets	Cymbidium Orchids	Cymbidium Orchid leaves are up to 1 m long, dried leaves are very strong and can be utilized for weaving baskets	Used for decorative purposes in domestic market, are very durable and give aesthetic look, are also degradable hence, reducing farm waste	Eco-friendly product, good for women empowerment, can earn up to Rs. 150–200/- per basket	D.R. Singh, Raj Kumar, D. Barmen, P. Ravi Kishor, ICAR-NRCO, Sikkim

10.5 STATUS OF WASTE VALORIZATION RESEARCH AT THE GLOBAL SCALE

Waste creation differs greatly among locations and is impacted by economic growth as well as time. As a result, making the ideal waste treatment choices is critical in order to fulfill the demands of all successful methods. Technological advancements, enhanced pollution control systems, political incentives and strict standards urge all nations to engage in the investigation of waste management options. Emerging innovations have recently been invented to manufacture value-added goods from various forms of trash, such as agricultural leftovers and food scraps, as well as waste from the dairy sector. Many items have been recovered from various sorts of garbage across the world. The development of numerous important products such as bioplastic, bioethanol, salts and other nutrients is aided by various integrated and holistic techniques for organic waste usage and food industry waste valorization technologies. Some global advancements in the valorization of food waste and organic waste are mentioned in Table 10.3.

TABLE 10.3
Various Strategies of Waste Valorization Reported in Different Parts of the World

Waste Material	Processing	Application/Uses/Products	References
Spent coffee grounds	Mechanical and thermal conversion	Production of bioplastic	[54]
Mixture of food waste from a local grocery store	Hydrothermal liquefaction (HTL)	Energy-dense biocrude oil	[55]
Pineapple waste	Fermentation and saccharification processes	Bioethanol production and bromelain extraction	[56]
Sugarcane bagasse	Fermentation and saccharification processes	Bioethanol production	[57]
Tapioca starch and sugarcane bagasse fiber (SBF)	Solution casting method	Bioplastics	[58]
Agricultural waste	Polyhydroxybutyrate (PHB) accumulating	Biodegradable plastic	[59]
Cheese whey waste	Fed-batch fermentation	D-lactic acid	[21]
Whey waste	Redox-mediated electrochemical desalination	Desalination and salt concentration for reuse, as well as the simultaneous recovery of highly purified protein contents	[60]
Delactosed whey permeate	Microbial fermentation (*Corynebacterium glutamicum*)	Production of amino acids and other important compounds	[61]

In the case of waste valorization, worldwide, sugarcane waste valorization is widely accepted [56]. Sugarcane processing generates a large quantity of trash, which may be valorized into biofuels and value-added chemicals using the circular bioeconomy idea. In this context, biomass integrated-gasifier/gas turbine combined cycle technology has continued its efforts for power generation using cane waste and bagasse, particularly in Brazil and Cuba [56]. Similarly, researchers are also working on the production of second-generation biofuels, such as lignocellulosic ethanol from cane waste in conjunction with bagasse. Another environmentally acceptable method of valorization is the conversion of sugarcane leaves into charcoal for use in cooking [57, 58].

Another case study of waste valorization in the production of different types of prebiotics, such as oligosaccharides; waste from the agro-industries has been converted from sucrose to fructo-oligosaccahrides with the help of enzymes called fructosyltransferases [61]. Similarly, waste from the pharmaceutical industry has been converted from sucrose to fructo-oligosaccahrides with the help of enzymes called fructosyltransferases [62]. The Xylo-oligosaccharides (XOS) are another important class of oligosaccharides which have been explored using agro-wastes via the enzyme Drieslase [61]. Similar efforts have been made to develop a continuous and cost-effective chemo-enzymatic process for the fractionation of holocellulose from agro-wastes into soluble arabino-XOS, soluble XOS and soluble cellooligosaccharide [62].

10.6 CONCLUSION AND FUTURE SCOPE

The creation of wealth from waste is a challenging area of research. However, this will provide a long-term solution to the problem of waste management. Apart from this, the technologies that have explored different areas of wealth creation utilizing various waste materials have generated a lot of valuable products in the form of energy, high-value chemicals and other raw materials. Some of these technologies have been transferred or commercialized; several of them are present at various stages of commercialization. So, in order to achieve more benefits in the waste valorization aspects, there is a need for further research along with the proper documentation of these technologies in the form of a compendium that will attract different stakeholders or investors. Further, the new business ventures and startups may also take the initiative to scale up technologies on a large scale that will ultimately serve as a stepping stone for employment opportunities and a clean environment.

REFERENCES

1. Kumar A and Agrawal A. 2020. Recent trends in solid waste management status, challenges, and potential for the future Indian cities – A review. *Current Research in Environmental Sustainability* 2, 100011.
2. Patel U and Ahluwalia I J. 2018. *Solid Waste Management in India: An Assessment of Resource Recovery and Environmental Impact*, Indian Council for research on International relations, **358**, 1–38.
3. Sabater C, Ruiz L, Delgado S, Ruas-Madiedo P and Margolles A. 2020. Valorization of vegetable food waste and by-products through fermentation processes. *Frontiers in Microbiology* **11**, 2604.

4. Arancon R A D, Lin C S K, Chan K M, Kwan T H and Luque R. 2013. Advances on waste valorization: New horizons for a more sustainable society. *Energy Science & Engineering* **1**, 53–71.

5. Otles S and Kartal C. 2018. 11 - Food Waste Valorization. *Sustainable Food Systems from Agriculture to Industry*, ed. C M Galanakis (Academic Press), pp. 371–99.

6. Mohan S V, Butti S K, Amulya K, Dahiya S and Modestra J A. 2016. Waste biorefinery: A new paradigm for a sustainable bioelectro economy. *Trends in Biotechnology* **34**, 852–5.

7. Rocha-Meneses L, Raud M, Orupõld K and Kikas T. 2017. Second-generation bioethanol production: A review of strategies for waste valorisation. *Agronomy Research* **15**, 830–47.

8. Sydney E B, Neto C J D, Novak A C, Medeiros A B P, Nouaille R, Larroche C and Soccol C R. 2016. Bioethanol Wastes: Economic Valorization. *Green Fuels Technology* (Springer), pp 255–89.

9. Jõgi K and Bhat R. 2020. Valorization of food processing wastes and by-products for bioplastic production. *Sustainable Chemistry and Pharmacy* **18**, 100326.

10. Tsang Y F, Kumar V, Samadar P, Yang Y, Lee J, Ok Y S, Song H, Kim K-H, Kwon E and Jeon Y J. 2019. Production of bioplastic through food waste valorization. *Environment International* **127**, 625–44.

11. Elkhalifa S, Al-Ansari T, Mackey H R and McKay G. 2019. Food waste to biochars through pyrolysis: A review. *Resources, Conservation and Recycling* **144**, 310–20.

12. Lin C S K, Pfaltzgraff L A, Herrero-Davila L, Mubofu E B, Abderrahim S, Clark J H, Koutinas A, Kopsahelis N, Stamatelatou K and Dickson F. 2013. Food waste as a valuable resource for the production of chemicals, materials and fuels. Current situation and global perspective. *Energy & Environmental Science* **6**, 426–64.

13. Perteghella A and Vaccari M. 2017. Organic waste valorization through composting process: A full-scale case study in Maxixe, Mozambique. *Environmental Engineering & Management Journal (EEMJ)* **16**, 1819–1826

14. Bhatnagar A, Kaczala F, Hogland W, Marques M, Paraskeva C A, Papadakis V G and Sillanpää M. 2014. Valorization of solid waste products from olive oil industry as potential adsorbents for water pollution control—a review. *Environmental Science and Pollution Research* **21**, 268–98.

15. Gullon P, Gullon B, Astray G, Carpena M, Fraga-Corral M, Lage M P and Simal-Gandara J. 2020. Valorization of by-products from olive oil industry and added-value applications for innovative functional foods. *Food Research International* **137**, 109683.

16. Yaashikaa P R, Kumar P S, Saravanan A, Varjani S and Ramamurthy R. 2020. Bioconversion of municipal solid waste into bio-based products: A review on valorisation and sustainable approach for circular bioeconomy. *Science of the Total Environment* **748**, 141312.

17. Xiong X, Yu I K M, Dutta S, Mašek O and Tsang D C W. 2021. Valorization of humins from food waste biorefinery for synthesis of biochar-supported Lewis acid catalysts. *Science of the Total Environment* **775**, 145851.

18. Khounani Z, Hosseinzadeh-Bandbafha H, Moustakas K, Talebi A F, Goli S A H, Rajaeifar M A, Khoshnevisan B, Salehi Jouzani G, Peng W, Kim K-H, Aghbashlo M, Tabatabaei M and Lam S. 2021. Environmental life cycle assessment of different biorefinery platforms valorizing olive wastes to biofuel, phosphate salts, natural antioxidant, and an oxygenated fuel additive (triacetin). *Journal of Cleaner Production* **278**, 123916.

19. Rajendran N, Gurunathan B, Han J, Krishna S, Ananth A, Venugopal K and Sherly Priyanka R B. 2021. Recent advances in valorization of organic municipal waste into energy using biorefinery approach, environment and economic analysis. *Bioresource Technology* **337**, 125498.

20. Moreno A D, Ballesteros M and Negro M J. 2020. 5 – Biorefineries for the Valorization of Food Processing Waste. *The Interaction of Food Industry and Environment*, ed. C Galanakis (Academic Press), pp. 155–90.
21. Liu P, Zheng Z, Xu Q, Qian Z, Liu J and Ouyang J. 2018. Valorization of dairy waste for enhanced D-lactic acid production at low cost. *Process Biochemistry* **71**, 18–22.
22. Arshad R N, Abdul-Malek Z, Roobab U, Qureshi M I, Khan N, Ahmad M H, Liu Z and Aadil R M. 2021. Effective valorization of food wastes and by-products through pulsed electric field: A systematic review. *Journal of Food Process Engineering* **44**, e13629
23. Sarkar O, Butti S K and Mohan S V. 2018. Chapter 6 – Acidogenic Biorefinery: Food Waste Valorization to Biogas and Platform Chemicals. *Waste Biorefinery*, eds. T Bhaskar, A Pandey, S V Mohan, D-J Lee and S K Khanal (Elsevier), pp. 203–18.
24. Gottardi D, Siroli L, Vannini L, Patrignani F and Lanciotti R. 2021. Recovery and valorization of agri-food wastes and by-products using the non-conventional yeast *Yarrowia lipolytica*. *Trends in Food Science & Technology* **115**, 74–86.
25. Barba F J. 2021. An integrated approach for the valorization of cheese whey. *Foods* **10**, 564.
26. Panghal A, Patidar R, Jaglan S, Chhikara N, Khatkar S K, Gat Y and Sindhu N. 2018. Whey valorization: Current options and future scenario–a critical review *Nutrition & Food Science* **48**(3), 520–535.
27. Gregg J S, Jürgens J, Happel M K, Strøm-Andersen N, Tanner A N, Bolwig S and Klitkou A. 2020. Valorization of bio-residuals in the food and forestry sectors in support of a circular bioeconomy – A Review. *Journal of Cleaner Production* **267**, 122093.
28. Yadav J S, Yan S, Pilli S, Kumar L, Tyagi R D and Surampalli R Y. 2015. Cheese whey: A potential resource to transform into bioprotein, functional/nutritional proteins and bioactive peptides. *Biotechnology Advances* **33**, 756–74.
29. Prazeres A R, Carvalho F and Rivas J. 2012. Cheese whey management: A review. *Journal of Environmental Management* **110**, 48–68.
30. Zotta T, Solieri L, Iacumin L, Picozzi C and Gullo M. 2020. Valorization of cheese whey using microbial fermentations. *Applied Microbiology and Biotechnology* **104**, 2749–64.
31. Lin J-C, Mariuzza D, Volpe M, Fiori L, Ceylan S and Goldfarb J L. 2021. Integrated thermochemical conversion process for valorizing mixed agricultural and dairy waste to nutrient-enriched biochars and biofuels. *Bioresource Technology* **328**, 124765.
32. Kwon D, Yi S, Jung S and Kwon E. 2021. Valorization of synthetic textile waste using CO_2 as a raw material in the catalytic pyrolysis process. *Environmental Pollution* **268**, 115916.
33. Santiago B, Moreira M T and González-García F G. 2020. Identification of environmental aspects of citrus waste valorization into D-limonene from a biorefinery approach. *Biomass and Bioenergy* **143**, 105844.
34. Leal C, Gouvinhas I, Santos R A, Rosa E, Silva A M, Saavedra M J and Barros A I R N A. 2020. Potential application of grape (*Vitis vinifera* L.) stem extracts in the cosmetic and pharmaceutical industries: Valorization of a by-product. *Industrial Crops and Products* **154**, 112675.
35. Babaei M, Tsapekos P, Alvarado-Morales M, Hosseini M, Ebrahimi S, Niaei A and Angelidaki I. 2019. Valorization of organic waste with simultaneous biogas upgrading for the production of succinic acid. *Biochemical Engineering Journal* **147**, 136–45.
36. Bian H, Gao Y, Luo J, Jiao L, Wu W, Fang G and Dai H. 2019. Lignocellulosic nanofibrils produced using wheat straw and their pulping solid residue: From agricultural waste to cellulose nanomaterials. *Waste Management* **91**, 1–8.
37. Majee S, Halder G and Mandal T. 2019. Formulating nitrogen-phosphorous-potassium enriched organic manure from solid waste: A novel approach of waste valorization. *Process Safety and Environmental Protection* **132**, 160–8.

38. Wongsirichot P, Gonzalez-Miquel M and Winterburn J. 2019. Holistic valorization of rapeseed meal utilizing green solvents extraction and biopolymer production with pseudomonas putida. *Journal of Cleaner Production* **230**, 420–9.
39. Swain B, Shin D, Joo S Y, Ahn N K, Lee C G and Yoon J-H. 2017. Selective recovery of silver from waste low-temperature co-fired ceramic and valorization through silver nanoparticle synthesis. *Waste Management* **69**, 79–87.
40. Nath A, Verasztó B, Basak S, Koris A, Kovács Z and Vatai G. 2016. Synthesis of lactose-derived nutraceuticals from dairy waste whey—A review. *Food and Bioprocess Technology* **9**, 16–48.
41. Kwan T H, Pleissner D, Lau K Y, Venus J and Lin P A. 2015. Techno-economic analysis of a food waste valorization process via microalgae cultivation and co-production of plasticizer, lactic acid and animal feed from algal biomass and food waste. *Bioresource Technology* **198**, 292–9.
42. Chojnacka K, Górecka H, Michalak I and Górecki H. 2011. A review: Valorization of keratinous materials. *Waste Biomass Valor* **2**, 317–21.
43. Marić S, Jocić A, Krstić A, Momčilović M, Ignjatović L and Dimitrijević A. 2021. Poloxamer-based aqueous biphasic systems in designing an integrated extraction platform for the valorization of pharmaceutical waste. *Separation and Purification Technology* **275**, 119101.
44. Papoutsis K, Grasso S, Menon A, Brunton N P, Lyng J G, Jacquier J-C and Bhuyan D J. 2020. Recovery of ergosterol and vitamin D2 from mushroom waste – Potential valorization by food and pharmaceutical industries. *Trends in Food Science & Technology* **99**, 351–66.
45. García-Pérez P, Lozano-Milo E and Gallego L M. 2020. From ethnomedicine to plant biotechnology and machine learning: The valorization of the medicinal plant *Bryophyllum* sp. *Pharmaceuticals* **13**, 444.
46. Lopes C, Antelo L T, Franco-Uría A, Alonso A A and Pérez-Martín R. 2015. Valorisation of fish by-products against waste management treatments – Comparison of environmental impacts. *Waste Management* **46**, 103–12.
47. Caruso G, Floris R, Serangeli C and Di Paola L. 2020. Fishery wastes as a yet undiscovered treasure from the sea: Biomolecules sources, extraction methods and valorization. *Marine Drugs* **18**, 622.
48. Duque A F, Castro P M L and Reis M A M. 2016. Treatment and valorization of cellulosic fraction of brewery waste. *New Biotechnology* **33**, S69.
49. Rivera P, Leos S, Solis M D and Domínguez V E. 2021. Recent trends on the valorization of winemaking industry wastes. *Current Opinion in Green and Sustainable Chemistry* **27**, 100415.
50. Silveira N C, Oliveira G H D, Damianovic M H R Z and Foresti E. 2021. Two-stage partial nitrification-anammox process for nitrogen removal from slaughterhouse wastewater: Evaluation of the nitrogen loading rate and microbial community analysis. *Journal of Environmental Management* **296**, 113214.
51. Bacelo H, Vieira B R C, Santos S C R, Boaventura R A R and Botelho C M S. 2018. Recovery and valorization of tannins from a forest waste as an adsorbent for antimony uptake. *Journal of Cleaner Production* **198**, 1324–35.
52. Braghiroli F L and Passarini L. 2020. Valorization of biomass residues from forest operations and wood manufacturing presents a wide range of sustainable and innovative possibilities. *Current Forestry Reports* **6**, 172–83.
53. ICAR. 2020. *Creating Wealth from Agricultural Waste* (India: Indian Council of Agricultural Research, New Delhi).
54. Kourmentza C, Economou Ch N, Tsafrakidou P and Kornaros M. 2018. Spent coffee grounds make much more than waste: Exploring recent advances and future exploitation strategies for the valorization of an emerging food waste stream. *Journal of Cleaner Production* **172**, 980–92.

55. Motavaf B and Savage P E. 2021. Effect of process variables on food waste valorization via hydrothermal liquefaction. *ACS ES&T Engineering* **1**, 363–74.

56. Gil S and Maupoey F. 2018. An integrated approach for pineapple waste valorisation. Bioethanol production and bromelain extraction from pineapple residues. *Journal of Cleaner Production* **172**, 1224–31.

57. Jugwanth Y, Sewsynker-Sukai Y and Kana G. 2020. Valorization of sugarcane bagasse for bioethanol production through simultaneous saccharification and fermentation: Optimization and kinetic studies. *Fuel* **262**, 116552.

58. Asrofi M, Sapuan S M, Ilyas R A and Ramesh. 2021. Characteristic of composite bioplastics from tapioca starch and sugarcane bagasse fiber: Effect of time duration of ultrasonication (Bath-type). *Materials Today: Proceedings* **46**, 1626–30.

59. Getachew A and Woldesenbet F. 2016. Production of biodegradable plastic by polyhydroxybutyrate (PHB) accumulating bacteria using low cost agricultural waste material. *BMC Res Notes* **9**, 509.

60. Kim N, Jeon J, Elbert J and Su K C. 2022. Redox-mediated electrochemical desalination for waste valorization in dairy production. *Chemical Engineering Journal* **428**, 131082.

61. Shen J, Chen J, Jensen P R and Solem C. 2019. Development of a novel, robust and cost-efficient process for valorizing dairy waste exemplified by ethanol production. *Microbial Cell Factories* **18**, 51.

62. Samanta A K, Jayapal N, Kolte A P, Senani S, Sridhar M, Dhali A, Suresh K P, Jayaram C and Prasad C S 2015. Process for enzymatic production of xylooligosaccharides from the xylan of corn cobs. *Journal of Food Processing and Preservation* **39**(6), 729–36.

11 Eco-Friendly Cities and Villages with Sustainability
Futuristic Perspectives

Bhima Sridevi[1], Shital Patel[1], and Egharevba Godshelp Osas[2]

[1]Department of Pharmacy, Bharat Institute of Technology, Mangalpally, Ibrahimpatnam, Telangana, India
[2]Industrial Chemistry Programme, Department of Physical Sciences, Landmark University, Omu-Aran, Kwara State, Nigeria

CONTENTS

11.1 Introduction 238
    11.1.1 Brief History of Eco-Friendly Cities and Villages 238
    11.1.2 Definition of Eco-Friendly Cities and Villages 239
    11.1.3 Sustainability for 21st-Century Dwellers 239
11.2 Principles of Building Eco-Friendly Urban Settlements 240
    11.2.1 Challenges of Building Sustainable Eco-Friendly Cities
        and Villages 241
    11.2.2 Principles with Measures...245
        11.2.2.1 Principle 1: Renewable Energy for Zero CO_2
                Emissions 247
        11.2.2.2 Principle 2: Zero-Waste City...247
        11.2.2.3 Principle 3: Climate and Context....................................247
        11.2.2.4 Principle 4: Landscape, Gardens, and Urban Biodiversity......247
        11.2.2.5 Principle 5: Water..248
        11.2.2.6 Principle 6: Sustainable Transport and Good Public
                Space 248
        11.2.2.7 Principle 7: Local and Sustainable Materials 249
        11.2.2.8 Principle 8: Density and Retrofitting for Accessible
                Districts...249
        11.2.2.9 Principle 9: Green Buildings and Districts......................249
        11.2.2.10 Principle 10: Livability, Healthy Communities, and
                Mixed-Use Programs...250
        11.2.2.11 Principle 11: Local Food and Short Supply Chains.........251

DOI: 10.1201/9781003369554-11

11.1 INTRODUCTION

Over the past few years, the global population increase and technological progress sig-
nify that we are altering the way we live our lives; as a result, the sustainability of our
planet is in greater danger than ever before, endangering the lives of many plants and
animal species, as well as humans, due to the utilization of automobiles, industrialization, urbanization, technologies, plastic use, artificial intelligence, etc. By the end of
the 21st century, developing nations are predicted to have tripled their aggregate built-
up with urban areas, accounting for about 90% of global urban growth. In 1950, only
30% of the world's population lived in urban areas, which grew up to 55% by 2018.
Approximately two-thirds of the population in Europe and Central Asia live in cities. Important causes for urbanization are industrialization, better social and economic
life and better resources of education, modernization and change in mode of living.
Urbanization offers positive benefits in terms of GDP growth of nations, monetary
growth, expanding business activities, social-cultural incorporation, resourceful services,
etc. Issues caused by rapid urbanization are overcrowding, unemployment, slum area,
sewage problem, trash disposal, urban crimes, traffic issues, increasing energy demand,
environmental issues, etc., which are escalating.

11.1.1 BRIEF HISTORY OF ECO-FRIENDLY CITIES AND VILLAGES

So, this unprecedented city, governments and the global development community
pose enormous problems as a result of urban expansion. It confronts us with a unique
opportunity to organize, design, develop and manage communities that are both environmentally and economically viable. The term "Climate Change" was introduced
in the year 2000. It encompasses not only global warming but also extreme weather
effects like air pollution, land pollution, water pollution, droughts, floods, deforestation, shifting wildlife populations and habitats, rising seas and a range of other
consequences like every year, 6 million hectares of productive land converts into
useless deserts and more than 11 million hectares of forests are destroyed. Climate
change, economic transformation and urbanization have risen drastically for future
life on the planet and are overwhelming, but the question of what to do about it
remains controversial. Even governments have collectively pledged to slow global
warming in contemporary eco-city research. Various vital international agreements
have also been made to tackle climate change issues: *Montreal Protocol, 1987, UN*

Framework Convention on Climate Change (UNFCCC), 1992, Kyoto Protocol, *2005, Paris Agreement, 2015.* A UN body established in 1988, Intergovernmental Panel on Climate Change (IPCC), regularly assesses the latest climate science and produces consensus-based reports for countries [1].

More than 35,000 million metric tons of CO_2 gas were dispensed into the atmosphere in the year 2020. This "greenhouse effect" might increase by average temperatures high enough to change agricultural production zones, elevate sea levels high enough to flood coastal cities and destabilize national economies. Acidification burns forests and lakes throughout Europe, along with ruining the artistic and architectural legacy of nations. It may have acidified enormous expanses of territory beyond repair. The combustion of fossil fuels emits greenhouse gases into the atmosphere, triggering gradual global warming. Other industrial pollutants have the potential to severely erode the planet's protective ozone layer [2].

11.1.2 Definition of Eco-Friendly Cities and Villages

New approaches are needed to maintain and raise the standard of living in cities as a result of urbanization and climate change. Green Cities Europe is a framework that promotes public space greening by offering creative ideas, knowledge based on scientific research and technical knowledge. Its operations focus on issues, notably health, climate change, the economy, biodiversity and social cohesion [3]. Biodiversity, climate, wellness and air quality all benefit from public eco-friendly spaces. It is clear that, if urbanization increases, there should be many new fundamental approaches to address the queries: How could cities continue to be a part of urbanization's prospects for economic growth and poverty reduction while also addressing its drawbacks? How can cities achieve this rate and size of urbanization while still being aware of their own capacity constraints? How may ecological and economic concerns be reconciled so that they produce cumulative benefits?

The conversion of existing villages to green ones results in multiple benefits, including a reduction in demand for water (20%–30% saving), reduction in demand for power (30%–40% energy saving), better managed solid waste, access to basic health care, education, transport, recreational facilities and access to safe drinking water and hygienic sanitation. In order to achieve this motto, the Confederation of Indian Industry (CII) established the Indian Green Building Council (IGBC) in 2001. The IGBC offers an array of services including ratings and certifications to encourage a sustainable built environment. It has been working complicatedly toward converting existing villages into green and self-sustainable villages and established the IGBC Green Village Rating System, which also helps to identify green features that can be implemented to make a village green [4].

11.1.3 Sustainability for 21st-Century Dwellers

Sustainable cities strengthen citizen's and society's well-being by incorporating ecological processes into urban development and management, whereas by successfully employing all tangible and intangible assets, economic cities deliver support and prospects for citizens and enterprises, enabling productive, diverse and sustainable economic activity. Eco2 City, as the name suggests, is focused on the synergy and interconnection of ecological and economic sustainability, as well as their inherent potential to reinforce each other in a city setting with appropriate strategic approaches that enhance economically the resource efficiency and simultaneously decrease harmful pollution and unnecessary waste. This ensures a viable future comprised of enhanced quality of life, improves their economic competitiveness and resilience and strives to function harmoniously with natural systems on global ecosystems that can be realized by exceptional leadership, management, policies, laws, standards and criteria, strategic collaborations, urban design and society through integrated urban planning, which leverages the benefits of ecological systems by protecting and nurturing the assets for future generations. The main challenge to reconcile nature is to convert urban centers into resilient and sustainable communities that benefits inhabitants by lowering energy costs, enhancing service quality, decreasing waste, offering better urban landscapes and economic opportunities. A sustainable and green city is a community of inhabitants, residents, workers and visitors who help in maintaining a clean, healthy, and safe environment for all people in society, as well as subsequent generations. The term "green city" refers to a strategy for improving the long-term viability of metropolitan regions. In essence, it encompasses all of the urban conceptions' qualities. The green and sustainable city is "in balance with nature" where all forms of nature from living organisms to their habitats are maintained and extended for the benefit of city residents [5].

A holistic vision of eco-friendly cities and villages in futuristic approach has the following objectives:

- Identification of challenges behind the development of sustainable cities and villages;
- Description of principles and measures to combat challenges for Green Urbanism, which consider accessibility of increasing population, environmental issues, economic development and optimization of energy use;
- Enlightenment of eco-cities and villages concept with some case studies, which gives a basic idea regarding earlier scenario and after implementing the solutions to existed challenges for attaining the sustainable green growth.

This chapter is organized as follows: Section 11.2 contains the main text containing challenges, 15 principles and measures to combat the challenges to attain the Green Urbanism. Section 11.3 explains about the concept of Eco-Village. Section 11.4 includes the information about the selected Eco-cities and Eco-villages as case studies. Section 11.5 concludes the chapter and mentions the future scope of eco-friendly cities and villages.

11.2   PRINCIPLES OF BUILDING ECO-FRIENDLY URBAN SETTLEMENTS

For the past 35 years, an international debate on the eco-city notion has emerged as a crucial research subject in relation to the future of urbanism and the city core [6]. Almost up to 75% of global energy consumption and carbon emissions are the foremost "victims" for climate change. We have entered a "new era of uncertainty" according to Ulrich Beck, wherein energy, water and food supply are critical. We live in a world whereby non-calculable unpredictability is increasing at the same rate as technical advancements [7]. Integrated urban development with an emphasis on energy, water and urban microclimate plays a leading role, and policymakers must interact in order to achieve it. Mexico City, Sao Paulo, Tokyo, Shanghai, Calcutta, Mumbai and Beijing evolved as infinite metropolitan landscapes by the end of the 20th century. As a result, they are new kinds of megacities that show the impossibility of a well-planned city and strategic regulation.

11.2.1 CHALLENGES OF BUILDING SUSTAINABLE ECO-FRIENDLY CITIES AND VILLAGES

Meanwhile, a new planning paradigm has arisen from the concepts and theory of "Sustainability Science" [6]. Cities around the globe are grappling with the question of how to design a climate-friendly and energy-efficient city. The key concerns in the Asian region include rapid expansion and urbanization processes, as well as human migration. The United States, Germany and other European countries are looking into measures to lessen their unsustainable urban sprawl and massive reliance on autos, with a focus on energy-efficient building conversions and optimizing material and energy flows. Almost 40% of European cities are decreasing, whereas others remain static, a few well-known cities are developing and profiting. The integration and reintroduction of biodiversity into the urban built environment will require design approaches for new urban areas. Other considerations include lowering the risk of urban flooding, stormwater harvesting, novel public transportation and eco-mobility concepts, resource recovery through local trash recycling, biodiversity defense and the formation of ecological interconnections [8]. One of the greatest challenges of our day is preserving biodiversity in the face of urbanization, habitat fragmentation, environmental degradation and climate change. In order to have sustainable development, the following challenges are to be noted:

a. **Air pollution:** Air pollution is the presence of gases, particulates (pollens, dust, smoke, soil and chemicals) in the atmosphere that are harmful to the health of humans and other living beings and causes damage to the climate. According to the World Health Organization (WHO), air pollution is responsible for about 7 million deaths per year around the world. Carbon dioxide, methane, nitrous oxide, water vapor and fluorinated carbon are greenhouse gases. Greenhouse gases trap heat in the atmosphere and are responsible for global warming. Methane is emitted during the production, processing, storage, transmission and distribution of natural gas and crude oil. Coal mining is also a source of CH_4 emission. Nitrous oxide is a byproduct from the use of synthetic and organic fertilizers and burning of agricultural residues. Hydrofluorocarbons (HFCs) are used as refrigerants, aerosol propellants, foam-blowing agents, solvents and fire retardants. HFCs have 13 times more global warming potential than CO_2 [9]. Air pollutants are being increased by increasing transportation, industrialization, combustion of fossil fuels and burning of agricultural wastes.

b. **Water:** Water is also natural resource, crucial for life. An approximate 60 million people died of diarrheal diseases related to unsafe drinking water and malnutrition. Dumping of garbage from household and noxious chemicals from factories into water bodies cause water pollution. Hot water discharged by a thermal power plant into water bodies like a river or a local sea affects aquatic lives. Oil waste from industries, ships and machinery sometimes gains access to water bodies, make layer over water surface, causing the death of marine life. Lack of a proper waste management system allows the waste to drain in water bodies. It changes color and temperature of water and causes eutrophication and poses serious hazard to humans, animals and plants. Wastewater from household carry pathogens and also acts as breeding place for disease carriers by decreasing dissolved oxygen content which affects all living systems. Overusage of pesticides and synthetic fertilizers leach to ground water with rain water and contaminate the ground water.

c. **Land:** Land degradation is an emergent restraint to global sustainable development. It is increasing at a rate of 5–10 million hectares and directly impacts the health and livelihoods of an estimated 1.5 billion people globally [10]. The study of land degradation in the 21st century must recognize that urban activities emphasize many of the proximate and underlying causes of global land degradation. The urban environments have increased at the expense of forest, crop land, open areas and, to some extent, wetlands. The urban area expanded to peripheral areas at more rate than the population growth rate because of increasing demand, high rate of land, commercial area, industrial area and other developments. As the migration of people in urban area increases, pressure for land and other resources are increased and it leads to development of slum areas lacking basic amenities. Within countries, poverty has been aggravated by the imbalanced distribution of land and other resources. These factors, combined with growing demands for the commercial use of good land often to grow crops for exports have pushed many crofters onto poor land and robbed them of any hope of participating in their nation's GDP growth. Deforestation at large scale is another example of major menaces to the integrity of regional ecosystems. With the population growth, waste management system is not so updated which exacerbates land pollution and endanger urban forestry and public health [11].

d. **Transport:** It is the 29% contributor to greenhouse gases in the United States. Transport is one of the major contributors of air and noise pollution. Transportation serves a critical role in ensuring the daily work of urban life. Since goods and services in cities are located separately, transportation is required for moving goods and people to permit social and economic activities. Economic growth of a nation depends on an efficient transportation system. People who live and work in major cities are facing escalating levels of traffic, delays, total travel time, costs and frustration and road accidents. It should be a surprise that the increasing population in cities is stressing on urban transportation systems. More vehicles have expanded demand for parking place and impeded public space also.

e. **Energy:** Urbanization has increased energy consumption along key pathways by urban spatial expansion. Urban households consume 50% more energy than rural households per capita. Nonrenewable energy resources, such as fossil fuels and minerals, may lead to resource scarcity in the future. The rate of depletion should, in particular, be taken into account for resources' criticality, the availability of technology to reduce depletion and the chance of substitutes being accessible. In order to achieve sustainable development, the rate of depletion of nonrenewable resources must be minimized. A reliable and long-term energy supply is essential for long-term growth. In order for developing countries' energy consumption to catch up to that of industrialized countries by 2025, global energy consumption must rise. Energy becomes more expensive as demand rises, making it inaccessible to low-income individuals, hence increasing inequality.

f. **Waste management:** A dramatic population growth, changing consumption patterns, economic development and industrialization have resulted in increased urban waste generation. In many circumstances, the current procedures for disposing of toxic wastes entail unacceptable hazards. Radioactive waste from the nuclear power sector can last hundreds of years. Currently, solid waste landfills in and around cities are contributing to high levels of land pollution, water pollution and air pollution, which have a negative effect on the health of millions of people around the world. Clogging of waste in drainage and sewer is a precursor of flooding and water-borne diseases like typhoid, cholera and diarrhea. Open burning of the garbage is a cause of increased upper respiratory diseases. The degradation of organic material produces methane gas which can cause fires and explosions and contribute to global warming. An inadequate waste management infrastructure and routine lack of compliance with waste management rules aid to inefficiency of waste management system.

g. **Education:** Economic and social development can and should be mutually reinforcing. Money expended on education and health can raise human efficiency. Only the law is not enough to enforce the common interest. Human activities and lifestyles have made an imbalance to nature and their effects can be realized by education. Education is a key for achieving environmental and ethical awareness, values and attitudes, skills and behavior consistent with sustainable development and for effective public participation in decision-making. Both formal and non-formal educations are indispensable in changing people's attitudes so that they have the capacity to assess and address their sustainable development concerns [12]. The environment is best secured by decentralizing the management of resources upon which local communities defend, and giving these communities an effective say over the use of these resources. It will also require promoting citizens' initiatives, empowering peoples' organizations and strengthening local democracy. Education can improve agricultural productivity, enhance the status of women, reduce population growth rates, enhance environmental protection and generally raise the standard of living [13]. Only knowledge-oriented education will not change the system but the action is required to have sustainable development.

h. **Ecology:** Development does simplification of ecosystems and reduces the diversity of species. The loss of plant and animal species can highly affect the options for future generations, so sustainable development requires the conservation of plant and animal species. The planet's species are exhaustible. There is a growing scientific consensus that species are vanishing at rates never before witnessed on the planet, although there is also controversy over those rates and the risks they bring about. Yet there is still time to cease the simplification of ecosystem. The diversity of species is necessary for the normal functioning of ecosystems and the biosphere as a whole. The genetic material in wild species contributes greatly yearly to the world economy in the form of improved crop species, new drugs and medicines and raw materials for industry. But utility aside, there are also moral, ethical, aesthetic, cultural and purely scientific reasons for conserving wild beings. A first priority is to establish the problem of disappearing species and threatened ecosystems on political agendas as a major economic and resource issue. Education for sustainable development should raise critical thinking, self-awareness, integrated problem solution and strategic approaches [2].

i. **Policies/political issues:** The enforcement of common interest often suffers because areas of political jurisdiction and areas of impact do not coincide. Energy policies in one jurisdiction cause acid precipitation in another. No supranational authority exists to resolve issues and the common interest can only be articulated through international cooperation. When people have no alternatives, pressure on resources increases. Economic and ecological concerns are interdependent, not necessarily in opposition. For example, policies that conserve the quality of agricultural land and protect forests improve the long-term anticipations for agricultural development. An increase in the energy efficiency and material use presents ecological purposes but can also reduce costs. Sustainability requires the enforcement of wider responsibilities for the impacts of decisions. This requires change in the legal and institutional frameworks that will enforce the common interest [2].

j. **Population and human resources:** In major parts of the world, available environmental resources cannot sustain the growth rate of the population as the rate at which population is exceeding any reasonable expectations of improvements in housing, health care, food security, resources or energy supplies. The problem is how the population correlates with available resources. Thus, the "population problem" must be dealt with in part by attempts to get rid of mass poverty in order to ensure more unbiased access to resources and by providing education to improve human capacity to manage those available resources. Quick decisions are required to control higher rates of population growth. Human resource development is a significant requirement to build up technical knowledge and capabilities, as well as to impart new values to help individuals and nations manage with the swiftly changing environmental, social and developmental phenomenon.

Special attention has to be given to the tribal and indigenous community as their traditional lifestyles are being disrupted by the forces of economic development and their lifestyles can recommend many valuable lessons in the management of resources in complex ecosystems of forest, mountain and dry land to modern societies. Some are threatened with virtual eradication by insensitive development over which they have no control [2].

k. **Food security:** World cereal production has steadily come behind exceeded world population growth, hence there are many people in the world who do not get abundant food. Global agriculture is capable to accomplish the requirement of food for all, but often food is not accessible where it is needed. In industrialized countries, production has usually been highly subsidized and protected from international competition. These subsidies have boosted the overuse of soil and chemicals, the pollution of both water resources and foods with these fertilizers and pesticides and the farmland degradation. Financial burden over a nation's economy is also produced by the subsidy. Much of the subsidy has produced excess food and some of this excess has been sent at discount rates to the developing world, where it has eroded the farming policies of recipient nations. However, there is growing awareness in some countries for the environmental and economic consequences of such ways, and the significance of agricultural policies to motivate conservation. Many developing countries have suffered the opposite problem: farmers are not sufficiently supported. In some, a major breakthrough in food production has been produced by improved technology associated to price incentives and government services. But elsewhere, the food-growing small farmers have been neglected. Coping with often insufficient technology and few economic incitements, many farmers are pushed onto marginal land: too dry, too steep and nutritionally deficient. Forests are cleared and productive dry lands rendered sterile. Most developing nations need more functional incentive systems to encourage production, especially of food crops. Food security requires concentration to questions of distribution because hunger often arises from lack of purchasing power rather than lack of available food [2, 14].

So, the above mentioned are major challenges facing to attain sustainable development in green cities and green villages.

11.2.2 PRINCIPLES WITH MEASURES

Green Urbanism is a comprehensive idea of a sustainable approach to urban systems, which is focused on maintaining a healthy balance between cities. It is multidisciplinary, allowing for existence and evolution without negatively influencing the planet's life support systems and ecology. In addition to architects and urban designers, it demands the participation of landscape architects, urban planners, engineers, ecologists, sociologists, transport planners, physicists, psychologists, economists and other professionals. Green Urbanism emphasizes every effort to reduce energy, water and material usage.

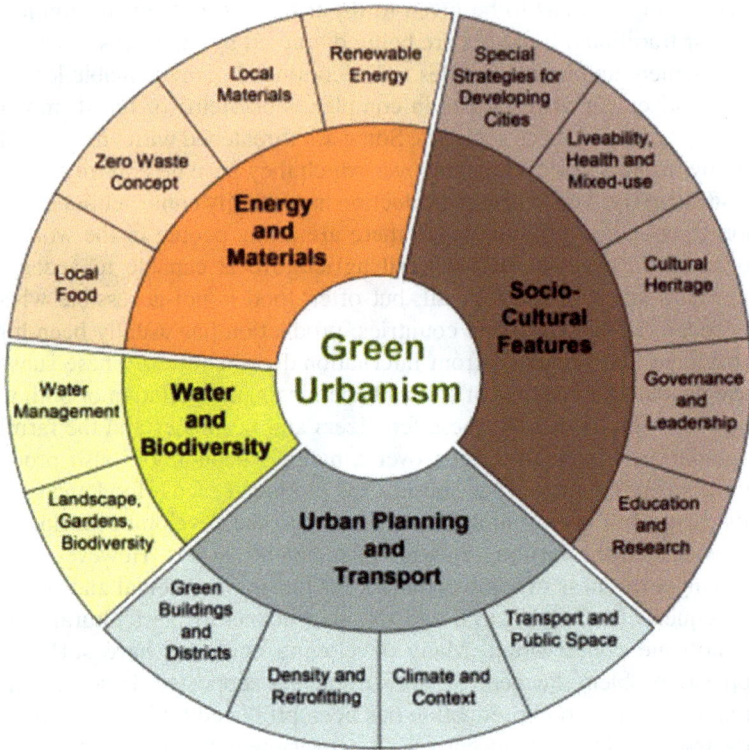

FIGURE 11.1 Principles of Green Urbanism as a conceptual model.

Fifteen Principles of Green Urbanism (Figure 11.1) are discussed in this section as just a conceptual model and a set of rules for how we might be able to tackle the enormous challenge of transforming the existing neighborhoods, districts and communities, as well as how we might be able to re-think the way we design, build and operate in the future. The goal of Green Urbanism is to achieve:

- Zero fossil-fuel energy use
- Zero waste
- Zero carbon emissions

Although the principles explain the techniques required for eco-villages, they must be customized to the location, context and scale of urban development. Some of the concepts may be difficult to implement at first, but they are all necessary in order to save money, improve livability and provide possibilities for residents to interact socially. It is important to emphasize that in order to enable sustainable urban development and assure the sustainability of eco-villages on many levels, all urban design components and sectors must cooperate together and cannot be evaluated independently. This necessitates a comprehensive perceptive of the accessible life-cycles connected with every development location, as well as the investigation of complicated energy/water/materials flow patterns. How these interconnections have been disrupted or destroyed, the planning process must include regenerative components like native species re-establishment.

The following major principles and measures provide a set of basic aspects that further describe the Eco2 Framework and also serve as an even more holistic approach to urban development.

11.2.2.1 Principle 1: Renewable Energy for Zero CO_2 Emissions

Energy supply infrastructure and energy efficiency use were elevated with renewable power perhaps natural gas as a transition fuel in the energy mix, always quickly changes away from heavy fossil fuels such as coal and oil, transforms with local solutions for renewable and the infusion of renewable into the energy mix, the municipal district transforms from an energy consumer to an energy producer. Decentralized energy delivery via a distributed network based on local renewable energy sources is also a must-achieve aim. With solar PV, solar thermal, wind biomass and geothermal energy, city villages could be transformed into local renewable energy power plants and other emerging technologies. Increasing solar power to decarbonize the power supply to 10% of the energy, installing smart grids and regulating solar hot water provide at least 50% of the energy on-site from renewable sources such as precinct-scale wind turbines and small-scale biogas plants.

11.2.2.2 Principle 2: Zero-Waste City

A zero-waste city has to be a closed-loop, circular ecosystem. Turning garbage into a resource is what sustainable waste management entails. Nature's zero-waste management approach should be adopted by all cities. Reducing, recycling, reusing and composting garbage to generate energy is a part of zero-waste urban planning. As a result, by executing "zero waste city" ideas and plans, we will be able to boost resource recovery rates to 100% and eliminate landfilling. All building proposals should strive for a minimum of 50% on-site renewable energy generation, with the energy supply coming from decentralized energy generation and having to consider locally accessible resources as well as the cost and availability of the technology matter.

11.2.2.3 Principle 3: Climate and Context

In terms of orientation, solar radiation, rain, humidity, prevailing wind direction, topography, sheltering, lighting, noise, air pollution and so on, each site or location has its own distinct set of characteristics. The following are some of the features of this principle: climatic conditions are considered as the basic influence for form generation in the development of any project; recognizing the site and its context is critical at the start of any sustainable design project. In addition, each strategic planning process should begin with a map of the city's location and an inventory of its assets. Accordingly, develop a strategic framework that takes into account the urban climate and bioregional context.

11.2.2.4 Principle 4: Landscape, Gardens, and Urban Biodiversity

Many beautiful parks and public gardens of a sustainable city are a source of great delight. This pleasure can best be generated by focusing on local biodiversity, habitat and ecology, as well as wildlife rehabilitation, forest conservation and the preservation of regional features. Easy access to all of these public parks, gardens and public spaces with leisure and recreation options is a crucial component. Biodiversity loss can be reversed by enhancing the natural environment. A healthy and resilient city is one that retains and optimizes its open spaces, natural landscapes and recreational activities. All urban design projects in a sustainable city should include inner-city gardens, urban farming/agriculture and green roofs. It is required to boost the eco-resilience systems by creating urban landscapes that reduce the "urban heat island" (UHI). Furthermore, road shortening reduces traffic congestion and reduces the UHI effect, allowing for additional tree planting. Preserving green space, gardens and farmland, as well as maintaining a green belt surrounding the city and planting trees to absorb CO_2, is a critical objective. Conservation of natural resources, respect for natural energy streams, restoration of stream and river banks and conservation of species dive are all important. Extend tree-planting projects and expand tree-planting in public areas. Wetland restoration will boost landscape and biodiversity while purifying and recycling gray water.

11.2.2.5 Principle 5: Water

By wastewater treatment and sensitive urban water management, closed urban water management and high-water quality enable sustainable water management to provide water security. Reducing energy use, finding more effective uses for water resources, guaranteeing high water quality and protecting aquatic environments are all facets of this idea. Storm water and flood management strategies must be implemented as part of urban drainage and wastewater treatment upgradation. The eco-district must assure the delivery of safe water and sanitation as part of its adequate and inexpensive health care provisions. Fresh water consumption is kept to 125 liters per person per day by using solar-powered desalination and recycling wastewater. To filter and recycle gray water, establish wetlands. In place of a water treatment facility, green infrastructure exploits the natural water cycle to provide access to clean water. For instance, restoring wetlands and planting trees can lessen the need for man-made floods.

11.2.2.6 Principle 6: Sustainable Transport and Good Public Space

Access to basic transportation services is critical since it helps to lessen reliance on automobiles, hence reducing the need to travel. Sustainable cities must address transportation by including eco-mobility principles and smart infrastructure such as electric vehicles, integrated transportation systems (bus transit, light rail, bike stations), increased public space networks and connectivity and a focus on transport-oriented development (TOD). Public transit that is both affordable and easily available gets automobiles off the road, lowering the harmful emissions emitted by regular driving commutes and errands. Investing in public transportation, expanding tramlines and introducing free hybrid buses are all possibilities. Provide multimodal public transportation with a large number of links and options, regular trams and buses, and safe pedestrian and cycling networks. Pedestrians and cyclists should be given higher precedence on city streets. Incentivizing alternative travel by designating some lanes for buses, electric cars and carpooling would be another wonderful method to improve public transportation and encourage greener practices. With innovative technology like hyperloop and smart highways, the destiny of transportation is tackling greener public transit. "Green TODs are designed that will allow for the creation of a variety of medium-density dwelling typologies as well as a number of transit options, resulting in a balance of residents and employment."

11.2.2.7   Principle 7: Local and Sustainable Materials

Construction of cities utilizing local regional materials with lower embodied energy and prefabricated modular systems is essential. Prefabrication has come and gone in modern architecture before, but this time, the focus will be on sustainability, owing to greater collaboration with makers of construction systems and building components throughout the design phase. We must encourage innovation and be conscious of sustainable production and consumption, as well as the embodied energy of materials and the flow of energy in closing life cycles. We must stress green manufacturing and waste-reduction technologies. Lightweight constructions, enclosures and local materials with lower embodied energy are more environmentally beneficial. Improved material and system specifications, endorsed by new material research and technological innovation, as well as reduced material diversity in multi-component products, are needed to aid in resource recovery, disassembly, value retention, and the possibility of reusing entire building components is highly required. Buildings' long-term durability will be improved and costs will be reduced as a result of success in this field.

11.2.2.8   Principle 8: Density and Retrofitting for Accessible Districts

The various aspects of this principle include encouraging the densification of the city center through mixed-use urban infill, center regeneration and green TODs, raising sustainability by developing business prospects around green transit-oriented developments, maximizing the interaction between urban planning and transportation networks, renovating inefficient building stock and systematically lowering the urban district's carbon footprint. Small- and medium-sized towns require special tactics due to their unique environment, and inventive approaches are required to address the vulnerabilities of Small Island States or coastal communities. People will return to the center of the city as public space is improved through urban renewal schemes. Continue to improve the streets and add bike lanes. Increase the number of pedestrian crossings to improve pedestrian connectivity. Natural components suited for active living can be added to public spaces to make them more usable.

11.2.2.9   Principle 9: Green Buildings and Districts

Novel design typologies must be developed at a minimal cost, and buildings must be operationally neutral and survive longer. We should use facade technology (Figure 11.2) with responsive buildings for bio-climatic architecture to stay ahead of cooling breezes and natural cross-ventilation, maximizing cross-ventilation, day-lighting and opportunities for night-flush cooling; we need to concentrate on the exterior of the structure. We need to focus on the low consumption of resources and materials, including the reuse of building elements and design for disassembly. It is critical to revitalize the city with energy-efficient green design, resulting in more adaptable

FIGURE 11.2   Greenery integrated in facades for buildings in Singapore and Seoul.

and long-lasting structures. Employing passive design principles, the city should implement deep green building design techniques and provide solar access to all new structures. It is critical to revitalize the city with energy-efficient green design, resulting in more adaptable and long-lasting structures. Buildings with more flexibility in their plans have a longer lifespan. Encourage the use of energy-efficient building designs and complete home insulation.

11.2.2.10   Principle 10: Livability, Healthy Communities, and Mixed-Use Programs

Affordable housing, mixed-use program and a healthy community are all priorities for the city. The key to long-term viability is land use development patterns. A mixed-use, mixed-income city promotes social sustainability and inclusiveness while also helping to repopulate the city center. Demographic shifts are an important concern for urban planners. The importance of mixed land uses is particularly important. All private developments should be required under master plans to include 40%–50% public (social) housing that is integrated into private housing. Green TODs should be the focus of higher densities. Essentially, these reforms will try to bring more sustainable lifestyle choices, with jobs, shopping, housing, and a city campus, all being close by, IT and teleporting from home assisting to dramatically reduce redundancy. Every project should incorporate at least 25% affordable housing and use prefabricated modular construction systems. Houses in the city should be taxed less. Encourage the retrofitting and adaptive reuse of historic structures.

11.2.2.11   Principle 11: Local Food and Short Supply Chains

Local food production, localized supply and a concentration on urban farming and agriculture, especially "eat local" and "slow food" initiatives, are all parts of this principle. The sustainable city provides appropriate acreage for food production in the city, a return to communal and allotment gardens of the past, where roof gardens are transformed into urban market gardens. Residents of the eco-city should cultivate and farm collectively, sharing food and composting kitchen waste and garden clippings, as well as growing "community" vegetables. It will be necessary to acquire and consume locally in order to eliminate the demand for gasoline-based transportation. The expense of food processing as well as paper recycling and reusing paper bags and glass containers will all have to be reviewed. Perhaps half of our food will have to be supplied organically, without the use of oil-based fertilizers or pesticides and farmed on local allotments. At least 20% of public parks should be used for urban farming. Make a communal garden. Maintain the urban hinterland as a source of food. Within the urban ecology, re-establish linkages to natural ecosystems.

11.2.2.12   Principle 12: Cultural Heritages, Identity, and Sense of Place

All sustainable cities strive for improved air quality, improved health and less pollution, as well as the development of resilient communities, robust public space networks and modern community facilities. Every city, on the other hand, has its own distinctive environment, whether it is by the sea, a river, in the desert or on a mountain; hence, whether the climate is tropical, arid or temperate, each site has its own particular habitat. All of these variables, including materials, history, and population demands, will be considered in the city's design. The emergence of grassroots strategy is the essence of a location. Creative ideas necessitate studio space in ancient buildings and warehouses that are both affordable and flexible. Cities will develop in accordance with the specific characteristics and characteristics of locations, the demographic characteristics of the population and the inventiveness of the officials and residents.

11.2.2.13   Principle 13: Urban Governance, Leadership, and Best Practice

If we aim to transform current cities into sustainable compact communities, good urban administration is critical. It must provide efficient public transportation, good public space and cheap housing as well as acceptable standards of urban administration, and change will not occur without political support. For their urban goals to be realized, city councils require strong administration and political support. A city on the road to sustainable practices is one that leads and designs holistically, implements change harmoniously and shares decision-making and accountability with the empowered citizenry. In order to ensure people-sensitive urban design and to encourage grassroots participation in balancing community needs with growth, public consultation exercises and grassroots engagement are crucial. Subsidize the clean-tech sector by innovative governmental policies, incentives and subsidies. To support the change from public to private partnership involves community organizations and non-governmental organizations.

11.2.2.14   Principle 14: Education, Research, and Knowledge

Technical training and up-skilling, research, sharing of experiences and knowledge dissemination through research papers concerning ecological city theory and sustainable design are all parts of this approach. Waste recycling water efficiency and sustainability are all subjects that need to be taught in primary and secondary schools. Changes in attitude and personal lifestyles will be necessary. The city is a center of institutions where information can be disseminated such as galleries, libraries and museums. We should ensure that citizens have ample access to educational and training opportunities. It is possible to establish research centers for sustainable urban development policies and best practices in eco-city design, as well as assessment methods to monitor environmental performance.

11.2.2.15   Principle 15: Strategies in Developing Countries

Cities in emerging countries need specific sustainability policies to balance the effects of increasing urbanization and globalization. Developing and emerging countries have unique requirements that necessitate specific tactics, technological transfers and financial mechanisms. Those in poor countries cannot adopt the same strategies or engage in the same discussions as cities in affluent countries. Cities in emerging countries should build specific sustainability plans are being developed to counteract the effects of increasing urbanization and globalization. Cities must adapt strategies suited to the developing world, such as low-cost construction and mass dwelling typologies, in order to achieve fast urbanization [15].

   Most of the requirements imply that design should be on the community and district scales with projects on urban infill or restoration (brown field) sites near existing built areas and transportation connections. Our cities will continue to use vast amounts of raw resources and produce large amounts of garbage, requiring more land and energy. Around the same moment, the economies of scale afforded by cities allow us to make renewable energy sources more economically viable. Since cities consume vast amounts of energy, transportation systems must be rapidly altered so that they might run on local renewable energy sources often as feasible – but at least 50% of the time. An intelligent power network often known as a smart grid can transfer and distribute electricity generated and stored locally. As a result, zero-emission city precincts will shift from electricity consumers to power producers [16]. They would become local power plants using photovoltaic, solar heat generation and refrigeration, wind energy, biomass and other renewable energy sources.

11.3   TELESCOPING THE BLISS OF ECO-FRIENDLY SUSTAINABLE CITIES AND VILLAGES

The rate at which urbanization grows is a growing concern at the international level as it is threatening safe drinking water supplies, waste management, environmental issues and pressure over urban land and public health. The international community has anxiety that the damage to ecosystem may be irreversible, which has noticed the sustainable model in eco-villages. To define eco-village in a strict manner is difficult. The Global Network of Eco-villages defines an eco-village as "an intentional or traditional community using local participatory processes to holistically integrate ecological, economic, social and cultural dimensions of sustainability in order to regenerate social and natural environments" [17]. It is a small and self-sustained group who is mostly from rural areas and lives for natural resources. Eco-village is a well-created place where ecological and economical values are balanced. Here renewable natural resources are used to satisfy the needs of people like use of solar and wind for energy production. Communities try to minimize environmental pollution by reducing, reusing, recycling and regenerating natural resources. They use regenerative technologies such as organic farming which reduce land and water pollution, bioclimatic constructions and reforestation. It is planned in such a way that all the basic facilities such as safe drinking water, transport, health care system, education, digitalization, self-sustainable economic growth and public services like bank are available. If communities' needs are satisfied at eco-village, then it can be assumed that the migration rate to the urban area will decrease as well as it prevents environment pollution and protects ecosystem.

So, conversion of existing villages to eco-village or establishment of new eco-village is promoted at international levels. In 2001, CII established the IGBC, which offers ratings and certifications to encourage a sustainable built environment on the assessment of village infrastructure, water conservation, energy efficiency and green innovation to encourage transformation of existing village to green eco-villages.

11.4   CASE STUDIES

Case 1: Curitiba [18–20]
Curitiba has a rich and glorious history. Curitiba, the capital of Brazil's state Parana, has long been a cultural and economic center in the region, having evolved from a "sleeping city" surrounded by farmland to a draw for European immigration in the late 19th century. Curitiba saw a watershed moment in the 1940s with the influx of agricultural laborers. The city's population has more than doubled in the last 20 years. Jaime Lerner, mayor of Curitiba city introduced a plan for sustainable city in 1972.
   Sustainable Solutions:

1. **Bus Rapid Transit System:** It was innovated with a new system of raised platforms that reduce time for passenger inflow and outflow, longer buses to increase capacity and a pre-payment system that eliminates the need for bus drivers to issue tickets and collect money on the go, as well as roads with express lanes for buses. It has aided Curitiba in maintaining a rapid, low-cost and low-emission public transportation system. The city's designating streets are for pedestrians only and specific bike lanes have also aided transportation system.
2. **Green space:** Curitiba had only one park in 1971, today it has more than 50 square meters of green space per person. Since the 1970s, 1.5 million trees have been planted and 28 public parks were developed in it. To combat flooding, Curitiba surrounded the urban area with fields of grass, saving itself the cost and environmental expense of dams. To maintain the fields,

rather than machines, the city uses sheep, saving energy and providing manure for farmers and wool.

3. **Waste management:** Today, around 70% of garbage is recycled in Curitiba. Under the Green Exchange program, residents trade trash for tokens. It allows for the exchange of notebooks, bus tokens and food in return for recycling. Today, 90% of the city participates in its recycling program. This protects the environment, as well as uplifts education, increases food access and expedites transport for the poor.

4. **Education:** Curitiba has a free University for the Environment, which teaches the city's poor about sustainability and empowers them. Information about the city's green design is provided to citizens. Encouragement for a culture of pride around sustainability and provision of knowledge helps to maintain the city's greenness.

Curitiba is a city with a very high Human Development Index (0.856) and in 2010, it was awarded the Globe Sustainable City Award. Curitiba's imaginative, innovative and practical urban planning aided it to be the gold standard in sustainable urban planning, making it one of the world's foremost green cities while also reducing poverty and increasing population. Its per-capita income is 66% greater, and its economic growth rate for 30 years is 3.1% higher than the national average. It has presented that in sustainable urban planning and development, resource constraints are not the barrier. It has revealed that sustainable urban planning is in fact a valuable asset in the future.

Case 2: Vancouver city [21–23]
Vancouver is encircled by beautiful natural resources, such as mountains and seashores. It has a long and interesting history. In the 1870s, it was a particular challenging community populated by Americans, Europeans and Scots. A devastating fire devastated the city in less than an hour, just two months after it was incorporated as a city in April 1886. In 1914, it was resurrected as a thriving port. It was Canada's largest Pacific coast port for trade and the industrial, commercial and financial center after World War II, and it grew into the country's main corporate hub. Since World War II, the city has been a favorite destination for immigrants from other parts of Canada and Asian countries, and it has grown swiftly. The city's population had topped 1 million by the mid-1960s. Protests against freeways triggered the commencement of sustainability reforms in the early 1970s. The creation of a sustainable city is a constant process. The city of Vancouver revealed its plans in 2011 to become the world's greenest city by 2020 by implementing the strategy known as the "Greenest City Action Plan."

Vancouver's green initiatives:

Clean energy: In Vancouver, around 85% of electricity is supplied by hydroelectricity and 10% by solar and wind energy. No use of fossil fuels for electricity generation is proposed in the greenest city action plan.

Green transportation: Approximately 40% of Vancouver's carbon emissions originate from transport, so reducing emissions in this sector is important. Roadways were redeveloped for walkers, cyclists and made it bike-friendly, so that the number of car rides can be reduced. Mass public transit options in Vancouver include extensive public bus systems, trolley buses, light rapid transit rail and commuter rail lines (Sky Train and West Coast Express); and Sea Bus. All of the mass transit options utilize clean energy technologies like electric or hybrid vehicles.

Urban planning: To oppose urban sprawl, vertical buildings are preferred. In addition to carbon-neutral buildings, the integration of energy, sewers, parks and water utilities are developed for highly efficient gains. Public green places were constructed such that all residents have access within a 5-minute walk.

Zero waste: Zero waste by 2040 plan of Vancouver includes the collection of all types of scrap and waste, recovering of new resources and recycling while simultaneously evaluating what changes to infrastructure will be required, as well as recognizing technology gaps and providing robust public education about waste management and diversion issues. For that, zero waste centers were established. Apart from it, the reduction in Single-Use Item Strategy was implemented, which includes actions to reduce waste from polystyrene foam cups and take-out containers, plastic and paper shopping bags, disposable hot and cold drink cups and disposable straws and utensils by 2025.

Different policies got success to drop carbon emissions in the city by about 12% between 2007 levels and 2020. Since 2008, when the goal of zero waste was first announced as a target, the amount of waste sent to landfill by the city has reduced by 27% (as of 2015). However, achieving a reduction of carbon emission and waste to landfill is very impressive, and a continuing trend would put the city well on its way to make it the greenest city in the world. Vancouver has exhibited how a set of planning principles of basic land use, combined with independent thinking at the local level, can help to create a highly livable region and to combat the forces that lead to urban expansion.

Case 3: Gaviotas: Eco-village [24, 25]

Gaviotas is an eco-village located in the Llanos of Colombia. In 1971, it was founded by Paolo Lugari, who brought together a group of engineers, artists and scientists in an effort to design a sustainable village. This region at that time was so remote and poor in soil quality, untouched even by the country's political instability.

Sustainable initiatives:

1. **Inventions and innovations:** Gaviotas has made it a goal to constantly find answers that are particular to the difficulties that the community will face. Gaviotas' electricity comes from a variety of sustainable sources, including solar panels, wind turbines and a seesaw-powered water pump. Gaviotas' inventions are frequently modest adjustments to manufacturing methods that result in high-quality items at low prices. Gaviotas' engineers created pumps that leave the piston in place and instead lift and lower a cheap, light PVC (polyvinyl chloride) sleeve around the piston, while pumps existed in the region raised and lowered a heavy piston in a pipe. They designed simple solar water-heating panels out of low-cost building materials. The eco-village also creates a distinctive form of quick-create brick made with the dirt of the region.

2. **Health:** Gaviotas' team created a self-driven, functional hospital in 1980, which treated a considerable number of local people who worked in the

village and outlying plains until in the 1990s, the Colombian government
established the hospital at Gaviotas. While the hospital is no longer functional, the availability of clean water has increased in the surrounding region
because of the application of Gaviotas pump technology and is an effective
form of disease prevention as long as the locals maintain the pumps.

3. **Environmental impact:** Gaviotas' villagers planted over 1.5 million trees
   in the area and the groundcover began hosting tropical rainforest species
   which are indigenous to the region. As a result, forest area expanded to
   20000 acres. The presence of the forest has provided tropical climate by
   generating an additional 10% rainfall. Resin harvested from the planted
   pine trees has provided sustainable source of income to Gaviotas.

4. **Self-reliant:** It regularly uses fuel made from turpentine and plant oil or
   recycled cooking oil to run all diesel engines. Currently, a Bio-Diesel project is going on for market and local use. When funding began to wane in
   the 1990s, the villagers figured out that their pine forest was sustainable
   producers of valuable pine resin used in the production of turpentine and
   violins, which they could export.

Every family enjoys free housing, community meals and schooling. There are no
weapons, no police and no jail. There is no mayor. The United Nations named the
village a model of sustainable development.

11.5 CONCLUSION AND FUTURE SCOPE

Environmental challenges including air, water, land pollution, food security, inequality in access to resources, growing waste volumes, dwindling nonrenewable energy
sources and ecosystem threats became unavoidable and harder to manage as nations
developed in terms of GDP growth. Eco-friendly communities and villages are a
viable option to combat existing problems. In addition to the usual urban components, it also had specific features like full-fledged Urban Governance, Leadership,
Education, Research and Knowledge, Cultural Heritage, Identity and Sense of Place
with Sustainable Transportation and Good Public Spaces etc., then that would be
beneficial to accomplish sustainability against the increase in population, pollution and urban expansion. This chapter aims to provide solutions with the accessible resources and infrastructures to attain the challenges that arose with these
newer development and technologies for both ecological and economical sustainable
growth of cities at the international level.

It is expected that with the implementation of given principles, the future generations might lead a quality life with green ecosystems and city sustainability.
This chapter aids to achieve some of the goals such as good health and well-being,
quality education, clean water and sanitation, affordable and clean energy, sustainable cities and communities, responsible consumption and production, climate
action and life and land as declared in the 17 sustainable development goals at
the UN Sustainable Development Summit in September 2015 as a 2030 Agenda for
Sustainable Development.

REFERENCES

1. Lindsay Maizland. Global Climate Agreements: Successes and Failures. Available at: https://www.cfr.org/backgrounder/paris-global-climate-change-agreements [accessed July 20, 2021].
2. Our Common Future, From One Earth to One World: A Report by the World Commission on Environment and Development, 1987. Available at https://sustainabledevelopment .un.org/content/documents/5987our-common-future.pdf [accessed on June 30, 2021].
3. Green Cities Newsletters. The Positive Effect of European Green Cities. Available at https://thegreencities.eu/ [accessed July 15, 2021].
4. Suzuki H., Dastur A., Moffatt S., Yabuki N. and Maruyama H. Eco2 cities: Ecological cities as economic cities. (World Bank) 2010, xvii–xviii. Available at: https:// openknowledge.worldbank.org/handle/10986/2453 [accessed July 1, 2021].
5. Zaid Ahmed. Green City. Available at: https://www.slideshare.net/ah16/green-city-75843883 [accessed on June 15, 2021].
6. Lehmann Steffen. (2011). "What is green urbanism? Holistic principles to transform cities for sustainability" in Climate change-research and technology for adaptation and mitigation, ed. Juan Blanco and Houshang Kheradmand. (Intech Open). doi: 10.5772/23957
7. Ulrich Beck. Risk society. Towards a new modernity. (Sage, London), 2000.
8. Lehmann Steffen. Low carbon districts: Mitigating the urban heat island with green roof infrastructure. City. Culture and Society. 2014; 5(1): 1–8. doi: 10.1016/j.ccs.2014.02.002
9. Overview of greenhouse gases. Available at official website of United States Environmental Protection Agency: https://www.epa.gov/ghgemissions/overview-greenhouse-gases [accessed June 10, 2021]
10. Gichuki L., Brouwer R., Davies J., Vidal A., Kuzee M., Magero C., Walter S., Lara P., Oragbade C. and Gilbey B. Reviving land and restoring landscapes: Policy convergence between forest landscape restoration and land degradation neutrality. (Gland, Switzerland: IUCN), 1, 2019. doi: 10.2305/IUCN.CH.2019.11.en.
11. Nathanson Jerry A. Land pollution. Encyclopedia Britannica. Available at https://www. britannica.com/science/land-pollution [accessed August 2, 2021].
12. Education for sustainable development. Available at: https://en.reset.org/knowledge/ advancing-sustainable-development-through-education-india [accessed August 4, 2021].
13. Yadav Archana. Role of education in sustainable development of modern India. Annals of Education. 2016; 2(2): 80–84.
14. The future of food and agriculture: Trends and challenges. Food and Agriculture Organization of the United Nations. Rome. 2017. 116–140.
15. Tirla L., Gabriela M., Vijulie I., Matei E. and Octavian C. Green cities-urban planning models of the future in cities in the globalizing world and Turkey: A theoretical and empirical perspective, ed. Recep Efe, Neslihan Sam, Rıza Sam, Eduardas Spıriajevas, Elena Galay. (St. KlimentOhridski University Press Sofia), 2014, 462–479, doi: 10.13140/2.1.4143.6487
16. Lehmann Steffen. Transforming the city for sustainability – The principles of green urbanism. Journal of Green Building, 2011; 6(1): 104–113. doi: 10.3992/jgb.6.1.104
17. Ecovillage and traditional village: A comparison in down to earth. Available at https:// www.downtoearth.org.in/blog/governance/ecovillage-and-traditional-village-some-misconceptions-57599 [accessed June 20, 2021]
18. Suzuki H., Dastur A., Moffatt S., Yabuki N. and Maruyama H. Eco2 cities: Ecological cities as economic cities. (World Bank) 2010, 169–181. https://openknowledge. worldbank.org/handle/10986/2453 [accessed on July 1, 2021]

19. Story of cities #37: How radical ideas turned Curitiba into Brazil's 'green capital (2016) The guardian for 200 years, International edition, available at: https://www.theguardian .com/cities/2016/may/06/story-of-cities-37-mayor-jaime-lerner-curitiba-brazil-green-capital-global-icon [accessed on August 6, 2021]

20. The Brazilian city Curitiba Awarded the Globe Sustainable City Award 2010. Globe Award: Leading Sustainability Awards. globeforum.com. Available at https://web. archive.org/web/20140714212832/http://globeaward.org/winner-city-2010.html [accessed September 10, 2021]

21. Story of cities #38: Vancouver dumps its freeway plan for a more beautiful future (2016) The guardian for 200 years, International edition, Available athttps://www. theguardian.com/cities/2016/may/09/story-cities-38-vancouver-canada-freeway-protest-liveable-city [accessed on August 6, 2021]

22. https://www.greencitytimes.com/vancouver/ [accessed on August 6, 2021]

23. Vancouver – Greenest city. 2020. Green City Times. https://www.greencitytimes.com/vancouver-greenest-city-2020/ [accessed on August 6, 2021]

24. Friends of Gaviotas, Available at: http://www.friendsofgaviotas.org/ [accessed on August 6, 2021]

25. Weisman Alan. Gaviotas – A village to reinvent the world, 2nd edition (Chelsea Green Publishing), 2008, 93, 112.

Index

For Product Safety Concerns and Information please contact our EU
representative GPSR@taylorandfrancis.com
Taylor & Francis Verlag GmbH, Kaufingerstraße 24, 80331 München, Germany

* 9 7 8 1 0 3 2 4 3 9 5 0 1 *